# STUDENT
# SELF-STUDY
# GUIDE

# STUDENT SELF-STUDY GUIDE

## Donald W. Shive
## Louise E. Shive

Muhlenberg College

MORTIMER

# CHEMISTRY
## A CONCEPTUAL APPROACH

### Third Edition

**D. Van Nostrand Company**
New York • Cincinnati • Toronto • London • Melbourne

D. Van Nostrand Company Regional Offices:
New York      Cincinnati      Millbrae

D. Van Nostrand Company International Offices:
London      Toronto      Melbourne

Published by D. Van Nostrand Company
450 West 33rd Street, New York, N. Y.  10001

Published simultaneously in Canada by
Van Nostrand Reinhold Ltd.

10 9 8 7 6 5 4

# Preface

The *Self-Study Guide* is designed to accompany the
Third Edition of *Chemistry: A Conceptual Approach*,
by Charles E. Mortimer. The organization of the
*Guide* follows that of the text, each chapter corresponding to one in the text. The *Guide* may be used
with little or no assistance from the instructor.
It should be used, however, after the student has
attended lectures and read the text.

Each chapter is divided into four parts: Objectives,
Exercises, Answers to Exercises, and Self-Test. The
first part provides a list of explicit objectives;
the second contains verbal and mathematic exercises
designed to help the student meet these objectives.

Exercises should be done sequentially and answers
checked immediately upon completion.  The answers to
all exercises and detailed solutions to most problems
are given in the third part.  The answer to an exer-
cise is generally given in the left column and the
detailed solution to the right.  The solutions include many hints, tables, and figures.  Text refer-
ences [in brackets] are given with many answers.  The
last part is a self-test and should be completed
after studying lecture, text, and study-guide material.  The test is to be completed within a pre-
scribed time, and answers are given without comment
in the back of the *Guide*.  The self-test is repre-
sentative of test questions given at Muhlenburg College and is not intended to be comprehensive or
indicative of the type of questions that should be on
an exam.

We hope that this *Guide* is helpful and makes the
study of chemistry more enjoyable.


Donald and Louise Shive

# Contents

# STUDENT
# SELF-STUDY
# GUIDE

# 1

# Introduction

OBJECTIVES          (a) You should be able to demonstrate your knowledge
                        of the following terms by defining them, describing
                        them, or giving specific examples of them:

                            Celsius temperature scale [1.2]
                            centi- [1.2]
                            chemical change [1.1]
                            compound [1.1]
                            element [1.1]
                            energy [1.1]
                            Fahrenheit temperature scale [1.2]
                            International System of Units and SI units [1.2]
                            kilo- [1.2]
                            matter [1.1]

1

> micro-  [1.2]
> milli-  [1.2]
> phase [1.1]
> physical change [1.1]
> pure substance [1.1]

(b) You should be able to write numbers in scientific notation.

(c) You should be able to determine and work with the proper number of significant figures.

**EXERCISES**    I.   Write the following numbers in scientific notation:

_____  1. 751                  _____  5. 0.000745
_____  2. 781,000              _____  6. one million
_____  3. 781,000.0            _____  7. one-millionth
_____  4. 0.050                _____  8. one-third

II.  How many significant figures are contained in each of the following numbers?

_____  1. 2.57                      _____  7. $6.022 \times 10^{23}$

_____  2. 0.0057                    _____  8. 900,000

_____  3. 0.570                     _____  9. 1.00

_____  4. $5.7 \times 10^{-3}$      _____  10. 0.75

_____  5. $2.9979 \times 10^{10}$   _____  11. 1.75

_____  6. 0.0821                    _____  12. 1.750

III. Perform the following calculations and report the answer and the appropriate number of significant figures:

1. 6.0 + 297 + 8.75 =

2. 7.41 + 0.02 =

3. 6.9 + 0.001 =

4. $182 - 99.2 =$

5. $(7.10 \times 10^{21}) + (7.10 \times 10^{20}) =$

6. $(6.4 \times 10^{-1}) - (4.21 \times 10^{-2}) =$

7. $7.1/9.64 =$

8. $(6.022 \times 10^{23})(1.70) =$

9. $(1.074 \times 10^{-4})(9.9) =$

10. $(7.45 \times 6.1)/2.45 =$

**ANSWERS TO EXERCISES**

I.  Scientific notation

It is often inconvenient to work with the standard form of very large or very small numbers. For example, the multiplication involving the numbers 701,000 and 0.00000077 can easily be done incorrectly if we lose track of the zeros. An exponential notation, commonly referred to as scientific notation, is used to simplify calculations involving such numbers. The conversion from standard notation to scientific notation is quite simple, and a method for performing this conversion is summarized as follows:

(a) Move the decimal point so that there is a single digit (not zero) to its left, for example,

0.0000007⊙7

(b) Count the number of digits between the original and new decimal point positions to determine the magnitude of the exponent of 10, for example,

0.0000007⊙7
    1234567

$10^7$

(c) Determine the sign of the exponent of 10 by the direction from the original to the new decimal point position (right is negative, left is positive), for example,

0.0000007⊙7
⟶

The direction indicates a negative sign.

Thus, 0.00000077 equals $7.7 \times 10^{-7}$.

Properly used scientific notation clearly indicates the number of significant figures.  When we expressed 0.00000077 in scientific notation, we retained only the two sevens.  The zeros preceding the sevens were expressed by the magnitude of the exponent. Thus, the zeros are not significant. Only the two sevens are significant. We can summarize the general rules concerning zeros in a number as follows:

(a) Zeros to the left of any digit other than zero are not significant.

(b) Zeros to the right of any digit other than zero are significant when a decimal point is included in the number.  If a decimal point is not included in the number, the zeros to the right of any digit other than zero may or may not be significant.

(c) Zeros between digits other than zero are significant.

In the examples of this and the following section, we will see specific cases in which these general rules apply.

1. $7.51 \times 10^{2}$    3 significant figures

2.    The number of significant figures to which the measurement was made is not clear in this example.  The number can be written therefore in several forms, depending upon the precision of the measurement.

$7.81 \times 10^{5}$    3 significant figures

$7.810 \times 10^{5}$    4 significant figures

$7.8100 \times 10^{5}$    5 significant figures

$7.81000 \times 10^{5}$      6 significant figures

3. $7.810000 \times 10^{5}$      7 significant figures

4. $5.0 \times 10^{-2}$      2 significant figures

5. $7.45 \times 10^{-4}$      3 significant figures

6. $1 \times 10^{6}$      When a number is written in word form, we determine the number of significant figures by writing the number in numerical form and applying the general rules. If the number of significant figures is still unclear, we choose the lowest number. Thus, one million is 1,000,000 and has one significant figure. Occassionally, integers are given in word form and are meant to have an infinite number of significant figures. (See Table 1.1 of the study guide.)

7. $1 \times 10^{-6}$      1 significant figure

8. $3 \times 10^{-1}$      1 significant figure

Table 1.1 Significant figures of numbers written in word form

| NUMBER IN WORD FORM | NUMERICAL FORM | NUMBER OF SIGNIFICANT FIGURES |
|---|---|---|
| one thousand | 1000 | 1 |
| one and five one-hundredth | 1.05 | 3 |
| fifty-five | 55 | 2 |
| one hundred and ten | 110 | 2 |
| two thousand and twenty | 2020 | 3 |
| five million and twenty-four | 5,000,024 | 7 |
| six million and two thousand | 6,002,000 | 4 |

II. Significant figures

1. 3

2. 2      The zeros before and after the decimal point only show the position of the decimal point. It is better to write this number in scientific notation as $5.7 \times 10^{-3}$. *Zeros to the left of any digit other than zero are not significant.*

3. 3      *Zeros to the right of any digit other than zero are significant if a decimal point is included in the number.*

4. 2       When a number is written in scientific notation, the number of significant figures is determined by the whole number.  The exponent shows the position of the decimal.

5. 5

6. 3

7. 4       *Zeros between digits other than zero are significant.*

8. 1,2,3,4,5,6   The number of significant figures depends upon the precision of the measurement.

9. 3       Rule (b)

10. 2       Rule (a)

11. 3

12. 4       Rule (b)


     III. Calculations

1. $3.12 \times 10^{2}$   After the addition is performed, the answer, 311.75, is rounded off to the correct number of significant digits as determined by the smallest number of digits to the right of the decimal point in any of the numbers that are being added.  The number 279 has no digits to the right of the decimal.  The sum, therefore, is reported as 312, or preferably as $3.12 \times 10^{2}$.  This procedure should be followed whenever addition or subtraction is performed.

2. 7.43

3. 6.9

4. 83      Notice that the number 82.8 is rounded off to 83. The answer has fewer significant figures than either number used in the calculation.

5. $7.81 \times 10^{21}$

6. $6.0 \times 10^{-1}$

7.  0.74

In both multiplication and division the answer is usually rounded off to the smallest number of significant figures contained in any of the numbers involved in the calculation.  This rule normally works and your instructor may prefer that you use it exclusively; however, there are some exceptions.  A more rigorous approach is described in example 9 of this section of the study guide.

8.  $1.02 \times 10^{24}$

9.  $1.06 \times 10^{-3}$

Notice that in this calculation, $(1.074 \times 10^{-4})$ $(9.9)$, three significant figures are reported in the answer rather than two significant figures as in the number 9.9.  The number 9.9 is precise to about one part in 99, and the number 106 is precise to about one part in 106 (see following paragraph).  Reporting the answer as $1.1 \times 10^{-3}$ would reflect a precision of about one part in 11; therefore, the answer would be less precise than the number 9.9.  (See Section 1.3 of your text.)

Normally it is assumed that the last digit of any experimentally determined number is known only within plus or minus one.  For example, reporting a number as 78.6 indicates that the number is known to be within 78.5 and 78.7.  This range in value may be indicated as $78.6 \pm 0.1$.  Thus, the number is known to be precise to $\pm 0.1$ in 78.5, or $0.1/78.5 =$ 1/785, or 1 part in 785.  If this number is used in the calculation $(9.00000)(78.5)$, the answer should be reported to a precision of 1 part in 785 since the other number, 9.00000, indicates a precision of 1 part in 900,000.  Possible values of the answer and the corresponding precisions are

    (a)  $7.0 \times 10^2$          1 part in 70

    (b)  $7.06 \times 10^2$        1 part in 706

    (c)  $7.065 \times 10^2$      1 part in 7,065

    (d)  $7.0650 \times 10^2$    1 part in 70,650

Answer (b) is chosen because its precision most closely reflects that of the least precise number in the calculation.  Normally, however, an answer will have the same number of digits as the least precise number in the calculation.

10. 18

The fewest number of significant figures in any
number involved in the calculation is 2, in the
number 6.1.  Note that the answer is reported as
18, not 19.  In the study guide the convention for
rounding off a digit followed by a five is

    (a) an even digit followed by a five remains un-
       changed

    (b) an odd digit followed by a five is raised in
       value by one unit

Thus, if two significant figures are desired and
the number 18.5 is obtained in a calculation, the
answer would be reported as 18; however, if the
number were 19.5, the answer would be reported as
20.  Notice that only the first digit after the
digit being rounded off is considered.  Subsequent
digits are not considered.

SELF-TEST

Complete the test in 15 minutes:

I.  Answer each of the following:

1. The process of water changing into steam is
   called
   a.  freezing
   b.  a chemical change
   c.  a physical change
   d.  fusion

2. The number 0.070020 has how many significant
   figures?
   a. 1      b.  2      c.  4      d. 5

3. The number 0.070020 should be written in scientific notation as
   a   $7.002 \times 10^2$          c.  $7.0020 \times 10^{-2}$
   b.  $7.002 \times 10^{-2}$       d.  $7.0020 \times 10^2$

4. The number 700 should be written in scientific
   notation as
   a.  $7 \times 10^2$          c.  $7.00 \times 10^2$
   b.  $7.0 \times 10^2$        d.  cannot be determined

_____ 5. One microgram is equal to

     a. $10^3$ g      c. $10^6$ g

     b. $10^{-3}$ g      d. $10^{-6}$ g

_____ 6. One kilometer equals

     a. $10^3$ m      c. $10^6$ m

     b. $10^{-3}$ m      d. $10^{-6}$ m

_____ 7. The number 700.0 should be written in scientific notation as

     a. $7 \times 10^2$      c. $7.000 \times 10^2$

     b. $7.00 \times 10^2$      d. $7.000 \times 10^3$

_____ 8. The process of gasoline burning in an automobile cylinder is called
     a. a physical change      c. vaporization
     b. a chemical change      d. boiling

_____ 9. The sum of the numbers 71.742, 6.0, and 21.3413 is
     a. 99.0833      c. 99.1

     b. 99.0      d. $1.0 \times 10^2$

_____ 10. The number 199.969, rounded off to three significant figures, should be written as

     a. 200      c. $2 \times 10^2$
     b. 199      d. $2.00 \times 10^2$

_____ 11. The number 0.074 has how many significant figures?
     a. 1      c. 3
     b. 2      d. 4

_____ 12. The physical state of matter that assumes the shape of its container only within the limit of the volume that the sample occupies is
     a. vapor      c. liquid
     b. gas      d. solid

_____ 13. SI units are
     a. units of the International System
     b. Standard International Units

c. Substituted International Units
d. Scientific International Units

_____ 14. Multiplication of the number 23.6 by the number $7.50 \times 10^3$ yields

a. $1.77 \times 10^3$     c. $1.7 \times 10^5$

b. $1.77 \times 10^5$     d. $177 \times 10^3$

_____ 15. A cow produces milk from ingested foodstuffs. This process can be called
a. a liquefaction        c. a physical change
b. a chemical change     d. a biological marvel

# 2

# Atomic Structure

OBJECTIVES

(a) You should be able to demonstrate your knowledge
of the following terms by defining them, describing
them, or giving specific examples of them:

    atom [2.1, 2.6]
    atomic number [2.7, 2.12]
    atomic weight [2.7, 2.9]
    Balmer lines [2.11]
    Bohr radius [2.11]
    Bohr theory [2.11]
    de Broglie wavelength [2.14]
    diamagnetism [2.16]
    electromagnetic radiation [2.10]
    electron [2.2, 2.3]

11

electronic configuration [2.13, 2.17]
excited state [2.11]
frequency [2.10]
ground state [2.11, 2.13]
Heisenberg uncertainty principle [2.14]
Hund's rule [2.16]
isobar [2.8]
isotope [2.8, 2.9]
neutron [2.2, 2.5]
nuclide [2.8]
paramagnetism [2.16]
Pauli exclusion principle [2.15]
periodic table [2.12, 2.13]
photon [2.10]
proton [2.2, 2.4]
quantum numbers [2.15, 2.16]
quantum theory [2.10, 2.11]
screening [2.17]
wavelength [2.10]

(b) You should be able to determine the number of protons, neutrons, and electrons of any isotope of an
element in the periodic table.

(c) You should be able to calculate atomic weights from
masses and relative abundances of isotopes.

(d) You should be able to determine an actual isotopic
mass in u from the binding energy of the isotope
and the rest masses of the proton, neutron, and
electron.

(e) You should understand the relationship between energy
and frequency, $E = h\nu$ or $E = hc/\lambda$, and be able to
work problems relating to these equations.

(f) You should understand the Bohr theory.

(g) You should be able to write a complete set of quantum numbers for each electron of any element in the
periodic table or of an ion of that element .

(h) You should be able to write the electronic configuration of each element in the periodic table and of
any monatomic ion.  You should also be able to predict the number of unpaired electrons and the magnetic properties of each of these species.

EXERCISES          I.    Major contributions in the development of atomic
                        theory were made by the following scientists.  If
                        your instructor emphasizes names in lecture, you
                        should take the time to match the name of each
                        scientist with his contribution.

_____        1. Lucretius, 100 B.C.        a. determined the charge of
                                                    the electron, e, with
                                                    his famous oil drop
_____        2. Dalton, 1803                  experiment.

                                                 b. discovered radioactivity.
_____        3. Thomson, 1897
                                                 c. first observed positive
                                                    rays, or canal rays.
_____        4. Goldstein, 1886
                                                 d. proposed a theory of the
                                                    electronic structure of
_____        5. Rutherford, 1911              atoms based on the emis-
                                                    sion spectrum of hydro-
                                                    gen.
_____        6. Planck, 1900
                                                 e. proposed the existence
                                                    of the neutron.
_____        7. Chadwick, 1932
                                                 f. presented evidence for
                                                    the existence of a nu-
_____        8. Millikan, 1909                cleus in an atom from
                                                    alpha particle scatter-
                                                    ing data.
_____        9. Moseley, 1913
                                                 g. proposed the quantum
                                                    theory of radiant energy.
_____       10. Becquerel, 1896
                                                 h. proposed that the atomic
                                                    number of an element is
_____       11. Bohr, 1913                    equal to the number of
                                                    units of positive charge
                                                    in the nucleus.
_____       12. Mendeleev, 1869
                                                 i. was a Roman philosopher
                                                    who described the nature
_____       13. de Broglie, 1925              of the universe.

                                                 j. proposed the periodic
_____       14. Schrödinger, 1926             classification of the
                                                    elements.

_____    15. Heisenberg, 1927        k. determined the ratio of
                                                   the electron's charge
                                                   to mass, *e/m*.

                                                l. proposed that position
                                                   and velocity simulta-
                                                   neously cannot be meas-
                                                   ured exactly.

                                                m. postulated the wave
                                                   nature of the electron.

                                                n. presented the keystone
                                                   to wave mechanics.

                                                o. first proposed an atom-
                                                   ic theory.


II.    Match the name of the element with the atomic symbol.
       These symbols are part of the language of chemistry
       and should be mastered early in your course.

_____        1. silver          a. Ag
_____        2. hydrogen        b. Al
_____        3. sodium          c. As
_____        4. calcium         d. Au
_____        5. potassium       e. B
_____        6. boron           f. Ba
_____        7. flourine        g. Br
_____        8. copper          h. Ca
_____        9. gold            i. Cl
_____       10. nitrogen        j. Cr
_____       11. bromine         k. Cu
_____       12. antimony        l. F
_____       13. chromium        m. H
_____       14. zinc            n. He
_____       15. arsenic         o. Hg
_____       16. tin             p. K
_____       17. chlorine        q. Mg
_____       18. helium          r. N
_____       19. tungsten        s. Na
_____       20. lead            t. Pb
_____       21. barium          u. S
_____       22. silicon         v. Sb
_____       23. magnesium       w. Si
_____       24. sulfur          x. Sn
_____       25. mercury         y. W
_____       26. aluminum        z. Zn

III. Answer each of the following with *true* or *false*.
     If a statement is false, correct it. Use this
     section as a guide for further study.

_____    1. The energy of a quantum of radiation is directly
                      proportional to its wavelength.  The proportion-
                      ality constant, *h*, is Planck's constant.

_____    2. The atomic mass, *A*, is the sum of the number of
                      neutrons and protons in the nucleus.

_____    3. The atomic number is equal to the total number of
                      neutrons in the nucleus.

_____    4. Isotopes of an element have the same atomic weight
                      but differ in atomic number.

_____    5. Fundamental subatomic particles include the pro-
                      ton, neutron, electron, and nucleus.

_____    6. The charge of a proton is equal to but opposite
                      in sign from that of the electron.

_____    7. A neutron has a positive charge.

_____    8. The nucleus is always a combination of protons
                      and neutrons.

_____    9. Isotopes are atoms of different chemical reactiv-
                      ity but equal masses.

_____   10. The atomic weight of an element is a weighted
                      average of the naturally occurring isotopes of
                      that element.

_____   11. A diamagnetic material has unpaired electrons.

_____   12. Electrons are distributed among orbitals of a
                      subshell such that the number of electrons that
                      are paired is a maximum.

_____   13. Paramagnetic materials interact with magnetic
                      fields.

_____   14. The frequency of a quantum of radiation is in-
                      versely proportional to its wavelength.

_____  15. Electrons associated with an atom can exist only in discrete energy levels.

_____  16. It is possible for an electron to have the set of quantum numbers $n = 1$, $l = 1$, $m = +1$, $s = +\frac{1}{2}$.

_____  17. It is possible for an electron to have the set of quantum numbers $n = 2$, $l = 0$, $m = 0$, $s = +\frac{1}{2}$.

_____  18. It is possible for an electron to have the set of quantum numbers $n = 2$, $l = 1$, $m = 1$, $s = +\frac{1}{2}$.

_____  19. It is possible for an electron to have the set of quantum numbers $n = 4$, $l = 3$, $m = -2$, $s = +\frac{1}{2}$.

_____  20. A $d$ subshell can contain a maximum of six electrons.

_____  21. A noble gas is unreactive chemically because the electronic configuration of its outer shell consists of completely filled $s$ and $p$ subshells.

IV. Using the periodic table in the back of the study guide, identify the element that is described by the statement:

_____  1. $Z = 13$

_____  2. The atomic weight is 74.9 u.

_____  3. The atomic number is 50.

_____  4. The mass number is 109 and the number of neutrons is 62.

_____  5. There are 16 electrons in the neutral atom.

_____  6. Two isotopes exist, one with a mass of 10.013 u and one with a mass of 11.009 u.

_____  7. There are 18 electrons in the ion that has a single negative charge, $X^-$.

_____  8. There are 22 electrons in the ion that has two positive charges.

_____  9. The electronic configuration is $1s^2\ 2s^2\ 2p^3$.

10. The electronic configuration is $1s^2\ 2s^2\ 2p^6\ 3s^2\ 3p^6\ 3d^8\ 4s^2$.

11. The electronic configuration is $1s^2\ 2s^2\ 2p^6\ 3s^2\ 3p^6\ 3d^{10}\ 4s^2\ 4p^6$.

12. The electronic configuration of the ion that has a single positive charge, $x^+$, is $1s^2\ 2s^2\ 2p^6$.

13. The electronic configuration of the ion that has a single positive charge, $x^+$, is $1s^2\ 2s^2\ 2p^6\ 3s^2\ 3p^6\ 3d^{10}$.

14. The quantum numbers of the last electron added according to the aufbau method are $n = 2$, $l = 1$, $m = 0$, and $s = +\frac{1}{2}$.

15. The quantum numbers of the last electron added according to the aufbau method are $n = 3$, $l = 1$, $m = 0$, and $s = -\frac{1}{2}$.

16. The quantum numbers of the last electron added according to the aufbau method are $n = 4$, $l = 2$, $m = 0$, and $s = -\frac{1}{2}$.

V.  Write the electronic configuration of the following:

1. Ca, calcium

2. $Zn^{2+}$, the zinc(II) ion

3. $Cs^+$, the cesium ion

4. Kr, krypton

5. $S^{2-}$, the sulfur anion

VI. Write the four quantum numbers of the last electron added to the element:

1. O, oxygen

2. K, potassium

3. V, vanadium

_____

_____

_____

4. As, arsenic

5. Ni, nickel

6. Au, gold

VII.   In Chapter 2 of your text you have learned to calculate atomic weights from the masses of isotopes,
to calculate binding energies, and to determine the
energy of electromagnetic radiation from either the
wavelength or frequency of that radiation.  Work
the following problems to test your expertise:

1. Boron consists of two naturally occurring isotopes.  One isotope, $^{10}B$, has a mass of 10.01294 u
and the other, $^{11}B$, has a mass of 11.00931 u.
If the atomic weight of boron is 10.811, what
percent of each of the two isotopes is naturally
occurring boron?

2. Chlorine consists of two naturally occurring
isotopes, $^{35}Cl$ and $^{37}Cl$.  The $^{35}Cl$ isotope
is more abundant than the $^{37}Cl$ isotope
(75.53% vs. 24.47%).  Calculate the approximate
atomic weight of naturally occurring chlorine.
Assume that the mass of $^{35}Cl$ is 35.00 u and
that of $^{37}Cl$ is 37.00 u.

3. Radio and TV antennae are designed so that the
length of a crossbar is approximately equal to
the wavelength of the signal received.  If a
crossbar is 1 m long, what frequency in megahertz is it designed to receive?

4. Which has the higher energy:  a 1.0 nm X ray or
a $3.0 \times 10^{10}$ Hz microwave?

ANSWERS TO
EXERCISES

I.  Contributions of scientists

1. i. 2. o. 3. k. 4. c. 5. f. 6. g. 7. e. 8. a.
9. h. 10. b. 11. d. 12. j. 13. m. 14. n. 15. l.

II.  Names and symbols of elements

After checking your answers, write the name of the
element next to the symbol:

1. Ag _____

2. H _____

3. Na _____

4. Ca _____

5. K _____

6. B _____

7. F _____

8. Cu _____

9. Au _____

10. N _____

11. Br _____

12. Sb _____

13. Cr _____

14. Zn _____

15. As _____

16. Sn _____

17. Cl _____

18. He _____

19. W _____

20. Pb _____

21. Ba _____

22. Si _____

23. Mg _____

24. S _____

25. Hg _____

26. Al _____

## III. Principles of atomic structure

1. False [2.10]    Energy, $E$, is directly proportional to frequency, $\nu$.

$$E = h\nu$$

The constant of proportionality is called Planck's constant, $h$. Energy, however, is inversely proportional to wavelength, $\lambda$, since

$$E = hc/\lambda$$

in which $c$ is the speed of light.

2. True [2.7]

3. False [2.7]    The atomic number, $Z$, is the total number of protons in the nucleus of an atom.

4. False [2.8]    Isotopes have different atomic weights but the same atomic number. Isobars have the same mass number but different atomic numbers.

5. False [2.2]    The fundamental particles are the proton, the neutron, and the electron. The nucleus consists of protons and, except for hydrogen, neutrons.

6. True [2.3, 2.4]

7. False [2.5]        The word *neutron* is thought to be derived from the same word as *neutral*.  The neutron has no charge.

8. False [2.6]        The hydrogen nucleus contains only a proton.  All other nuclei contain both neutrons and protons.

9. False [2.6]        Although the masses of isotopes differ, the chemical reactivities are the same.  Isotopes are atoms of the same element.

10. True [2.9]

11. False [2.16]      In diamagnetic materials all electrons are paired.

12. False [2.16]      The number of electrons that are unpaired is a maximum, as stated in Hund's rule of maximum multiplicity.

13. True [2.16]

14. True [2.10]       The relationship between frequency and wavelength is

$$\nu = c/\lambda$$

in which $\nu$ is the frequency in hertz or $\sec^{-1}$; $\lambda$ is the wavelength in cm; and $c$ is the speed of light in a vacuum, in units of cm $\sec^{-1}$.

15. True [2.11]

16. False [2.15]      If $n = 1$, $l$ can only be zero.

17. True [2.15]       For $n = 2$, the following sets of quantum numbers are possible:

$$l = 0, \quad m = 0, \quad s = +\tfrac{1}{2} \text{ or } -\tfrac{1}{2}$$
$$l = 1, \quad m = +1, \quad s = +\tfrac{1}{2} \text{ or } -\tfrac{1}{2}$$
$$l = 1, \quad m = 0, \quad s = +\tfrac{1}{2} \text{ or } -\tfrac{1}{2}$$
$$l = 1, \quad m = -1, \quad s = +\tfrac{1}{2} \text{ or } -\tfrac{1}{2}$$

In general, for any value of $n$ the possible values of $l$ and $m$ are

$$l = (n - 1), (n - 2), \ldots, 0$$
$$m = +1, +(l - 1), \ldots, 0, \ldots, (l - 1), -l$$

18. True [2.15]

19. True [2.15]

20. False [ 2.17 ]   A maximum of 10 electrons can be added to a *d* subshell.  The *d* orbitals are being filled in each
series of transition elements.  The possible quantum
numbers of electrons in *d* orbitals are

$$l = 2 \quad m = +2, +1, 0, -1, \text{ or } -2 \quad s = +\tfrac{1}{2} \text{ or } -\tfrac{1}{2}$$

21. True [ 2.19]

IV.   Identification of elements

1. Al, aluminum
   [ 2.7 ]

2. As, arsenic
   [ 2.9]

3. Sn, tin
   [ 2.7]   The symbol of atomic number is Z.

4. Ag, silver
   [ 2.7]   The atomic number is obtained by subtracting the
number of neutrons from the mass number, 109 - 62 =
47.  The atomic number is 47.

5. S, sulfur
   [ 2.7]   The number of electrons in a neutral atom equals the
number of protons in that atom; therefore, Z = 16.

6. B, boron
   [ 2.9]   Boron is the only element that has an atomic weight
lying between 10.013 u and 11.009 u.

7. Cl, chlorine
   [ 2.7, 2.17]   The ion has one more electron than the neutral atom;
therefore, the neutral atom has 17 electrons, Z = 17.

8. Cr, chromium
   [ 2.7, 2.17]   The ion has a 2+ charge, i.e., two electrons, each
of which has a negative charge, have been removed.
The neutral atom should have 24 electrons, Z = 24.

9. N, nitrogen
   [ 2.7, 2.17]   Each superscript indicates the number of electrons
in the subshell.  The 1s subshell has 2 electrons,
the 2s subshell has 2 electrons, and the 2p subshell
has 3 electrons.  A total of 7 electrons indicates
an atom with Z = 7, nitrogen.  Check the periodic
table in the back of the study guide.

10. Ni, nickel
    [ 2.7, 2.17]   The total number of electrons in the atom is 28;
Z = 28.  Observe the position of nickel in the periodic table of Figure 2.1 of the study guide.  The

$1s^2$
$2s^2$  $2p^6$
$3s^2$  $3p^6$
$4s^2$  $3d^8$

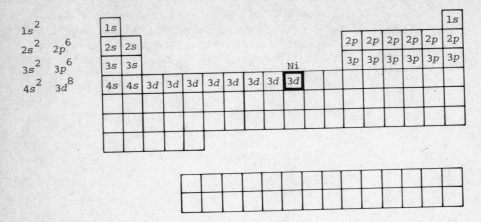

Figure 2.1 Periodic table

order of filling the atomic orbitals is shown up to Z = 28, and the corresponding spectroscopic notation is given to the left of the table.

11. Kr, krypton
    [2.7]

The sum of the superscripts of the electronic configuration is 36; Z = 36.

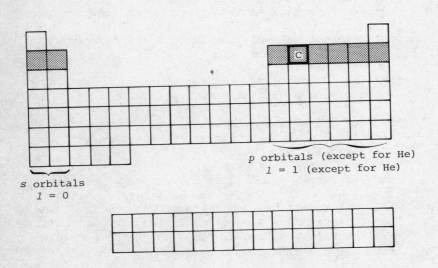

p orbitals (except for He)
l = 1 (except for He)

s orbitals
l = 0

Figure 2.2 Periodic table

Table 2.1 Order of filling the *p* orbitals

| ORDER | *m* | *s* | ELEMENT | | |
|---|---|---|---|---|---|
| | | | *n* = 2, *l* = 1 | *n* = 3, *l* = 1 | *n* = 4, *l* = 1 |
| 1 | +1 | +½ | B | Al | Ga |
| 2 | 0 | +½ | C | Si | Ge |
| 3 | −1 | +½ | N | P | As |
| 4 | +1 | −½ | O | S | Se |
| 5 | 0 | −½ | F | Cl | Br |
| 6 | −1 | −½ | Ne | At | Kr |

12. Na, sodium
    [ 2.7 ]                    $Z = 10$

13. Cu, copper
    [ 2.7 ]                    $Z = 29$

14. C, carbon          Since *n*, the principal quantum number, is 2, the
    [ 2.15 ]           element must be in the shaded area of the periodic
                       table in Figure 2.2 of the study guide.  This shaded
                       area is the second row of the periodic table.  Since
                       *l* = 1, the element is one in which the *p* subshell is
                       being filled.  The convention used in your text sug
                       gests that the *p* orbitals are filled in the order
                       given in Table 2.1 of the study guide.

15. Cl, chlorine       Elements in which the last electron is added to the
    [ 2.15 ]           *n* = 3 shell are indicated by the shaded area of the
                       periodic table of Figure 2.3 of the study guide.

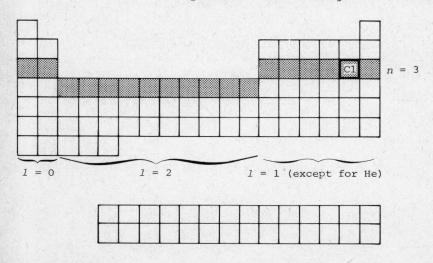

Figure 2.3 Periodic table

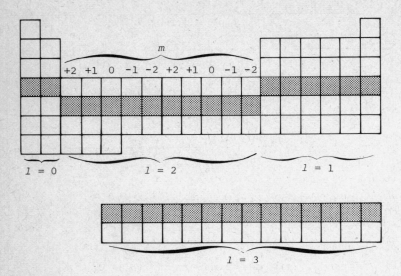

Figure 2.4   Periodic table

16. Pd, palladium
    [2.15]

Elements in which the last electron is added to the $n = 4$ shell are indicated by the shaded area of the periodic table of Figure 2.4 of the study guide. The convention used in your text suggests that the $d$ orbitals ($l = 2$) are filled in the order given in Table 2.2 of the study guide.

Table 2.2 Order of filling the $d$ orbitals

| ORDER | $m$ | $s$ | ELEMENT | |
| | | | $n=4,\ l=2$ | $n=3,\ l=2$ |
|---|---|---|---|---|
| 1 | +2 | $+\frac{1}{2}$ | Y | Sc |
| 2 | +1 | $+\frac{1}{2}$ | Zr | Ti |
| 3 | 0 | $+\frac{1}{2}$ | Nb | V |
| 4 | -1 | $+\frac{1}{2}$ | Mo | Cr |
| 5 | -2 | $+\frac{1}{2}$ | Tc | Mn |
| 6 | +2 | $-\frac{1}{2}$ | Ru | Fe |
| 7 | +1 | $-\frac{1}{2}$ | Rh | Co |
| 8 | 0 | $-\frac{1}{2}$ | Pd | Ni |
| 9 | -1 | $-\frac{1}{2}$ | Ag | Cu |
| 10 | -2 | $-\frac{1}{2}$ | Cd | Zn |

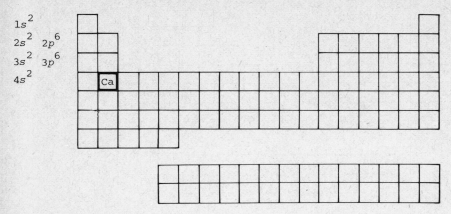

$1s^2$
$2s^2$ $2p^6$
$3s^2$ $3p^6$
$4s^2$

Figure 2.5   Periodic table

### V. Electronic configurations

1. Ca:
$1s^2$ $2s^2$ $2p^6$ $3s^2$
$3p^6$ $4s^2$

Reading the periodic table in Figure 2.5 of the study guide from left to right, we note that the electronic configuration of Ca is $1s^2$ $2s^2$ $2p^6$ $3s^2$ $3p^6$ $4s^2$. The arrangement of the periodic table corresponds to the aufbau filling order.

2. $Zn^{2+}$:
$1s^2$ $2s^2$ $2p^6$ $3s^2$
$3p^6$ $3d^{10}$

For $Zn^{2+}$ write the electronic configuration of elemental zinc first. Always write a configuration in order of increasing principal quantum number; be-- gin with the s subshell of a shell, then write the p subshell, the d subshell, and finally the f subshell of that shell. The configuration of Zn is $1s^2$ $2s^2$ $2p^6$ $3s^2$ $3p^6$ $3d^{10}$ $4s^2$ (see Figure 2.6 of the study guide).

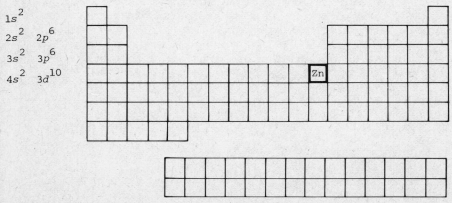

$1s^2$
$2s^2$ $2p^6$
$3s^2$ $3p^6$
$4s^2$ $3d^{10}$

Figure 2.6   Periodic table

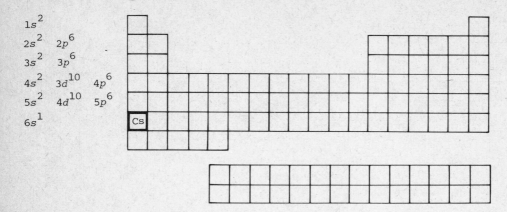

$1s^2$

$2s^2$  $2p^6$

$3s^2$  $3p^6$

$4s^2$  $3d^{10}$  $4p^6$

$5s^2$  $4d^{10}$  $5p^6$

$6s^1$

Cs

Figure 2.7  Periodic table

Form the ion by removing electrons from the subshells of the outermost shell. These subshells are the ones on the extreme right in the electronic configuration. The electronic configuration of $Zn^{2+}$ is $1s^2\ 2s^2\ 2p^6\ 3s^2\ 3p^6\ 3d^{10}$. Notice that the aufbau order of filling orbitals is not the same as the order of removing electrons for ionization. Remember that the aufbau method is only a convention. Electron addition is but a small part of the total change across a period from left to right: the elements are changing, protons and neutrons are being added to the nucleus, and Z is increasing with each addition of a proton. In the formation of an ion, on the other hand, only electrons are added to or subtracted from the atom, the element remains the same, the nucleus is not changed, and Z is not changed. *The order of adding or subtracting electrons to form ions is not and should not be expected to be the same as the aufbau order.* (See Section 2.17 of your text.)

3. $Cs^+$:
$1s^2\ 2s^2\ 2p^6\ 3s^2\ 3p^6$
$3d^{10}\ 4s^2\ 4p^6\ 4d^{10}$
$5s^2\ 5p^6$

The configuration of Cs is $1s^2\ 2s^2\ 2p^6\ 3s^2\ 3p^6\ 3d^{10}$ $4s^2\ 4p^6\ 4d^{10}\ 5s^2\ 5p^6\ 6s^1$ and that of the $Cs^+$ is $1s^2$ $2s^2\ 2p^6\ 3s^2\ 3p^6\ 3d^{10}\ 4s^2\ 4p^6\ 4d^{10}\ 5s^2\ 5p^6$. (See Figure 2.7 of the study guide.)

4. Kr:
$1s^2\ 2s^2\ 2p^6\ 3s^2\ 3p^6$
$3d^{10}\ 4s^2\ 4p^6$

5. $S^{2-}$:
$1s^2\ 2s^2\ 2p^6\ 3s^2\ 3p^6$

The electronic configuration of S is $1s^2\ 2s^2\ 2p^6\ 3s^2$ $3p^4$ and that of $S^{2-}$ is $1s^2\ 2s^2\ 2p^6\ 3s^2\ 3p^6$.

## VI. Quantum numbers

1. O:
$n = 2\ l = 1$
$m = +1\ s = -\frac{1}{2}$

The last electron added is the fourth one in the $2p$ subshell. Therefore, $n = 2$ and $l = 1$. The quantum numbers $m$ and $s$ by the convention used in your text are $+1$ and $-\frac{1}{2}$ respectively (see Table 2.1 of the study guide). The set of quantum numbers that corresponds to the last electron added is

$$n = 2, \quad l = 1, \quad m = +1, \quad s = -\tfrac{1}{2}$$

The periodic table of Figure 2.8 of the study guide may be helpful in the determination of the quantum numbers of the last electron added in the building of elements when the aufbau method is used.

2. K:
$n = 4\ l = 0\ m = 0\ s = +\frac{1}{2}$

3. V:
$n = 3\ l = 2\ m = 0\ s = +\frac{1}{2}$

4. As:
$n = 4\ l = 1\ m = -1\ s = +\frac{1}{2}$

5. Ni:
$n = 3\ l = 2\ m = 0\ s = -\frac{1}{2}$

6. Au:
$n = 5\ l = 2\ m = -1\ s = -\frac{1}{2}$

## VII. Mathematical problems

1. 19.903% $^{10}B$
   80.097% $^{11}B$
   [ 2.9 ]

Set $x$ = fraction of $^{10}B$, the isotope with a mass of 10.01294 u. Then $1 - x$ = fraction of $^{11}B$, the isotope with a mass of 11.00931 u. The atomic weight of naturally occurring boron is the weighted average of these isotopes; therefore,

$$10.01294(x) + 11.00931(1 - x) = 10.811$$

Multiplying terms, we obtain

$$10.01294x + 11.00931 - 11.00931x = 10.811$$

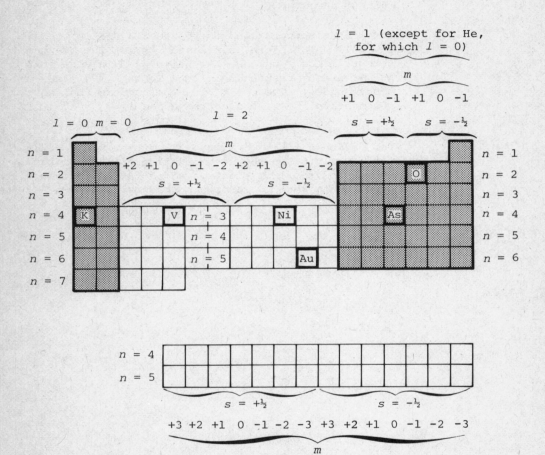

Figure 2.8  Periodic table

Combining terms containing $x$ and solving for $x$, we obtain

$$0.99637x = 0.19831$$

$$x = \frac{0.19831}{0.99637}$$

$$x = 0.19903$$

and thus

$$1 - x = 0.80097$$

Since a percentage is 100 times a corresponding fraction, the values expressed as percentages are 19.903% and 80.097%. Thus, naturally occurring boron is a mixture composed of 19.903% $^{10}$B and 80.097% $^{11}$B.

2. 35.49 u
   [ 2.9 ]

Using the information given in the problem, we obtain an answer of 35.49 u:

$$0.7553 \, (35.00 \text{ u}) + 0.2447 \, (37.00 \text{ u}) = 35.49 \text{ u}$$

This answer is very close to the actual atomic weight, 35.45 u.

3. $3 \times 10^2$ Mhz
   [ 2.10 ]

Since frequency is inversely proportional to wavelength, $\lambda$, and directly proportional to the velocity of light, $c$, we substitute values into the equation

$$\nu = c/\lambda$$

and find

$$\nu = (2.9979 \times 10^{10} \text{ cm sec}^{-1})/100 \text{ cm}$$
$$\nu = (3.0 \times 10^{10} \text{ cm sec}^{-1})/ \, (1 \times 10^2 \text{ cm})$$

Notice that units are included with all numbers and that the number of significant figures is adjusted according to the procedure outlined in Chapter 1 of your text. The frequency can be expressed in reciprocal seconds since the unit of length, the centimeter, can be canceled:

$$\nu = (3.0 \times 10^{10} \, \cancel{\text{cm}} \text{ sec}^{-1})/(1 \times 10^2 \, \cancel{\text{cm}})$$
$$\nu = 3 \times 10^8 \text{ sec}^{-1}$$

Notice that the frequency is expressed in one significant figure because the calculation involves a value, 100 cm, that has only one significant figure. Since a reciprocal second is a hertz and $10^6$ hertz is a megahertz, the frequency can be expressed as $3 \times 10^2$ Mhz:

$$\nu = (3 \times 10^8 \, \cancel{\text{hz}}) \, (1 \text{ Mhz}/10^6 \, \cancel{\text{hz}})$$
$$\nu = 3 \times 10^2 \text{ Mhz}$$

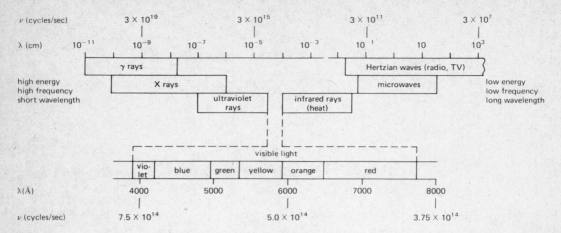

Figure 2.9   Electromagnetic radiation. Note that the approximate ranges of electromagnetic radiations are plotted on a logarithmic scale in the upper part of the diagram; the spectrum of visible light is not plotted in this way.

4. 1.0 nm X ray       See Figure 2.9 of the study guide.
   [2.10 ]            Combining

$$E = h\nu \qquad \text{and} \qquad \nu = \frac{c}{\lambda}$$

we obtain

$$E = \frac{hc}{\lambda}$$

Substituting values into the equation, we find that for the X ray

$$E = \frac{(6.63 \times 10^{-34} \text{ J sec}) \quad (3.0 \times 10^{10} \text{ cm sec}^{-1})}{1.0 \times 10^{-7} \text{ cm}} = 2.0 \times 10^{-16} \text{ J}$$

The energy of the microwave is calculated from the equation

$$E = h\nu$$

Substituting values into the preceding equation, we find that for the microwave

$$E = (6.63 \times 10^{-34} \text{ J sec})(3.0 \times 10^{10} \text{ sec}^{-1}) = 2.0 \times 10^{-23} \text{ J}$$

Thus, the X ray has the higher energy.

SELF-TEST             Complete the test in 45 minutes:

I.  Fill in the blank space with the word that most ap-
    propriately completes the statement or answers the
    question:

    1. _____ equals Planck's constant, $h$, times
       the frequency, $\nu$, of the electromagnetic radia-
       ion.

    2. The fundamental subatomic particles are the
       _____, _____, and _____.

    3. The fundamental subatomic particle that has no
       charge is the _____.

    4. Atoms with the same atomic number but different
       mass numbers are called _____.

    5. The maximum number of electrons that can exist in
       the $M$ shell of any atom is _____.

    6. A material is said to exhibit what type of mag-
       netic behavior if all electrons are paired?
       _____

    7. The electronic configuration of lithium is
       _____ and the value of $Z$ of the element
       is _____.

II. Do the following:

    1. Define the atomic mass unit.

    2. Write the quantum numbers of the last electron
       that is added to iridium, Ir, according to the
       Aufbau method.  Use a periodic table.

    3. Write the electronic configuration of Ir and that
       of $Ir^{2+}$.

    4. Determine the atomic weight of lithium.  Lithium
       is a mixture of two naturally occurring isotopes;
       7.40% of the mixture is $^6Li$, an isotope that has
       a mass of 6.0169 u, and 92.60% of the mixture is
       $^7Li$, an isotope that has a mass of 7.0182 u.

III. Answer each of the following:

_____   1. The differentiating electron of a sulfur atom has
which of the following sets of quantum numbers?
a. $n = 3$,   $l = 1$,   $m = +1$,   $s = +\frac{1}{2}$
b. $n = 3$,   $l = 1$,   $m = -1$,   $s = +\frac{1}{2}$
c. $n = 3$,   $l = 1$,   $m = +1$,   $s = -\frac{1}{2}$
d. $n = 3$,   $l = 1$,   $m = -1$,   $s = -\frac{1}{2}$

_____   2. Which of the following sets of quantum numbers
represents an impossible arrangement?
a. $n = 2$,   $l = 0$,   $m = 0$,   $s = -\frac{1}{2}$
b. $n = 7$,   $l = 4$,   $m = -1$,   $s = +\frac{1}{2}$
c. $n = 3$,   $l = 1$,   $m = +1$,   $s = -\frac{1}{2}$
d. $n = 3$,   $l = -1$,   $m = -1$,   $s = -\frac{1}{2}$

_____   3. The success of J.J. Thomson's experiment to meas-
ure the charge to mass ratio of the electron
depended upon
a. the behavior of an electron in an electrical
field.
b. the behavior of an electron in a magnetic
field.
c. the ability to detect the exact point of
impact of a stream of electrons with matter.
d. all of the preceding.

_____   4. The discovery and characterization of the neutron
occurred significantly later than that of either
the proton or the electron.  This can be attri-
buted to the fact that
a. the mass of the neutron is so similar to that
of the proton that the two particles are
essentially indistinguishable.
b. neutrons are buried deep within the nucleus
and are not easily accessible.
c. neutrons are uncharged particles.
d. neutrons have a very short mean lifetime.

_____   5. In which of the following sets do the nuclides
have an isotopic relationship to one another?

a. nuclide I   (7 protons, 6 neutrons)
   nuclide II  (6 protons, 7 neutrons)
b. nuclide I   (7 protons, 7 electrons)
   nuclide II  (6 protons, 6 electrons)
c. nuclide I   (7 protons, 6 neutrons)
   nuclide II  (7 protons, 7 neutrons)
d. nuclide I   (7 protons, 6 neutrons)
   nuclide II  (6 protons, 6 neutrons)

6. In which of the sets listed in problem 5 of the self-test are the nuclides isobars?

7. The wave mechanical approach to atomic structure permits the calculation of
   a. a volume about the nucleus in which an electron of specified energy will most probably be found.
   b. the most probable radius of an orbit that an electron of specified energy will follow.
   c. the most probable position of an electron of specified energy at a given time, $t$.
   d. the most probable spin value that will be associated with an electron of specified energy.

8. As electrons move from a ground state to an excited state,
   a. an emission spectrum results.
   b. energy is absorbed.
   c. heat is liberated.
   d. light is emitted.

9. The intensity of a spectral line observed in an atomic emission spectrum can be directly related to
   a. the difference in energy of the energy levels involved in the electron transition that gives rise to the line.
   b. the number of electrons undergoing the transition that gives rise to the line.
   c. the number of energy levels involved in the transition that gives rise to the line.
   d. the speed with which an electron undergoes a transition from one energy level to another.

_____ 10. The wavelength at which a spectral line is observed in an atomic emission spectrum is inversely related to
    a. the difference in energy of the energy levels involved in the electron transition that gives rise to the line.
    b. the number of electrons undergoing the transition that gives rise to the line.
    c. the number of energy levels involved in the transition that gives rise to the line.
    d. the speed with which an electron undergoes a transition from one energy level to another.

_____ 11. The quantum number that designates the spin of an electron is
    a. $n$   b. $l$   c. $m$   d. $s$

_____ 12. How do chemists know that the atomic weight of the $^{12}C$ nuclide is exactly 12.000?
    a. The $^{12}C$ nuclide weighs exactly 12 times as much as hydrogen.
    b. $6.02 \times 10^{23}$ atoms of $^{12}C$ weigh exactly 12 grams.
    c. The $^{12}C$ nuclide is composed of 6 protons and 6 neutrons, each weighing one unit.
    d. Chemists define it so.

_____ 13. Which of the following is the most paramagnetic?
    a. $K^+$   b. $Zn^{2+}$   c. $Cu^{2+}$   d. $Fe^{3+}$

_____ 14. Scandium
    a. is a transition element.
    b. has an atomic number of 45.
    c. has 21 electrons in the $Sc^{2+}$ ion.
    d. has a completely filled $n = 3$ subshell.

_____ 15. Which of the following is diamagnetic?
    a. K   b. $Na^+$   c. $Co^{2+}$   d. P

# 3

# Chemical Bonding

OBJECTIVES

(a) You should be able to demonstrate your knowledge
of the following terms by defining them, describing them, or giving specific examples of them:

anion [3.4]
atomic radius [3.1]
bond energy [3.9]
cation [3.4]
covalent bond [3.6]
dipole moment [3.8]
electron affinity [3.3]
electronegativity [3.9]
electrovalence number [3.4]
formal charge [3.7]

ionic charge [3.4]
ionic radius [3.5]
ionization potential [3.2]
isoelectronic [3.4]
lanthanide contraction [3.1]
lattice energy [3.4]
Lewis structure [3.6]
nomenclature [3.11]
oxidation number [3.10]

(b) Given the names or chemical formulas of common compounds, you should be able to write the corresponding chemical formulas or names.

(c) Using the periodic table of the elements, you should be able to predict relative sizes of atoms and ions and relative magnitudes of ionization potentials and electronegativities.

(d) You should be able to predict the common oxidation numbers of elements and to determine oxidation numbers of these elements in polyatomic molecules and ions.

(e) You should be able to draw Lewis structures of molecules and ions and to determine the formal charge of each atom in such structures.

(f) You should be able to predict which bonds in a given set have the most covalent character.

(g) You should be able to calculate the partial ionic character of a bond from dipole moment data and bond distance data. You should also be able to determine the polarity of a bond from the electronegativities of the elements that are bonded.

EXERCISES

I.  Write the formula of each of the following compounds in the space provided:

_____ 1. potassium sulfide      _____ 7. beryllium fluoride
_____ 2. sodium chloride        _____ 8. sodium sulfide
_____ 3. magnesium sulfide      _____ 9. ammonium iodide
_____ 4. calcium oxide          _____ 10. lithium hydroxide
_____ 5. barium sulfide         _____ 11. cesium iodide
_____ 6. strontium iodide       _____ 12. sodium cyanide

_____ 13. potassium dichromate
_____ 14. sodium bromide
_____ 15. calcium carbonate
_____ 16. potassium phosphate
_____ 17. sodium permanganate
_____ 18. hydrobromic acid
_____ 19. hypobromous acid
_____ 20. bromous acid
_____ 21. bromic acid
_____ 22. perbromic acid
_____ 23. potassium perbromate
_____ 24. barium sulfate
_____ 25. sodium nitride
_____ 26. carbon disulfide
_____ 27. diiodine pentoxide
_____ 28. nitrogen oxide
_____ 29. phosphorus
                    trifluoride
_____ 30. tetraarsenic
                    tetrasulfide
_____ 31. iodine trifluoride
_____ 32. dichlorine oxide
_____ 33. carbon dioxide
_____ 34. sulfur dioxide

_____ 35. dinitrogen
                    pentoxide
_____ 36. tin(II) bromide
_____ 37. tin(IV) bromide
_____ 38. mercury(I) sulfide
_____ 39. gold(III) chloride
_____ 40. diantimony trioxide
_____ 41. ferrous sulfate
_____ 42. ferric sulfate
_____ 43. copper(II) dihy-
                    drogen phosphate
_____ 44. titanium(IV)
                    chloride
_____ 45. tin(II) oxide
_____ 46. stannic oxide
_____ 47. calcium hydride
_____ 48. sodium peroxide
_____ 49. calcium phosphide
_____ 50. sodium monohydro-
                    gen phosphite
_____ 51. xenon hexafluoride
_____ 52. sodium bisulfate
_____ 53. barium arsenate
_____ 54. potassium arsenite

II. Write the name of each of the following compounds
    in the space provided:

_____ 1. $NF_3$

_____ 2. $NaCl$

_____ 3. $NaClO$

_____ 4. $NaClO_2$

_____ 5. $NaClO_3$

_____ 6. $NaClO_4$

_____ 7. $CCl_4$

_____ 8. $CaCrO_4$

_____ 9. $ZnHPO_4$

_____ 10. $Sn_3(PO_4)_2$

_____ 11. $N_2F_4$

_____ 12. $Cu(NO_3)_2$

_____ 13. $Ba_3N_2$

_____ 14. $TiCl_2$

_____ 15. $NH_3$

_____ 16. $MnO_2$

_____ 17. $N_2O_5$

_____ 18. $BaCr_2O_7$

III. Answer the following:

1. Which has the larger radius?

_____  a. Mg or Si
_____  b. Li or Cs
_____  c. Ni or Zn

_____  d. Ti or Cr
_____  e. Mg or $Mg^{2+}$
_____  f. Cl or $Cl^-$

2. Which has the larger electronegativity?

_____  a. K or Br

_____  b. O or Te

3. Which has the larger first ionization potential?

_____  a. Ba or Bi
_____  b. Al or Tl
_____  c. C or O

_____  d. Br or Kr
_____  e. P or O

4. Which has the larger first electron affinity?

_____  a. B or F

_____  b. Cl or I

5. Which bond is more polar?

_____  a. S—H or Se—H

_____  b. S—F or Se—F

6. Using the data in Table 3.1 of the study guide, predict whether a bond between the elements of each of the following pairs would be ionic or covalent. If you predict the bond to be covalent, estimate the degree of polarity of the bond.

_____  a. C, I
_____  b. Al, Cl
_____  c. K, Cl

_____  d. S, Cl
_____  e. Cs, S
_____  f. Ca, O

7. Using the concept of anion deformation, predict which one of the two compounds in each of the following pairs has the bond with the greater amount of covalent character:

_____  a. $SbCl_3$ or $BiCl_3$

_____  b. $CrCl_2$ or $CrCl_3$

_____  c. $HgCl_2$ or $HgI_2$

_____  d. BeO or BeS

_____  e. ZnS or CdS

_____  f. $Tl_2O$ or $Tl_2O_3$

_____  g. MgO or CO

_____  h. $SO_2$ or $SeO_2$

Table 3.1   Electronegativities of the elements[a]

| 1<br>H<br>2.1 | | | | | | | | | | | | | | | | | 2<br>He<br>— |
|---|---|---|---|---|---|---|---|---|---|---|---|---|---|---|---|---|---|
| 3<br>Li<br>1.0 | 4<br>Be<br>1.5 | | | | | | | | | | | 5<br>B<br>2.0 | 6<br>C<br>2.5 | 7<br>N<br>3.0 | 8<br>O<br>3.5 | 9<br>F<br>4.0 | 10<br>Ne<br>— |
| 11<br>Na<br>0.9 | 12<br>Mg<br>1.2 | | | | | | | | | | | 13<br>Al<br>1.5 | 14<br>Si<br>1.8 | 15<br>P<br>2.1 | 16<br>S<br>2.5 | 17<br>Cl<br>3.0 | 18<br>Ar<br>— |
| 19<br>K<br>0.8 | 20<br>Ca<br>1.0 | 21<br>Sc<br>1.3 | 22<br>Ti<br>1.5 | 23<br>V<br>1.6 | 24<br>Cr<br>1.6 | 25<br>Mn<br>1.5 | 26<br>Fe<br>1.8 | 27<br>Co<br>1.8 | 28<br>Ni<br>1.8 | 29<br>Cu<br>1.9 | 30<br>Zn<br>1.6 | 31<br>Ga<br>1.6 | 32<br>Ge<br>1.8 | 33<br>As<br>2.0 | 34<br>Se<br>2.4 | 35<br>Br<br>2.8 | 36<br>Kr<br>— |
| 37<br>Rb<br>0.8 | 38<br>Sr<br>1.0 | 39<br>Y<br>1.2 | 40<br>Zr<br>1.4 | 41<br>Nb<br>1.6 | 42<br>Mo<br>1.8 | 43<br>Tc<br>1.9 | 44<br>Ru<br>2.2 | 45<br>Rh<br>2.2 | 46<br>Pd<br>2.2 | 47<br>Ag<br>1.9 | 48<br>Cd<br>1.7 | 49<br>In<br>1.7 | 50<br>Sn<br>1.8 | 51<br>Sb<br>1.9 | 52<br>Te<br>2.1 | 53<br>I<br>2.5 | 54<br>Xe<br>— |
| 55<br>Cs<br>0.7 | 56<br>Ba<br>0.9 | 57–71<br>La–Lu<br>1.1–1.2 | 72<br>Hf<br>1.3 | 73<br>Ta<br>1.5 | 74<br>W<br>1.7 | 75<br>Re<br>1.9 | 76<br>Os<br>2.2 | 77<br>Ir<br>2.2 | 78<br>Pt<br>2.2 | 79<br>Au<br>2.4 | 80<br>Hg<br>1.9 | 81<br>Tl<br>1.8 | 82<br>Pb<br>1.8 | 83<br>Bi<br>1.9 | 84<br>Po<br>2.0 | 85<br>At<br>2.2 | 86<br>Rn<br>— |
| 87<br>Fr<br>0.7 | 88<br>Ra<br>0.9 | 89–<br>Ac–<br>1.1–1.7 | | | | | | | | | | | | | | | |

[a] Based on Linus Pauling, *The Nature of the Chemical Bond, Third Edition.* © 1960 by Cornell University Press. Used with the permission of Cornell University Press.

IV. Answer each of the following with *true* or *false*. If a statement is false, correct it.

_____  1. An atom of any Group V A element has five valence electrons.

_____  2. The oxidation number of each of the alkali metal ions is 2+.

_____  3. In all stable compounds there is a full octet of valence electrons surrounding each atom.

_____  4. The ionization potential of a noble gas is very high compared to that of any of the other elements in the same period.

_____  5. The electron affinity of a halogen is rather low compared to that of any of the other elements in the same period.

_____    6. Monatomic anions are named by replacing the
                      last part of the element's name with "ide."

_____    7. Ionization energy is the amount of energy re-
                      quired to add an electron to an isolated gase-
                      ous atom in its ground state.

_____    8. The element in the periodic table that has the
                      highest electronegativity is fluorine.

_____    9. $Na^+$ and Ne are isoelectronic species.

_____   10. Purely ionic bonds are formed by two atoms
                      sharing electrons.

_____   11. The second ionization potential of sodium should
                      be larger than the second ionization potential
                      of magnesium.

_____   12. The more polar the bond in a diatomic molecule
                      the lower the dipole moment of that molecule.


V. Draw the most probable Lewis structure of each of
   the following and include the formal charges. The
   formulas of the more complex, multi-element molecules are written to indicate the general atomic
   arrangement of the molecule. Thus, $N_2F_2$ is written
   as FNNF to show that the two nitrogen atoms are
   joined by at least one chemical bond and that each
   nitrogen atom forms at least one bond with a single
   fluorine atom.

_____  1. $O_2$                    _____  8. $C_2^{2-}$

_____  2. $O_3$                    _____  9. $BF_4^-$

_____  3. $CH_4$                   _____ 10. $NH_4^+$

_____  4. $N_3^-$                  _____ 11. $NO_3^-$

_____  5. $OF_2$                   _____ 12. $SO_4^{2-}$

_____  6. $H_2PO_2^-$              _____ 13. $N_2$

_____  7. FNNF                     _____ 14. $ClO^-$

_____  15. NNO                  _____  20. $SOF_2$

_____  16. $O_2SF_2$            _____  21. $O_2NONO_2$

_____  17. $H_2NNH_2$           _____  22. $ONNO^{2-}$

_____  18. $SnCl_4$             _____  23. $H_3COCH_3$

_____  19. $OH^-$               _____  24. $H_2CO$


VI. What is the oxidation number of

_____  1. B in $BF_4^-$        _____  7. Sn in $SnCl_4$

_____  2. N in $NH_4^+$        _____  8. Cr in $Cr_2O_7^{2-}$

_____  3. N in $NO_3^-$        _____  9. Cr in $CrO_4^{2-}$

_____  4. P in $H_2PO_2^-$     _____  10. Mn in $MnO_2$

_____  5. P in $PO_4^{3-}$     _____  11. Mn in $MnO_4^-$

_____  6. Cl in $ClO_4^-$      _____  12. Cl in $Cl_2$


ANSWERS TO
EXERCISES
                I. Formulas of compounds

1. $K_2S$         Each element of Group I A (see Figure 3.1 of the
2. NaCl           study guide for this and subsequent exercises)
                  has one electron in its valence shell and an oxi-
                  dation number of 1+ in its compounds.

3. MgS            Each element of Group II A has two electrons in its
4. CaO            valence shell and an oxidation number of 2+ in its
                  compounds.

5. BaS            Each element of Group VI A has 6 electrons in its
                  valence shell and usually an oxidation number of
                  2- in its compounds.

6. $SrI_2$        Each element of Group VII A has 7 electrons in its
7. $BeF_2$        valence shell and usually an oxidation number of
                  1- in its compounds.

8. $Na_2S$

9. $NH_4I$            An ammonium ion has a charge of 1+.

10. LiOH            A hydroxide ion has a charge of 1-.

11. CsI

12. NaCN            A cyanide ion has a charge of 1-.

13. $K_2Cr_2O_7$          A dichromate ion has a charge of 2-.

14. NaBr

15. $CaCO_3$

16. $K_3PO_4$

17. $NaMnO_4$            A permanganate ion has a charge of 1-.

18. HBr

19. HBrO

20. $HBrO_2$            These are the formulas of oxybromine acids. For-
                       mulas for oxyacids of other elements are given in
21. $HBrO_3$            Table 3.2 of the study guide.

22. $HBrO_4$

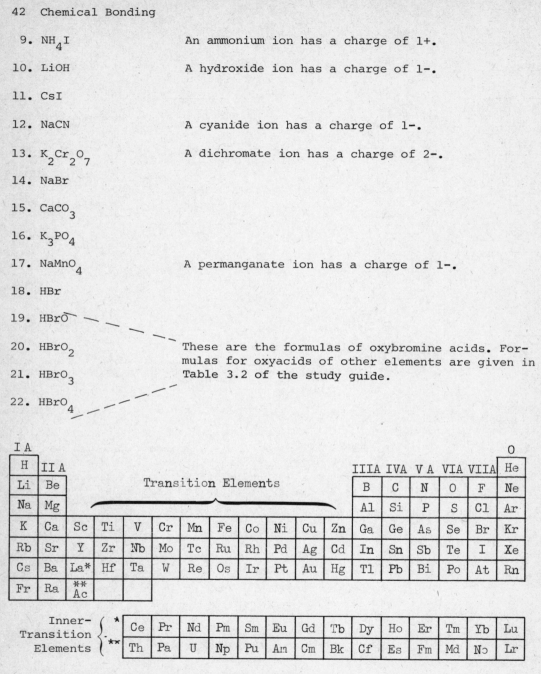

Figure 3.1  Periodic table of elements

23. $KBrO_4$            The perbromate ion has a charge of 1-.

24. $BaSO_4$          The sulfate ion has a charge of 2-.

25. $Na_3N$          In this compound the oxidation number of nitrogen is 3-.

26. $CS_2$

27. $I_2O_5$

Prefixes are used in the names of covalent compounds to indicate the number of atoms of each element in a molecule of that compound. In general, this type of nomenclature is used for any combination of non-metals. The shaded area of the periodic table in Figure 3.2 of the study guide indicates the elements that are nonmetals.

28. $NO$

29. $PF_3$

30. $As_4S_4$

31. $IF_3$

32. $Cl_2O$

33. $CO_2$

34. $SO_2$

35. $N_2O_5$

Table 3.2  Some common oxyacids

| NAME | FORMULA | NAME | FORMULA |
|------|---------|------|---------|
| sulfurous acid | $H_2SO_3$ | nitrous acid | $HNO_2$ |
| sulfuric acid | $H_2SO_4$ | nitric acid | $HNO_3$ |
| hypophosphorous acid | $H_3PO_2$ | hypoiodous acid | $HIO$ |
| phosphorous acid | $H_3PO_3$ | iodic acid | $HIO_3$ |
| phosphoric acid | $H_3PO_4$ | periodic acid | $HIO_4$ |

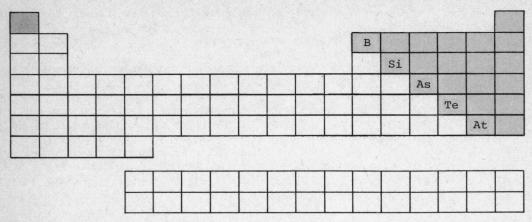

Figure 3.2  Elements that are nonmetals (indicated by shaded area)

36. $SnBr_2$

37. $SnBr_4$

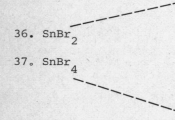

Metals that commonly display more than one oxidation state are indicated by the shaded area in the periodic table of Figure 3.3 of the study guide.  The oxidation number of such an element in a compound can be indicated in the name of that compound by Roman numerals in parentheses after the English name of that element.  Thus, $SnBr_2$ is named tin(II) bromide, and $SnBr_4$ is named tin(IV) bromide.

38. $Hg_2S$

39. $AuCl_3$

40. $Sb_2O_3$

41. $FeSO_4$

42. $Fe_2(SO_4)_3$

43. $Cu(H_2PO_4)_2$

44. $TiCl_4$

45. SnO

46. $SnO_2$

The suffixes *-ous* and *-ic* are often used to indicate the lower and the higher oxidation state of an element respectively.  Names such as this should be memorized.

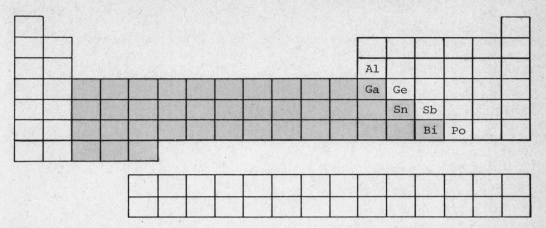

Figure 3.3   Metals that commonly display more than one oxidation state
(indicated by shaded area)

47. $CaH_2$   Hydrogen sometimes has an oxidation number of 1- in its compounds.

48. $NaO$   Oxygen sometimes has an oxidation number of 1- in its compounds.

49. $Ca_3P_2$

50. $Na_2HPO_3$

51. $XeF_6$

52. $NaHSO_4$

53. $Ba_3(AsO_4)_2$

54. $K_3AsO_3$


II. Names of compounds

   The preferred names are given for the compounds.

1. nitrogen
   trifluoride
2. sodium chloride

 3. sodium
    hypochlorite
 4. sodium chlorite
 5. sodium chlorate
 6. sodium perchlorate
 7. carbon tetra-
    chloride
 8. calcium chromate
 9. zinc(II) hydrogen-
    phosphate
10. tin(II) phosphate
11. dinitrogen tetra-
    fluoride
12. copper(II)nitrate
    or cupric nitrate
13. barium nitride
14. titanium(II)
    chloride
15. ammonia
16. manganese(IV)
    oxide
17  dinitrogen pent-
    oxide
18. barium dichromate

The names of anions of other common oxyacids are given in Table 3.3 of the study guide.

This common name is preferred.
The name "manganese dioxide" is frequently used, but such use should be discouraged.

Table 3.3   Anions of some common oxyacids

| NAME | FORMULA | NAME | FORMULA |
|---|---|---|---|
| sulfate ion | $SO_4^{2-}$ | nitrite ion | $NO_2^-$ |
| sulfite ion | $SO_3^{2-}$ | nitrate ion | $NO_3^-$ |
| hypophosphite ion | $PO_2^{3-}$ | hypoiodite ion | $IO^-$ |
| phosphite ion | $PO_3^{3-}$ | iodate ion | $IO_3^-$ |
| phosphate ion | $PO_4^{3-}$ | periodate ion | $IO_4^-$ |

III.    Atomic properties as related to chemical bonding

1. [3.1]  a. Mg      There is generally a **decrease** in atomic radius
                      across a period from left to right.

          b. Cs      There is generally an increase in atomic radius
                      in a group from top to bottom.

          c. Zn      In the transition series inner electrons of the
                      atoms screen the outer electrons from the nuclear
                      charge. Toward the end of the series the atomic ra-
                      dius actually increases with increasing atomic number.

          d. Ti      The screening effect of inner electrons is not so
                      pronounced in the beginning of a transition series,
                      and the atomic radius decreases with increasing
                      atomic number.

          e. Mg      The magnesium ion, $Mg^{2+}$, has the same nuclear
                      charge as the magnesium atom but two fewer elec-
                      trons. The radius of the magnesium ion is therefore
                      less than that of the magnesium atom.

          f. $Cl^-$   The chloride ion, $Cl^-$, has the same nuclear charge
                      as the chlorine atom but one additional electron.
                      The radius of the chloride ion is therefore larger
                      than that of the chlorine atom.

2. [3.9]  a. Br      There is generally an increase in electronega-
                      tivity across a period from left to right.

          b. O       There is generally a decrease in electronegativity
                      in a group from top to bottom.

3. [3.2]  a. Bi      There is generally an increase in ionization po-
                      tential across a period from left to right.

          b. Al      There is generally a decrease in ionization in a
                      group from top to bottom.

          c. O       See the explanation given for part (a) of this
                      question.

          d. Kr      Krypton is a noble gas and has a full outer shell
                      of electrons. A large amount of energy would be
                      required to remove an electron and form a less
                      stable ion.

        e. O          See the explanations given for parts (a) and (b)
                      of this question.

4. [3.3]  a. F        Electron affinity generally increases across a
                      period from left to right; however, there are a
                      few exceptions.  See Table 3.4 of the study guide.

          b. Cl       Electron affinity generally decreases in a group
                      from top to bottom; however, there are exceptions.
                      See Table 3.4 of the study guide.

5. [3.9]  a. S—H      Bond polarity is determined by the difference be-
                      tween the electronegativities of the atoms. Sulfur
                      is above selenium in the periodic table and there-
                      fore has a higher electronegativity. Thus, the dif-
                      ference in electronegativity between hydrogen and
                      sulfur is greater than that between hydrogen and
                      selenium.

          b. Se—F     The electronegativity difference between selenium
                      and fluorine is greater than that between sulfur
                      and fluorine.

6. [3.9]  a. covalent,   The electronegativity difference = 2.5 - 2.5 = 0,
             nonpolar    and the bond is predicted to be covalent and non-
                         polar.

### Table 3.4   Electron affinities (eV)

| H −0.77 | | | | | | | He +0.6[a] |
|---|---|---|---|---|---|---|---|
| Li −0.6[a] | Be +0.6[a] | B −0.2[a] | C −1.25 | N +0.3[a] | O −1.47 (+7.28)[b] | F −3.45 | Ne +1.0[a] |
| Na −0.2[a] | Mg +0.3[a] | Al −0.3[a] | Si −1.4[a] | P −0.6[a] | S −2.07 (+3.44)[b] | Cl −3.61 | |
| | | | | | | Br −3.36 | |
| | | | | | | I −3.06 | |

[a] Calculated values.
[b] Values in parentheses are calculated and pertain to the total energy effects for the addition of two electrons.

b. covalent, moderately polar

The electronegativity difference = $3.0 - 1.5 = 1.5$, and the bond is predicted to be covalent and moderately polar.

c. ionic

The electronegativity difference = $3.0 - 0.8 = 2.2$, and the bond is predicted to be ionic.

d. covalent, somewhat polar

The electronegativity difference = $3.0 - 2.5 = 0.5$, and the bond is predicted to be covalent and somewhat polar.

e. covalent, highly polar

The electronegativity difference = $2.5 - 0.7 = 1.8$, and the bond is predicted to be covalent and highly polar.

f. ionic

The electronegativity difference = $3.5 - 1.0 = 2.5$, and the bond is predicted to be ionic.

7. [3.8]

The following diagrams are drawn to an approximate scale so that they visually indicate the arguments used in predicting the molecule that has the bond with the greater degree of covalent character.

a. $SbCl_3$

Since $Sb^{3+}$ is smaller than $Bi^{3+}$, $Sb^{3+}$ has a larger charge density, concentration of charge, than $Bi^{3+}$. As a result, $Sb^{3+}$ distorts, i.e., attracts, the electron cloud of the chlorine ion more than the $Bi^{3+}$ does.

$Sb^{3+}$    $Cl^-$                $Bi^{3+}$    $Cl^-$

The larger the cation charge concentration the more distorted the anion and the more covalent the bond.

b. $CrCl_3$

$Cr^{3+}$ has a larger charge density than $Cr^{2+}$; there-fore, $Cr^{3+}$ distorts the anion more than $Cr^{2+}$ does and thus forms the more covalent bond.

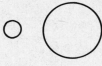

$Cr^{3+}$     $Cl^-$                $Cr^{2+}$     $Cl^-$

Larger cation charge            Smaller cation charge
  concentration                   concentration
More distorted anion            Less distorted anion
More covalent                   Less covalent

c. $HgI_2$

The larger $I^-$ ion can be more easily distorted than the smaller $Cl^-$ ion.

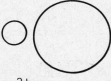

$Hg^{2+}$     $I^-$                 $Hg^{2+}$     $Cl^-$

Larger anion                    Smaller anion
More distortion of anion        Less distortion of anion
More covalent                   Less covalent

d. BeS

The sulfide anion is larger than the oxide anion and is therefore more easily distorted than the oxide anion.

$Be^{2+}$     $S^{2-}$              $Be^{2+}$     $O^{2-}$

Larger anion                    Smaller anion
More easily distorted           Less easily distorted
  anion                           anion
More covalent bond              Less covalent bond

e. ZnS

See the explanation given for part (a) of this problem.

f. $Tl_2O_3$

See the explanation given for part (b) of this problem.

g. MgO

See the explanation given for part (a) of this problem.

h. $SO_2$

See the explanation given for part (a) of this problem.

IV. Concepts of chemical bonding

1. True [3.10]

Valence electrons are those electrons in the outermost shell of an element. All elements of group V A (N, P, As, Sb, and Bi) have an $ns^2np^3$ electronic configuration.

2. False [3.10]

The oxidation number of each of the alkali metal ions ($Li^+$, $Na^+$, $K^+$, $Rb^+$, $Cs^+$, $Fr^+$) is 1+.

3. False [3.6]

Generally there are 8 electrons in the valence shell; however, there are numerous exceptions. For example, in $BF_3$

there are only 6 electrons in the boron valence shell. Also, in hydrogen compounds only 2 electrons are needed to fill the hydrogen valence shell. Many other atoms have fewer than or more than 8 electrons in the valence shell. Table 3.5 .of the study guide lists some typical examples.

4. True [3.3]

Since a noble gas has a full outer shell, $ns^2np^6$, a very large amount of energy is required to ionize an atom of that element, forming a less stable configuration.

Table 3.5   Atoms with fewer than or more than 8
            electrons in the valence shell

| MOLECULE | ATOM | NUMBER OF ELECTRONS IN ATOM'S VALENCE SHELL |
|---|---|---|
| $SF_6$ | S | 12 |
|  | F | 8 |
| $BrF_4^-$ | Br | 12 |
|  | F | 8 |
| $IF_5$ | I | 12 |
|  | F | 8 |
| $MgCl_2$ | Mg | 4 |
|  | Cl | 8 |
| $BeCl_2$ | Be | 4 |
|  | Cl | 8 |
| $PCl_5$ | P | 10 |
|  | Cl | 8 |

5. False [3.3]   The electron affinity of a halogen is the energy
                 evolved when an electron is added to an atom of
                 that element. Each halogen atom has an $ns^2np^5$ elec-
                 tronic configuration and with the addition of an
                 electron will have a very stable filled outer shell,
                 $ns^2np^6$, a noble gas configuration.

6. True [3.11]   See Table 3.6 of the study guide for the names of
                 anions of some common elements.

Table 3.6   Names of anions of some common elements

| NAME OF ELEMENT | SYMBOL | ANION | ANION NAME |
|---|---|---|---|
| fluorine | F | $F^-$ | fluoride |
| nitrogen | N | $N^{3-}$ | nitride |
| oxygen | O | $O^{2-}$ | oxide |
| carbon | C | $C^{4-}$ | carbide |
| sulfur | S | $S^{2-}$ | sulfide |
| bromine | Br | $Br^-$ | bromide |
| selenium | Se | $Se^{2-}$ | selenide |

7. False [3.2, 3.3]   Ionization energy is the amount of energy necessary to remove an electron from an isolated gas-
eous atom in its ground state, $X \rightarrow X^+ + e^-$.
Electron affinity is the energy effect accompany-
ing the process in which an electron is added to an
isolated atom in its ground state, $X + e^- \rightarrow X^-$.

8. True [3.9]   The general trend of electronegativity is an increase from left to right across a period and a
decrease from top to bottom within a group.

9. True [3.4]   The term *isoelectronic* is used to indicate that
species have the same electronic configuration.
For example,

$\text{Na}^+$ :  $1s^2 2s^2 2p^6$

$\text{Ne}$  :  $1s^2 2s^2 2p^6$

are said to be isoelectronic.

10. False [3.4]   Covalent bonds are those bonds formed by the shar-
ing of electrons. Ionic bonds are those bonds
formed by the attraction of oppositely charged ions.

11. True [3.2]   Singly ionized sodium has a stable configuration,
a noble gas structure, but magnesium only attains
this stable configuration after being doubly ionized.

12. False [3.8]   The more polar a bond, the larger its charge separation, and therefore the larger its dipole moment.

V. Lewis structures

1. $\ddot{\text{O}} = \ddot{\text{O}}$   Each oxygen has 6 electrons in its valence shell.
The total number of valence electrons to be dis-
tributed in $O_2$ is therefore

2  O atoms · (6 valence electrons/neutral O atom)
= 12 valence electrons

The number of valence electrons of an atom can be
obtained easily from the periodic table.  See Figure 3.4 of the study guide.  Note that the usual
number of valence electrons of an atom is identical
to the group number of the element, with the excep-
tion of the noble gases, the transition elements, and
the inner-transition elements.  Each of the noble gas

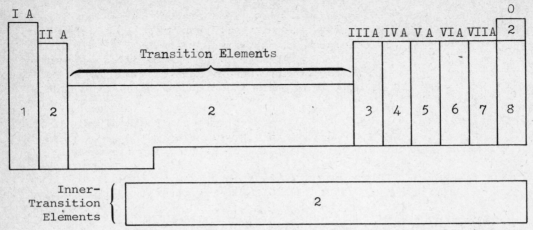

Figure 3.4  Usual number of valence electrons (_s_ and _p_ electrons) of an element in a group or in a series

elements has 8 valence electrons, except for helium which has 2 valence electrons. Each of the transition and the inner-transition elements has 2 valence electrons.

2. $\ddot{O} = \ddot{O} - \ddot{O}:$
       $\oplus$   $\ominus$

   or

   $:\ddot{O} - \ddot{O} = \ddot{O}$
       $\ominus$   $\oplus$

The total number of valence electrons to be distributed in $O_3$ is 18:

   3  O atoms  (6 valence electrons/neutral O atom) = 18 valence electrons

The following steps can be used to determine the formal charge of an atom in a molecule:

(a) Identify the number of valence electrons of the neutral atom; for example, the number of valence electrons of a neutral oxygen atom is 6.

(b) Determine the number of electrons surrounding each atom in the Lewis structure of the molecule. Assume that each bond is formed from a single electron of each of the bonded atoms. For example, for ozone the number of valence electrons surrounding each atom of oxygen is

$\ddot{O}$ :|: $\ddot{O}$ ·|· $\ddot{O}$ :

   6  |  5  |  7

(c) Subtract the number of electrons surrounding
the atom in the Lewis structure of the molecule (determined in Step c) from the number
of electrons in the valence shell of a neutral
atom of the element (determined in Step a).
This difference is the formal charge of the
atom in the molecule.

$$\overbrace{6-6=0}\ \overbrace{6-7=-1}$$

$$\ddot{O} = \ddot{O} - \ddot{O}:\qquad \text{or}\qquad \overset{\oplus}{\ddot{O}} = \overset{\ominus}{\ddot{O}} - \ddot{O}:$$

$$\underbrace{6-5=+1}$$

When drawing Lewis structures, always try to mini-
mize the formal charges of the atoms and to avoid
charges of the same sign on adjacent atoms.

3.

The total number of valence electrons is 8:

   1 C atom   (4 valence electrons/neutral C atom)
                                 = 4 valence electrons
   4 H atoms   (1 valence electron/neutral H atom)
                                 = 4 valence electrons
                 Total = 8 valence electrons

For neutral molecules the sum of the formal charges
is always zero. For charged molecules the sum of
the formal charges equals the charge of the molecule.

4. $\left[\ddot{N} = N = \ddot{N}\right]^{-}$
    $\overset{\ominus}{}$  $\overset{\oplus}{}$  $\overset{\ominus}{}$

The total number of valence electrons is 16:

   3 N atoms   (5 valence electrons/neutral N atom)
                            = 15 valence electrons
   Electrons needed to account for the 1- charge
           of the molecule = 1 valence electron
               Total = 16 valence electrons

The sum of the formal charges of the atoms equals
the charge of the molecule.

The structure

$$\left[:\ddot{N} - N \equiv N:\right]^{-}$$
$$\overset{2-}{}\quad\overset{+}{}$$

is improbable because it has a large formal charge
on a single atom.

5. 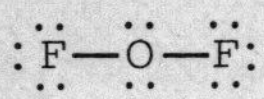 :F̈—Ö—F̈:   The total number of valence electrons is 20:

    1  O atom   (6 valence electrons/neutral O atom)
                          =   6 valence electrons
    2  F atoms  (7 valence electrons/neutral F atom)
                          = 14 valence electrons
              Total = 20 valence electrons

or in an abbreviated form

    O :  1 x 6 =  6
    F :  2 x 7 = 14
        Total = 20

6. 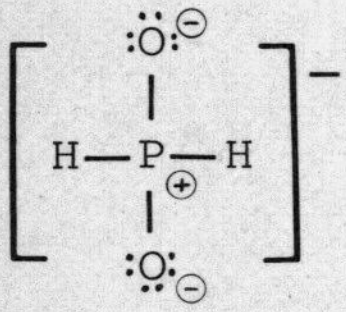 The total number of valence electrons is 20:

    P :  1 x 5 =  5
    O :  2 x 6 = 12
    H :  2 x 1 =  2
  charge (1-) :        1
        Total = 20

The structure

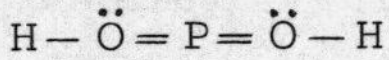

    H—Ö=P=Ö—H

is incorrect because there are too few electrons. The structure

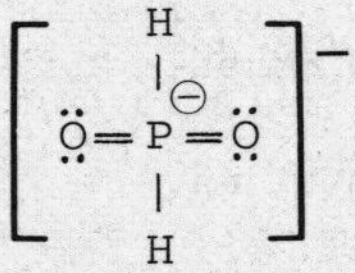

is doubtful because there are too many electrons in the phosphorus valence shell.

7. 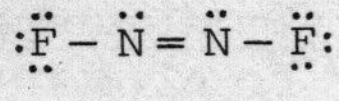 :F̈—N̈=N̈—F̈:   The total number of valence electrons is 24:

    F :  2 x 7 = 14
    N :  2 x 5 = 10
        Total = 24

The structures

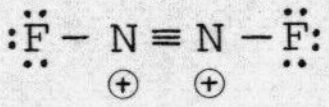

    :F̈—N≡N—F̈:   and   F̈=N̈—N̈=F̈

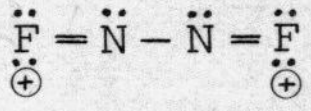

are incorrect because there are too few electrons.

The structure

$$\overset{..}{\underset{\oplus}{F}}=\overset{..}{N}-\overset{..}{\underset{\ominus}{N}}-\overset{..}{F}:\quad\quad \text{or} \quad\quad :\overset{..}{\underset{\ominus}{F}}-\overset{..}{N}-\overset{..}{N}=\overset{..}{\underset{\oplus}{F}}$$

is improbable because the molecule can be drawn so that all formal charges are zero; that is, each atom contributes an electron to the bond between it and another atom.

8.  $\left[:C\equiv \underset{\ominus}{C}:\right]^{2-}$  The total number of valence electrons is 10:

$$
\begin{array}{rl}
\text{C :} & 2 \times 4 = \phantom{0}8 \\
\text{charge (2-) :} & \phantom{2 \times 4 =}\underline{\phantom{0}2} \\
& \text{Total} = 10
\end{array}
$$

The structure

$$\left[\overset{\textcircled{\tiny 2-}}{\overset{..}{C}}=\overset{\textcircled{\tiny 2-}}{\overset{..}{C}}\right]4-$$

in incorrect because there are too many electrons.

9.

$$\left[\begin{array}{c} :\overset{..}{F}: \\ | \\ :\overset{..}{F}-B-\overset{..}{\underset{\ominus}{F}}: \\ | \\ :\overset{..}{F}: \end{array}\right]^{-}$$

The total number of valence electrons is 32:

$$
\begin{array}{rl}
\text{B :} & 1 \times 3 = \phantom{0}3 \\
\text{F :} & 4 \times 7 = 28 \\
\text{charge (1-) :} & \phantom{4 \times 7 =}\underline{\phantom{0}1} \\
& \text{Total} = 32
\end{array}
$$

10.

$$\left[\begin{array}{c} H \\ | \\ H-\overset{\oplus}{N}-H \\ | \\ H \end{array}\right]^{+}$$

The total number of electrons is 8:

$$
\begin{array}{rl}
\text{N :} & 1 \times 5 = 5 \\
\text{H :} & 4 \times 1 = 4 \\
\text{charge (1+) :} & \phantom{4 \times 1 =}\underline{-1} \\
& \text{Total} = \phantom{0}8
\end{array}
$$

Note that the charge of the ion is 1+. One electron has been removed from the valence shells.

11.

$$\left[\begin{array}{c} :\ddot{O}: \\ \| \\ \overset{\ominus}{:}\ddot{O}: \quad \overset{\oplus}{N} \quad :\ddot{O}:\overset{\ominus}{} \end{array}\right]^{-}$$

or

$$\left[\begin{array}{c} :\ddot{O}:\overset{\ominus}{} \\ | \\ :\ddot{O}: \quad \overset{\oplus}{N} \quad :\ddot{O}:\overset{\ominus}{} \end{array}\right]^{-}$$

or

$$\left[\begin{array}{c} :\ddot{O}:\overset{\ominus}{} \\ | \\ \overset{\ominus}{:}\ddot{O}: \quad \overset{\oplus}{N} \quad :\ddot{O}: \end{array}\right]^{-}$$

The total number of valence electrons is 24:

$$
\begin{array}{rl}
N : & 1 \times 5 = \phantom{0}5 \\
O : & 3 \times 6 = 18 \\
\text{charge } (1-) : & \phantom{1 \times 6 = }\underline{\phantom{0}1} \\
& \text{Total} = 24
\end{array}
$$

The structure

is doubtful because there are only 6 electrons in the valence shell of nitrogen and the formal charge of nitrogen is larger than necessary.

12.

$$\left[\begin{array}{c} :\ddot{O}:\overset{\ominus}{} \\ | \\ \overset{\ominus}{:}\ddot{O} - S - \ddot{O}:\overset{\ominus}{} \\ |^{(2+)} \\ :\ddot{O}:\overset{\ominus}{} \end{array}\right]^{2-}$$

The total number of valence electrons is 32:

$$
\begin{array}{rl}
S : & 1 \times 6 = \phantom{0}6 \\
O : & 4 \times 6 = 24 \\
\text{charge } (2-) : & \phantom{1 \times 6 = }\underline{\phantom{0}2} \\
& \text{Total} = 32
\end{array}
$$

13. $:N\equiv N:$

The total number of valence electrons is 10:

$$
\begin{array}{rl}
N : & 2 \times 5 = 10 \\
& \text{Total} = 10
\end{array}
$$

The structure

$$\dot{:}N - N\dot{:}$$

is improbable because there are too few electrons in the nitrogen valence shell, 6 instead of 8.

14. $\left[\overset{\ominus}{:}\ddot{O} - \ddot{C}l:\right]^{-}$

The total number of valence electrons is 14:

$$
\begin{array}{rl}
Cl : & 1 \times 7 = 7 \\
O : & 1 \times 6 = 6 \\
\text{charge } (1-) : & \phantom{1 \times 6 = }\underline{\phantom{0}1} \\
& \text{Total} = 14
\end{array}
$$

15. $:N \equiv N - \overset{\cdot\cdot}{\underset{\cdot\cdot}{O}}:$
       ⊕   ⊖

The total number of valence electrons is 16.  The structure

$\overset{\cdot\cdot}{N} = N = \overset{\cdot\cdot}{O}$
⊖    ⊕

is improbable because there are formal charges of opposite sign on adjacent atoms of the same element.

The structure

$:\overset{\cdot\cdot}{N} - N \equiv O:$
②⊖  ⊕  ⊕

is improbable because there are large formal charges on adjacent atoms.

The structure

$\overset{\cdot\cdot}{\underset{\cdot\cdot}{N}} - \overset{\cdot\cdot}{\underset{\cdot\cdot}{N}} - \overset{\cdot\cdot}{\underset{\cdot\cdot}{O}}:$
    ⊕  ⊖

is improbable.  The formal charge of each atom in the structure is acceptable.  Note, however, that the valence shell of each nitrogen atom has only 6 electrons, not 8 electrons.

16.
$:\overset{\cdot\cdot}{O}:$ ⊖
|
$:\overset{\cdot\cdot}{\underset{\cdot\cdot}{F}} - S - \overset{\cdot\cdot}{\underset{\cdot\cdot}{F}}:$
| ②⁺
$:\overset{\cdot\cdot}{O}:$ ⊖

The total number of valence electrons is 32.

17.
H  H
|   |
$:N - \overset{\cdot\cdot}{N}:$
|   |
H  H

The total number of valence electrons is 14.

18.
$:\overset{\cdot\cdot}{Cl}:$
|
$:\overset{\cdot\cdot}{\underset{\cdot\cdot}{Cl}} - Sn - \overset{\cdot\cdot}{\underset{\cdot\cdot}{Cl}}:$
|
$:\overset{\cdot\cdot}{Cl}:$

The total number of valence electrons is 32.

19. $\left[ :\overset{\cdot\cdot}{\underset{\cdot\cdot}{O}} - H \right]^{-}$ ⊖

The total number of valence electrons is 8.

20.

or

The total number of valence electrons is 40. The structures

and

are improbable because in each there are positive formal charges on three adjacent atoms.

21. $\left[\; :\ddot{O} - \ddot{N} = \ddot{N} - \ddot{O}: \;\right]^{2-}$

The total number of valence electrons is 24. The structure

$$\left[\; :\ddot{\underset{\ominus}{O}} - \ddot{\underset{\ominus}{N}} - \ddot{N} = \ddot{O} \;\right]^{2-}$$

is improbable because the formal charges are localized at one end of the molecule.

22.

The total number of valence electrons is 20.

VI. Oxidation numbers

1. 3+

The sum of the oxidation numbers of the atoms in a polyatomic ion always equals the charge of that ion. For $BF_4^-$ the sum of the oxidation numbers is 1- since there are 4 fluorine atoms, each of which has an oxidation number of 1-, and 1 boron atom, which has an oxidation state of 3+. In abbreviated notation this sum can be indicated as

```
4   F atoms (1- per F atom) = 4-
1   B atom (3+ per B atom) = 3+
                     sum = 1-
```

2. 3-

For $NH_4^+$

```
4   H atoms (1+ per H atom) = 4+
1   N atom (3- per N atom) = 3-
                     sum = 1+
```

and the charge of the ion is 1+.

3. 5+

For $NO_3^-$

$$
\begin{array}{llr}
3 & \text{O atoms (2- per O atom)} & = 6- \\
1 & \text{N atom (5+ per N atom)} & = \underline{5+} \\
 & \text{sum} & = \overline{1-}
\end{array}
$$

and the charge of the ion is 1-.

4. 1+

For $H_2PO_2^-$

$$
\begin{array}{llr}
2 & \text{O atoms (2- per O atom)} & = 4- \\
2 & \text{H atoms (1+ per H atom)} & = 2+ \\
1 & \text{P atom (1+ per P atom)} & = \underline{1+} \\
 & \text{sum} & = \overline{1-}
\end{array}
$$

and the charge of the ion is 1-.

5. 5+

For $PO_4^{3-}$

$$
\begin{array}{llr}
4 & \text{O atoms (2- per O atom)} & = 8- \\
1 & \text{P atom (5+ per P atom)} & = \underline{5+} \\
 & \text{sum} & = \overline{3-}
\end{array}
$$

and the charge of the ion is 3-.

6. 7+

For $ClO_4^-$

$$
\begin{array}{llr}
4 & \text{O atoms (2- per O atom)} & = 8- \\
1 & \text{Cl atom (7+ per Cl atom)} & = \underline{7+} \\
 & \text{sum} & = \overline{1-}
\end{array}
$$

and the charge of the ion is 1-.

7. 4+

For $SnCl_4$ the sum of the oxidation numbers is zero.

$$
\begin{array}{llr}
4 & \text{Cl atoms (1- per Cl atom)} & = 4- \\
1 & \text{Sn atom (4+ per Sn atom)} & = \underline{4+} \\
 & \text{sum} & = \overline{0}
\end{array}
$$

8. 6+

For $Cr_2O_7^{2-}$

$$
\begin{array}{llr}
7 & \text{O atoms (2- per O atom)} & = 14- \\
2 & \text{Cr atoms (6+ per Cr atom)} & = \underline{12+} \\
 & \text{sum} & = \overline{2-}
\end{array}
$$

and the charge of the ion is 2-.

9. 6+

For $CrO_4^{2-}$

$$
\begin{array}{llr}
4 & \text{O atoms (2- per O atom)} & = 8- \\
1 & \text{Cr atom (6+ per Cr atom)} & = \underline{6+} \\
 & \text{sum} & = \overline{2-}
\end{array}
$$

and the charge of the ion is 2-.

10. 4+                    For $MnO_2$ the sum of the oxidation numbers is 0.

$$
\begin{array}{ll}
2 \text{ O atoms (2- per O atom)} & = 4- \\
1 \text{ Mn atom (4+ per Mn atom)} & = \underline{4+} \\
\text{sum} & = 0
\end{array}
$$

11. 7+                    For $MnO_4^-$

$$
\begin{array}{ll}
4 \text{ O atoms (2- per O atom)} & = 8- \\
1 \text{ Mn atom (7+ per Mn atom)} & = \underline{7+} \\
\text{sum} & = 1-
\end{array}
$$

and the charge of the ion is -1.

12. 0                     Any atom in a molecule of an element has an oxidation number of zero.

# SELF-TEST

Complete the test in 45 minutes:

I. Complete the following table:

| FORMULA OF COMPOUND | NAME OF COMPOUND | SYMBOL OF ELEMENT | OXIDATION NUMBER OF ELEMENT |
|---|---|---|---|
| $SnCl_4$ | _____ | Sn | _____ |
| $As_2O_3$ | _____ | As | _____ |
| $Cu(NO_3)_2$ | _____ | N | _____ |
| _____ | manganese (IV) oxide | Mn | _____ |
| $Cr_2O_3$ | _____ | Cr | _____ |
| $NaC_2H_3O$ | _____ | Na | _____ |
| $N_2O_4$ | _____ | _____ | 4+ |
| $H_2SO_4$ | sulfuric acid | S | _____ |
| _____ | sodium chlorite | _____ | 3+ |

II. Answer each of the following:

_____ 1. Which of the following elements has the greatest electron affinity?

a. Na        c. I

b. O         d. Cl

_____ 2. Which of the following elements has the lowest first ionization potential?

a. Na        c. I

b. O         d. Cl

_____ 3. Which of the following ions has the largest ionic radius?

a. $Na^+$      c. $Ca^{2+}$

b. $K^+$       d. $Ga^{3+}$

_____ 4. Which of the following elements has the greatest electronegativity?

a. Na        c. Sn

b. Ca        d. F

_____ 5. Which of the following elements has the smallest electronegativity?

a. P         c. As

b. S         d. Ge

_____ 6. Which of the following compounds contains an element with an oxidation state of 5+?

a. HClO        c. $HClO_3$

b. $HNO_2$      d. $HBrO_4$

_____ 7. Which of the following compounds contains bonds with the most covalent character?

a. $SbCl_3$      c. $SbI_3$

b. $BiCl_3$      d. $BiI_3$

_____ 8. Which of the following statements is the most accurate description of the observed trends in atomic radii?

a. Atomic radii are approximately constant for elements of a group of the periodic table and increase with increasing atomic number across a period of the periodic table.

b.  Atomic radii decrease with increasing atomic number across a period of the periodic table and increase with increasing atomic number of the elements of a group in the periodic table.

c.  Atomic radii increase with increasing atomic number, Z.

d.  Atomic radii increase with increasing atomic number, but the transition elements show a change in this trend due to screening by inner electrons.

_____  9. Which of the following compounds contains bonds with the most ionic character?

a.  $MgBr_2$                c.  $SiBr_4$

b.  $AlBr_3$                d.  $PBr_5$

_____  10. The ion $MnO_4^-$ can best be described as

a.  a Lewis structure       c.  a transition metal
b.  a monatomic ion            ion
                            d.  an anion

III. Draw Lewis structures of $PCl_3$, HCCH, $SOF_2$, and $H_2CO$.

# 4.

# Molecular Geometry and the Covalent Bond

OBJECTIVES

(a) You should be able to demonstrate your knowledge of the following terms by defining them, describing them, or giving specific examples of them:

antibonding orbitals [4.4]
atomic (also called network) crystals [4.8]
band [4.7]
bond order [4.4]
degeneracy [4.7]
delocalization [4.1, 4.5]
dipole moment [4.6]
electron pair repulsions [4.3]
hybrid orbitals [4.2]
ionic crystals [4.8]

London forces [4.6]
metallic crystals [4.7, 4.8]
molecular crystals [4.8]
molecular geometry [4.3]
molecular orbitals [4.4, 4.5]
resonance [4.1]
resonance form [4.1]

(b) You should be able to draw resonance structures of molecules and ions.

(c) You should be able to use the concepts of orbital hybridization and electron pair repulsion to predict molecular shapes.

(d) You should be able to use the concepts of resonance, hybrid orbitals, electron pair repulsion, molecular orbitals, intermolecular forces, and metallic bonding as needed to rationalize such information about molecules as bond length, bond order, bond angles, dipole moments, magnetic susceptibility, boiling points, and melting points.

(e) You should be familiar with the properties and structures of the various types of crystalline solids.

**EXERCISES**

I. Draw resonance structures of a molecule of each of the following compounds.  Include formal charges.

1. $COCl_2$ This compound, called phosgene or carbonyl chloride, is a poisonous gas.  Carbon is the central atom in the compound.

2. $SO_3$ Sulfur trioxide is an atmospheric contaminant that is responsible for imparting the characteristic blue color to smoke.  It is the acid anhydride of sulfuric acid, i.e., $SO_3$ reacts with water to form $H_2SO_4$:

$$SO_3(aq) + H_2O(aq) \longrightarrow H_2SO_4(aq)$$

3. $C_3O_2$ Little chemistry is reported for this rare gas, which is called tricarbon dioxide.

4. $ClCN$ Chlorine cyanide or cyanogen chloride is the name of this poisonous, volatile liquid.

5. $NO_3^-$     This extremely common anion is the nitrate ion.  Nitrogen is the central atom in the structure of this ion.

6. $O_3$     This compound, ozone, is a major reactant in photochemical smog production and is formed by the action of ultraviolet light on oxygen in the upper atmosphere.

7. $N_2O$     The chemical name of this compound is dinitrogen monoxide or nitrous oxide.  The compound is the anesthetic commonly referred to as "laughing gas."  It is also used as a propellant gas in whipped cream cans.

II. Draw the most probable Lewis structure of a molecule of each of the following compounds.  Predict the hybrid orbitals of the central atom in each molecule and the geometry of the molecule.

1. $BF_3$     Boron trifluoride is a colorless, pungent gas used extensively in organic reactions.

2. $PH_3$     This compound, which is called phosphine or hydrogen phosphide, is a poisonous, colorless gas that is spontaneously inflammable.

3. $BrF_5$     Bromine pentafluoride is a reactive gas.

4. $XeF_2$     Xenon difluoride is a noble gas fluoride.

5. $XeF_4$     Xenon tetrafluoride is also a noble gas fluoride.

6. $H_2CO$     This organic compound, which is commonly used to preserve animal specimens, is called formaldehyde.

III. Use the concept of electron pair repulsions to predict the geometric configuration of a molecule of each of the following compounds:

1. $XeF_4$, xenon tetrafluoride

2. $AsF_5$, arsenic pentafluoride

3. $SnF_4$, tin tetrafluoride or tin(IV) fluoride

4. $IF_2^+$, the iodine difluoride cation

IV.   Write the electronic configuration of each of the following homopolar, diatomic molecules and ions. Determine the bond order of each and determine whether the molecule or ion is paramagnetic or diamagnetic.  Also draw the Lewis structure of each.

1. NO          2. $CN^-$          3. $O_2^+$

V.   Predict the geometric configuration of a molecule of each of the following compounds.  Describe the bonding between all atoms in the molecule.  Remember that the molecular shapes will only be approximations based on the theories with which you are now familiar.

1. $H_2CO$, formaldehyde

2. $C_6H_6$, a six-carbon ring compound, benzene

3. $CCl_4$, carbon tetrachloride

4. HCCH, acetylene

5. $SO_2$, sulfur dioxide

6. $CO_2$, carbon dioxide

7. HF, hydrogen fluoride

VI.   From the following list of molecules and atoms (described by their Lewis structures) choose the ones that answer the question. A molecule or atom may be used to answer more than one question.

a. 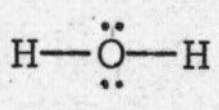    H—Ö—H               c. 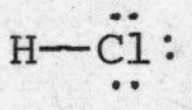    H—Cl:

b. 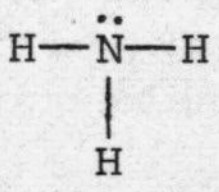    H—N̈—H               d. 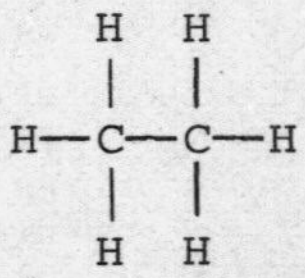         H   H
           |                              |   |
           H                         H—C—C—H
                                          |   |
                                          H   H

e.

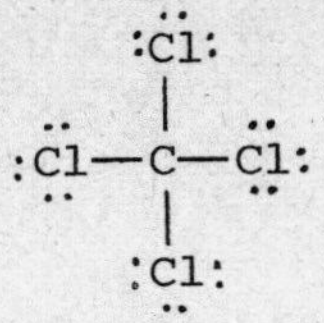

$$:\overset{\cdot\cdot}{Cl}:$$
$$:\overset{\cdot\cdot}{Cl}-C-\overset{\cdot\cdot}{Cl}:$$
$$:\overset{\cdot\cdot}{Cl}:$$

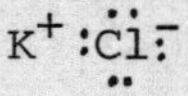

f.

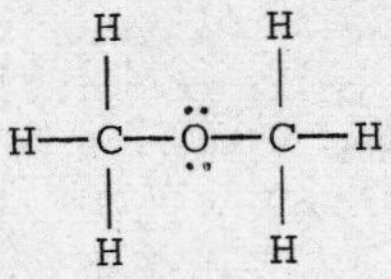

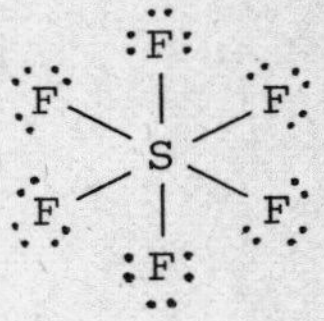

$$H-\overset{H}{\underset{H}{C}}-\overset{\cdot\cdot}{O}-\overset{H}{\underset{H}{C}}-H$$

g.   Na·

h. 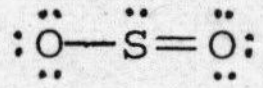   :Ö—S̈=Ö:

i.   ·C̈·

j.   K⁺ :C̈l:⁻

k.

S surrounded by six F atoms (SF₆)

l.

:F̈:
|
:Ï—F̈:
|
:F̈:

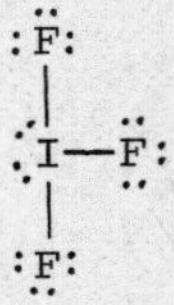

m.

Xe surrounded by six F atoms (XeF₆)

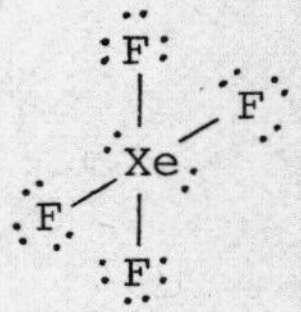

1. Which molecule(s) is polar?
2. Which molecule(s) is nonpolar?
3. Which molecule(s) contains one or more atoms that probably use $sp^3$ hybrid orbitals in bonding?
4. Which molecule(s) contains π bonds?
5. Which atom(s) forms metallic bonds?
6. Which molecule(s) forms ionic crystals?
7. Which molecule(s) contains 7 σ bonds?
8. Which molecule(s) or atom(s) contains unpaired electrons?
9. Which molecule(s) is planar?
10. Which molecule(s) does not contain nonbonding pairs of electrons?

**ANSWERS TO EXERCISES**

I. Resonance structures
See Section 4.1 of your text.

1.

(a)                    (b)                    (c)

Three resonance structures (a, b, and c) can be
drawn.  The structure of a $COCl_2$ molecule can be
described as a resonance hybrid, or a weighted average, of the three structures.  Structure (a) is more
important than either (b) or (c), i.e., the actual
structure of a $COCl_2$ molecule contains more of the
expected character of structure (a) than of that of
either (b) or (c).  If you had been asked to draw a
single Lewis structure to describe the molecule,
structure (a) would be correct because it most closely approximates the structure of the molecule.

2.

(a)                    (b)                    (c)

In sulfur trioxide all S—O bonds are of equal length.
We may therefore conclude that the actual structure
is an average of the three resonance forms drawn.
Each of the structures contributes equally to the
actual structure of the $SO_3$ molecule.  Each S—O
bond in·$SO_3$ is equivalent and has a strength that
is between that of a single bond and that of a double bond.  There is no reason to expect any one of
the structures to contribute more to the actual
structure than either of the other two.

3.

(a)                    (b)                    (c)

Structure (a) is the main contributor to the actual
structure of tricarbon dioxide. Structures (b) and
(c) are assumed to be important because the bond
lengths in the molecule are found to be shorter than
those expected if only structure (a) were important.

4.   :C̈l—C≡N:   ⟷   C̈l=C=N̈  (with ⊕ over Cl, ⊖ over N)
        (a)                    (b)

Structure (a) is a more important contributor to the actual structure than structure (b) because of the formal charges in (b).

5.

All three structures contribute equally to the actual structure of the nitrate ion.

        (a)              (b)              (c)

6.

        (a)         (b)         (c)         (d)

Four structures can be drawn for the ozone molecule. Structures (a) and (b) are expected to contribute most to the actual structure. Structures (c) and (d) do not obey the octet rule:  the valence shell of oxygen is not filled.  You would not have known *a priori* that (c) and (d) are important.

7.   N̈=N=Ö   ⟷   :N≡N—Ö:
        (a)              (b)

Both are important contributors to the actual structure of nitrous oxide.  A third structure

:N—N≡O:

does not contribute appreciably.

II.  Hybrid orbitals and molecular geometry
     See Section 4.2 of your text.

1.   :F̈—B—F̈:       *sp²* hybrid orbitals       triangular planar
        |                                          geometry
       :F̈:

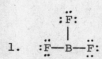

120°

Since experimental evidence shows the molecule to be triangular planar, boron is assumed to use $sp^2$ hybrid orbitals in its bonding.  If we wish to predict the shape of a polyatomic molecule before sufficient experimental facts are known, however, we can use the following scheme, which we shall designate as Method 1:

(a) Draw a Lewis structure.

(b) Add the number of nonbonding electron pairs that are in the valence shell of the central atom in the Lewis structure to the number of atoms bonded directly to the central atom in that structure.

(c) Recall from Table 4.1 of the study guide the correspondence between the sum determined in part (b) and the type of hybrid orbitals used by the central atom and the geometry of these orbitals.

Table 4.1   Geometry and hybridization of atomic orbitals

| NUMBER OF NONBONDING ELECTRON PAIRS + NUMBER OF ATOMS BONDED TO THE CENTRAL ATOM | TYPE OF HYBRID ORBITALS USED BY THE CENTRAL ATOM | GEOMETRY OF HYBRID ORBITALS ABOUT THE CENTER OF THE CENTRAL ATOM |
|---|---|---|
| 2 | $sp$ | linear |
| 3 | $sp^2$ | triangular planar |
| 4 | $sp^3$ | tetrahedral |
| 4 | $dsp^2$ | square planar[a] |
| 5 | $dsp^3$ | trigonal pyramidal |
| 6 | $d^2sp^3$ | octahedral |

[a] Occasionally the $d$ orbitals of the shell next to the outer shell are involved in bonding of molecules containing Rh(I), Ir(I), Pt(II), Pd(II), Au(III), Ni(II), and Cu(II).  Most commonly, however, the $s$, $p$, and $d$ orbitals of the outer shell are used in the bonding of molecules.  (See Section 4.2 of your text for further discussion.)

Using Method 1 to predict the shape of the $BF_3$ molecule, we find

$$\ddot{\text{:F:}}$$

(a)   $:\ddot{\text{F}}—\text{B}—\ddot{\text{F}}:$

(b) number of nonbonding pairs
    in valence shell of boron  =   0
    number of atoms directly
    bonded to boron               =   3
    ____

                          sum =   3

(c) From Table 4.1 of the study guide we note
    that the number 3 corresponds to $sp^2$ hybrid
    orbitals of the central atom and a triangular
    planar geometry of these orbitals about the
    center of the central atom.  Boron uses these
    three orbitals to bond with the fluorine atoms.
    We therefore predict the shape of the $BF_3$ mole-
    cule to be triangular planar.

An alternate method of predicting molecular geometry, which we shall designate as Method 2, can be
used.  This method consists of the following steps:

(a) Count the number of atoms bonded directly to
    the central atom.

(b) Diagram the valence shell of the central atom
    according to the relative energies of the sub-
    shells.

(c) Hybridize the orbitals of the valence subshells
    that are necessary for the bonding of the cen-
    tral atom to the peripheral atoms.[1]  Assume
    that each bond is formed with one electron
    from the central atom and one from a peripheral
    atom.  If a π bond is formed between the cen-
    tral atom and a peripheral atom, leave one un-
    hybridized $p$ orbital to allow for the formation
    of such a bond.

(d) From the previous information and the informa-
    tion in Table 4.1 of the study guide, predict
    the molecular geometry of the molecule.

_____

1. Occasionally the $d$ orbitals of the shell next to the outer shell are involved in the bonding of molecules, i.e., those containing Rh(I), Ir(I),
   Pt(II), Pd(II), Au(III), and Ni(II).  Most commonly, however, the $s$, $p$,
   and $d$ orbitals of the outer shell are used in the bonding of molecules.
   (See Section 4.2 of your text for further discussion.)

Using Method 2 to predict the molecular geometry of $BF_3$, we find

(a) There are three fluorine atoms bonded directly to boron.

(b) and (c)

$2p$ $\underline{\;1\;}$ $\underline{\quad}$ $\underline{\quad}$          $2p$ $\underline{\quad}$

$sp^2$ $\underline{\;1\;}$ $\underline{\;1\;}$ $\underline{\;1\;}$

$2s$ $\underline{\;1\!\!\downarrow\;}$

B (ground state)      B (hybrid state)

(d) Boron uses $sp^2$ hybrid orbitals in its bonding in $BF_3$. From Table 4.1 of the study guide we note that the geometry of $sp^2$ orbitals is tri-angular planar. We therefore predict that the molecule has a triangular planar geometry.

2. H—P̈—H          $sp^3$ hybrid              trigonal pyramidal geometry
   |                    orbitals
   H

According to Method 1, we predict the molecular shape to be that of a trigonal pyramid:

(a) The Lewis structure is

H—P̈—H
   |
   H

(b) There are three bonded atoms and one nonbonding pair of electrons, a numerical total of 4.

(c) According to the information in Table 4.1 of the study guide, phosphorus should use $sp^3$ hybrid orbitals in its bonding and these orbitals should have a tetrahedral geometry about the center of the phosphorus atom.

(d) We predict the molecular shape of the molecule to be that of a trigonal pyramid.

P
H       H
H

According to Method 2, we predict the molecule to have a trigonal pyramidal shape:

(a) There are 3 hydrogen atoms bonded to the central phosphorus atom in $PH_3$.

(b) and (c)

$3p$  $\underline{\uparrow}$  $\underline{\uparrow}$  $\underline{\uparrow}$

$sp^3$  $\underline{\uparrow\downarrow}$  $\underline{\uparrow}$  $\underline{\uparrow}$  $\underline{\uparrow}$

$3s$  $\underline{\uparrow\downarrow}$

P (ground state)          P (hybrid state)

The bond angles in $PH_3$ (as in $NH_3$, see Section 4.2 of your text) are found to be closer to the tetrahedral angle than to $90^\circ$. Thus, we say that $sp^3$ hybrid orbitals are used in the bonding, not the $3p$ orbitals.

(d) We predict the molecule to have a trigonal pyramidal shape.

Your text discusses other molecules in which the central atom is thought to use $sp^3$ hybrid orbitals in bonding: $CH_4$, $NH_3$, and $H_2O$. Methane is a tetrahedral molecule, ammonia is a trigonal pyramidal molecule (a nonbonding pair of electrons occupies one of the $sp^3$ hybrid orbitals), and water is an angular molecule (nonbonding pairs of electrons occupy two of the $sp^3$ hybrid orbitals). Note the position of the central atom of each molecule in the periodic table (see Figure 4.1 of the study guide) and recall that a hydrogen atom, like an atom of each of the halogens, needs only one electron to fill its valence shell.

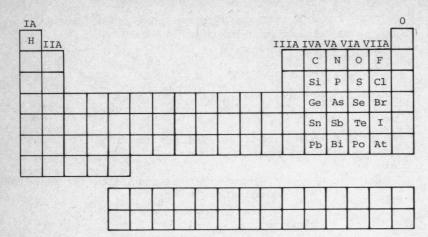

Figure 4.1 Periodic table

A molecule composed of an atom of an element of
group IV A bonded to four atoms of an element of
group VII A or to four atoms of hydrogen has a tetrahedral shape.   The central atom of such a molecule
is thought to use $sp^3$ hybrid orbitals in bonding.
In addition, a molecule that is composed of any combination of atoms of an element of group VII A and
hydrogen bonded to an atom of a group IV A element is
predicted to have a tetrahedral shape, and the central atom of such a molecule is thought to use $sp^3$
hybrid orbitals in bonding.   Examples of such mole-
cules include $CH_4$, $CF_4$, $CCl_4$, $CBr_4$, $CI_4$, $CH_3F$, $CH_3Cl$,
$CH_3Br$, $CH_3I$, $CH_2F_2$, $CH_2Cl_2$, $CH_2Br_2$, $CH_2I_2$, $CHF_3$,
$CHCl_3$, $CHBr_3$, $CHI_3$; $SiH_4$, $SiF_4$, ...; $GeH_4$, $GeF_4$, ...;
$SnH_4$, $SnF_4$, ...; and $PbH_4$, $PbF_4$, ....

A molecule composed of an atom of an element of
group V A bonded to three atoms of an element of
group VII A or to three hydrogen atoms is predicted
to have a trigonal pyramidal shape.   The central
atom of such a molecule is thought to use $sp^3$ hybrid
orbitals in bonding.   A nonbonding pair of electrons
occupies one of the $sp^3$ hybrid orbitals.   Examples
of such molecules include $NH_3$, $NF_3$, $NCl_3$, $NBr_3$, $NI_3$;
$PH_3$, $PF_3$, $PCl_3$, $PBr_3$, $PI_3$; $AsH_3$, $AsF_3$, ...; $SbH_3$,
$SbF_3$, ...; and $BiH_3$, $BiF_3$, ....

Any single atom of an element of group VI A bonded directly to two atoms of an element of group VII A is thought to use $sp^3$ hybrid orbitals in bonding. Nonbonding pairs of electrons occupy two of the $sp^3$ hybrid orbitals. A few examples of such molecules are $H_2O$, $H_2S$, $OF_2$, and $OCl_2$. Write the formulas of a few more molecules of this type.

3.

$d^2sp^3$ hybrid orbitals

square pyramidal geometry

Method 2:

    (a) There are 5 fluorine atoms bonded to the bromine atom in $BrF_5$.

    (b) and (c)

$4d$ _ _ _ _ _       $4d$ _ _ _

$d^2sp^3$ ⇅ ↑ ↑ ↑ ↑ ↑

$4p$ ⇅ ⇅ ↑

$4s$ ⇅

Br (ground state)   Br (hybrid state)

    (d) We predict the molecular shape to be that of a square pyramid. One of the $d^2sp^3$ orbitals is occupied by a nonbonding pair of electrons.

4.

:F:
|
:Xe:
|
:F:

$dsp^3$ hybrid orbitals

It has been shown experimentally that $XeF_2$ is linear. At this time you may not have been able to successfully predict the shape.

5.   F⟍ :: ⟋F
     ⟍Xe⟍
     F⟋ :: ⟍F          $d^2sp^3$ hybrid orbitals          square planar geometry

6.        :O:
          ‖
     H—C—H              $sp^2$ hybrid orbitals            triangular planar

Method 1:

(a)  The Lewis structure is

              :O:
              ‖
         H—C—H

(b)  number of nonbonding pairs
     in valence shell of carbon   =  0
     number of atoms directly
     bonded to carbon                    =  3
                             sum =   3

(c)  From the previous information and the informa-
     tion in Table 4.1 of the study guide, we pre-
     dict the molecule to have a triangular planar
     shape.

Method 2:

(a)  There are 3 atoms bonded directly to carbon.

(b)  and (c)

     2p ⎯1⎯ ⎯1⎯ ⎯            2p ⎯1⎯
                             $sp^2$ ⎯1⎯ ⎯1⎯ ⎯1⎯



     2s ⎯1↓⎯

     C (ground state)       C (hybrid state)

Note that one of the half-filled $2p$ orbitals
is involved in π bond formation and is not
hybridized.

(d) From the previous information and the information in Table 4.1 of the study guide we predict the molecule to have a triangular planar
shape.  Indeed, it is triangular planar.

III.  Electron pair repulsions and molecular geometry
See Section 4.3 of you text.

1. square planar
configuration

To predict the geometry of a molecule from a consideration of electron pair repulsions, we can use the
following method if the molecule contains no double
bonds:

Electron-pair repulsion method:

(a) Determine the number of electrons in the valence shell of the neutral central atom.  If
the molecule is an ion, account for its charge
by adding or subtracting electrons in the valence shell of the central atom.

(b) Determine the total number of electrons contributed to the central atom's valence shell
by the atoms directly bonded to the central
atom.

(c) Divide the sum of (a) and (b) by 2 to determine the number of electron pairs.

(d) Distribute the electron pairs in the valence
shell of the central atom such that they are
as far away from each other as possible.  Such
orientations of electron pairs are given in
Table 4.2 of the study guide.

(e) Bond the periphereal atoms to the central atom
such that minimum repulsion occurs.  Molecular
shapes predicted from a consideration of minimum electron pair repulsions are given in Table 4.3 of the study guide.

Table 4.2   Orientations of electrons pairs about the center of the central atom of a molecule

| NUMBER OF ELECTRON PAIRS | ORIENTATION |
|---|---|
| 2 | linear |
| 3 | triangular planar |
| 4 | tetrahedral |
| 5 | trigonal bipyramidal |
| 6 | octahedral |

Table 4.3   Number of electron pairs in the valence shell of the central atom and molecular shape

| NUMBER OF ELECTRON PAIRS | | | | |
|---|---|---|---|---|
| TOTAL | BONDING | NONBONDING | SHAPE OF MOLECULE OR ION | EXAMPLES |
| 2 | 2 | 0 | linear | $HgCl_2$, $CuCl_2^-$ |
| 3 | 3 | 0 | triangular planar | $BF_3$, $HgCl_3^-$ |
| 3 | 2 | 1 | angular | $SnCl_2$, $NO_2^-$ |
| 4 | 4 | 0 | tetrahedral | $CH_4$, $BF_4^-$ |
| 4 | 3 | 1 | trigonal pyramidal | $NH_3$, $PF_3$ |
| 4 | 2 | 2 | angular | $H_2O$, $ICl_2^+$ |
| 5 | 5 | 0 | trigonal bipyramidal | $PCl_5$, $SnCl_5^-$ |
| 5 | 4 | 1 | distorted tetrahedral | $TeCl_4$, $IF_4^+$ |
| 5 | 3 | 2 | T shaped | $ClF_3$ |
| 5 | 2 | 3 | linear | $XeF_2$, $ICl_2^-$ |
| 6 | 6 | 0 | octahedral | $SF_6$, $PF_6^-$ |
| 6 | 5 | 1 | square pyramidal | $IF_5$, $SbF_5^{2-}$ |
| 6 | 4 | 2 | square planar | $BrF_4^-$, $XeF_4$ |

Using the electron-pair repulsion method to predict
the geometry of $XeF_4$, we find

(a) The total number of valence electrons in a
neutral atom of xenon is 8.  Since $XeF_4$ is uncharged, no electrons must be subtracted from
or added to the valence shell of xenon.

(b) One electron is donated to the valence shell
of xenon by each fluorine atom.  The total number of electrons donated is therefore 4.

(c) 8 electrons + 4 electrons = 12 electrons
12 electrons/2 = 6 electron pairs

(d) According to the information in Table 4.2 of
the study guide, we predict the electron pairs
to have an octahedral orientation about the
center of the xenon atom.

(e) According to the information in Table 4.3 of
the study guide, we predict the molecule to
have a square planar shape.

2. trigonal
bipyramidal

Electron-pair repulsion method

(a) Five valence electrons are in the valence shell
of arsenic.

(b) Five electrons are contributed by the peripheral atoms, one by each fluorine atom.

(c) There is a total of 5 electron pairs.

(5 + 5)/ 2 = 5

(d) Electron pairs are distributed such that they
have a trigonal bipyramidal orientation.

(e) We predict the shape of the molecule is that
of a trigonal bipyramid.

3. tetrahedral

$$
\begin{array}{c}
\text{F}\\
|\\
\text{F}-\text{Sn}-\text{F}\\
|\\
\text{F}
\end{array}
$$

Electron-pair repulsion method

(a) Four electrons are in the valence shell of tin.

(b) Four electrons are contributed by the peripheral atoms, one by each fluorine atom.

(c) There is a total of 4 electron pairs.

(d) Electron pairs are distributed such that they have a tetrahedral orientation.

(e) We predict the shape of the moleucle to be that of a tetrahedron.

4. angular

(a) There are seven electrons in the valence shell of a neutral iodine atom.  Since the molecule has a 1+ charge, we must subtract one electron from the valence shell of the iodine atom.  Thus, the valence shell of the iodine atom has 6 electrons.

(b) Two electrons are contributed by the peripheral atoms, one by each fluorine atom.

(c) There is a total of 4 electron pairs.

(d) Electron pairs are distributed such that they have a tetrahedral orientation.

(e) We predict the molecule to have an angular shape.

IV.  Molecular orbitals

It is imperative that you read your class notes very meticulously to ascertain the depth to which your instructor wishes you to master this material.

1. In an NO Molecule 15 electrons are distributed as described in your text:

$$(\sigma_{1s})^2 \ (\sigma_{1s}^*)^2 \ (\sigma_{2s})^2 \ (\sigma_{2s}^*)^2 \ (\sigma_{2p})^2 \ (\pi_{2p})^4 \ (\pi_{2p}^*)^1$$

The bond order is $2\frac{1}{2}$:

$$\text{bond order} = \frac{10 \text{ bonding electrons } - \text{ 5 antibonding electrons}}{2} = 2\frac{1}{2}$$

A molecular orbital energy-level diagram for the higher-energy molecular orbitals of NO can be drawn:

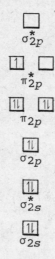

From this diagram we readily see that there is one unpaired electron in an NO molecule.  Thus, an NO molecule is paramagnetic.  Lewis structures do not give an adequate description of the NO molecule:

$$:\overset{\cdot}{N} = \overset{\cdot\cdot}{O}: \longleftrightarrow :\overset{\cdot\cdot}{N} = \overset{\cdot\cdot}{O}: \longleftrightarrow :\overset{\cdot}{N} = \overset{\cdot}{O}:$$

2. In a $CN^-$ ion there are 14 electrons to be distributed in molecular orbitals:

$$(\sigma_{1s})^2 \ (\sigma_{1s}^*)^2 \ (\sigma_{2s})^2 \ (\sigma_{2s}^*)^2 \ (\pi_{2p})^4 \ (\sigma_{2p})^2$$

The bond order is 3:

$$\text{bond order} = \frac{10 \text{ bonding electrons} \ - \ 4 \text{ antibonding electrons}}{2} = 3$$

A molecular orbital energy-level diagram shows that there are no unpaired electrons in a $CN^-$ ion.  A $CN^-$ ion is therefore diamagnetic.

The Lewis structure

$$[:C \equiv N:]^-$$

gives an adequate description of the bond order of the molecule.

3. In an $O_2^+$ ion there are 15 electrons to be distributed in molecular orbitals:

$$(\sigma_{1s})^2 \; (\sigma_{1s}^*)^2 \; (\sigma_{2s})^2 \; (\sigma_{2s}^*)^2 \; (\sigma_{2p})^2 \; (\pi_{2p})^4 \; (\pi_{2p}^*)^1$$

The bond order is $2\frac{1}{2}$:

$$\text{bond order} = \frac{10 \text{ bonding electrons } - \; 5 \text{ antibonding electrons}}{2} = 2\frac{1}{2}$$

A molecular orbital diagram shows that there is one unpaired electron in the $O_2^+$ ion. An $O_2^+$ ion therefore should be paramagnetic.

Lewis structures do not give an adequate description of the bond order of the ion:

$$:\!\overset{..}{O} = \overset{..}{O}: \;\longleftrightarrow\; :\!\overset{..}{O} = \overset{..}{O}: \;\longleftrightarrow\; :\!\overset{..}{O} \equiv O:$$

V.   Geometric configurations of molecules

1. triangular planar

The Lewis structure is

$$\begin{array}{c} :O: \\ \| \\ H\!-\!C\!-\!H \end{array}$$

According to the concept of electron pair repulsions, we predict the $\sigma$ bonding pairs of electrons to be distributed about the center of the carbon atom in the configuration of a plane triangle:

$$\begin{array}{c} O \\ \| \\ H \overset{C}{\underset{120^\circ}{\leftrightarrow}} H \end{array} \; 120^\circ$$

The central carbon atom is said to be $sp^2$ hybridized, i.e., carbon uses $sp^2$ hybrid orbitals in its bonding. The $2s$ orbital and two of the $2p$ orbitals of carbon form the $sp^2$ hybrid orbitals of carbon while the remaining $p$ orbital of carbon overlaps a $p$ orbital of oxygen to form a $\pi$ bond.

2. hexagonal
   planar

There are two resonance forms of benzene:

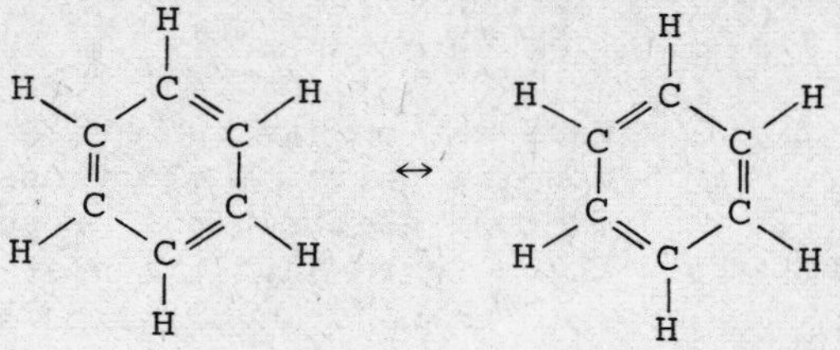

$\leftrightarrow$

According to the concept of electron pair repulsions, we predict the σ bonding pairs of electrons to be distributed about the centers of the appropriate carbon atoms in the configuration of a plane triangle:

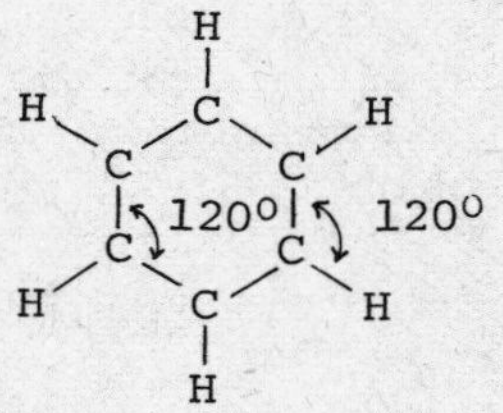

Each carbon is said to be $sp^2$ hybridized. The $2s$ orbital and two of the $2p$ orbitals of carbon form the $sp^2$ hybrid orbitals of carbon while the remaining $p$ orbital of carbon enters into π bonding.

3. tetrahedral

The Lewis structure is

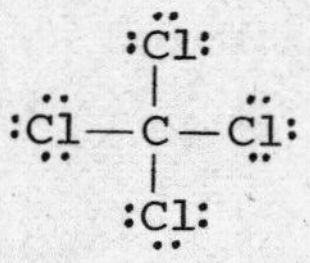

According to the concept of electron pair repulsions, we predict a tetrahedral geometry:

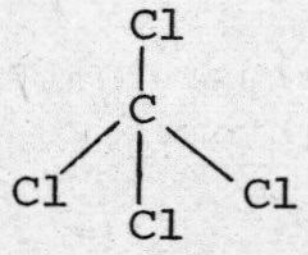

The carbon is said to be $sp^3$ hybridized.

4. linear

The Lewis structure is

$$H-C \equiv C-H$$

Each carbon has two atoms bonded to it and no non-bonding pairs of electrons.  According to the concept of electron pair repulsions, we predict the molecule to have a linear shape.  Each carbon is said to be *sp* hybridized and forms two σ bonds.

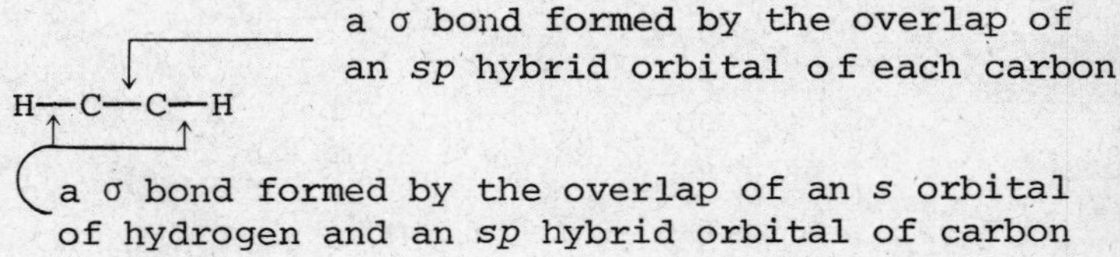

a σ bond formed by the overlap of an *sp* hybrid orbital of each carbon

a σ bond formed by the overlap of an *s* orbital of hydrogen and an *sp* hybrid orbital of carbon

Each carbon also forms two π bonds.  These bonds are formed by the overlap of the unhybridized *p* orbitals of each carbon.

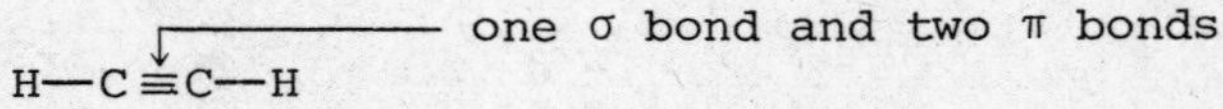

one σ bond and two π bonds

5. angular

The resonance forms of $SO_2$ are

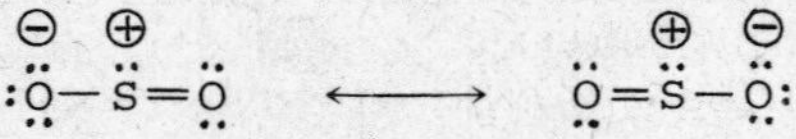

Sulfur is bonded to two atoms and has one nonbonding pair of electrons.  According to electron pair repulsions, we predict the σ and nonbonding electron pairs to be distributed about the center of the sulfur atom in the configuration of a plane triangle:

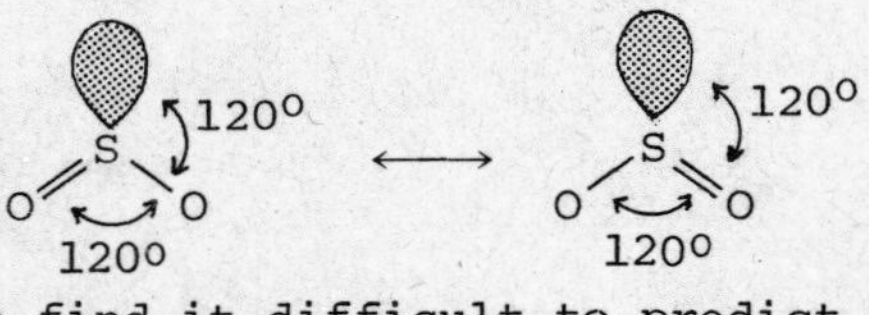

We find it difficult to predict the bonding of the oxygen atoms because the π bond is not localized between any two atoms:

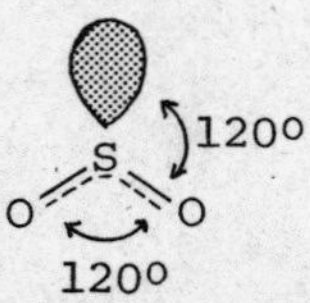

6. linear

The resonance forms of $CO_2$ are

$$:\overset{..}{\underset{..}{O}}-C=\overset{..}{\underset{..}{O}} \quad \longleftrightarrow \quad \overset{..}{\underset{..}{O}}=C-\overset{..}{\underset{..}{O}}:$$

Since carbon is bonded to two atoms and has no non-bonding pairs of electrons, we predict a linear molecular structure. The $\pi$ bond is not localized between any two atoms.

7. linear

The Lewis structure is

$$H:\overset{..}{\underset{..}{F}}:$$

The fluorine is said to be $sp^3$ hybridized. A tetrahedral arrangement of the bonding pair of electrons and the nonbonding pairs of electrons provides the most effective separation of electron pairs.

VI.   Structure of molecules

1. a, b, c, f,
   h, j, l

The water molecule is angular. The individual dipoles combine to make the molecule a dipole. The direction of an individual dipole is indicated by an arrow, the arrow pointing toward the negative end of the dipole.

Ammonia is triangular pyramidal. The individual dipoles combine to make the molecule a dipole.

The individual dipoles cancel in a molecule that is symmetrical about a central point:

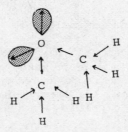

(nonbonding          (nonbonding
electron pairs       electron pairs
of Cl omitted        of F omitted for
for clarity)         clarity)

A $(CH_3)_2O$ molecule, like water, is angular.  The individual dipoles combine to make the molecule a dipole.

Because the molecule $SO_2$ is angular, the individual dipoles do not cancel:

The $IF_3$ molecule is T-shaped and the individual dipoles do not cancel:

   (nonbonding electron pairs of
    fluorines omitted for clarity)

The individual bond dipoles of $XeF_4$ totally cancel each other and the molecule is nonpolar.

F, F
  Xe
F  F

(nonbonding electron pairs of
   fluorines omitted for clarity)

2. d, e, k, m            See Section 4.6 of your text.

3. a, b, c, d, e, f      See Section 4.5 of your text.

a.  The oxygen atom probably uses $sp^3$ hybrid
    orbitals.

b.  The nitrogen atom probably uses $sp^3$ hybrid
    orbitals.

c.  The chlorine atom *possibly* uses $sp^3$ hybrid
    orbitals.

d.  Each carbon atom probably uses $sp^3$ hybrid
    orbitals.

e.  The carbon atom probably uses $sp^3$ hybrid
    orbitals, and the chlorine atom *possibly* uses
    $sp^3$ hybrid orbitals.

f.  The oxygen atom and the carbon atoms each
    probably use $sp^3$ hybrid orbitals.

4. h            See Sections 4.4 and 4.5 of your text.

5. g            See Section 4.9 of your text.

6. j            See Section 4.9 of your text.

7. d            See Sections 4.4 and 4.5 of your text.

8. g, i         See Section 4.4 of your text.

g.  An atom of sodium has a single electron in the
    2s subshell.

   i.  An atom of carbon has 2 unpaired electrons in
       the 2p subshell.

9. a, c, h, j, l, m     See Section 4.3 of your text.

10. d


SELF-TEST              Complete the test in 30 minutes:

           1.  Complete the table:

| FORMULA | NUMBER OF ELECTRON PAIRS | | SHAPE OF MOLECULE OR ION |
| | BONDING | NONBONDING | |
| --- | --- | --- | --- |
| $SCl_2$ | _____ | _____ | _____ |
| $XeF_4$ | _____ | _____ | _____ |
| $AlH_4^-$ | _____ | _____ | _____ |
| $TeCl_4$ | _____ | _____ | _____ |
| $SeF_5^-$ | _____ | _____ | _____ |

           2.  Crystalline $PCl_5$ conists of an ionic lattice of
               $PCl_4^+$ and $PCl_6^-$ ions.  In the vapor and liquid
               states, however, the compound exists as $PCl_5$ mole-
               cules.  What type of hybrid orbitals does P employ
               in each of these species, and what is the geometry
               of the molecule or ion?

|  | HYBRID ORBITALS | GEOMETRIC SHAPE |
| --- | --- | --- |
| $PCl_5$ | _____ | _____ |
| $PCl_4^+$ | _____ | _____ |
| $PCl_6^-$ | _____ | _____ |

           3.  Complete the following table.  The σ orbitals in-
               clude only the σ2s and σ2p orbitals, and the σ*
               orbitals include only the σ*2s and σ*2p orbitals.

| MOLECULE | TOTAL NUMBER OF ELECTRONS IN | | | | BOND ORDER | NUMBER OF UNPAIRED ELECTRONS |
|---|---|---|---|---|---|---|
| | $\sigma$ ORBITALS | $\sigma^*$ ORBITALS | $\pi$ ORBITALS | $\pi^*$ ORBITALS | | |
| $Be_2$ | _____ | _____ | _____ | _____ | _____ | _____ |
| $B_2$ | _____ | _____ | _____ | _____ | _____ | _____ |
| $N_2$ | _____ | _____ | _____ | _____ | _____ | _____ |
| $O_2$ | _____ | _____ | _____ | _____ | _____ | _____ |
| $NO^+$ | _____ | _____ | _____ | _____ | _____ | _____ |

4. The thiocyanate ion, SCN⁻, is linear and the atoms are arranged in the order given. Draw Lewis structures for the three resonance forms of the ion *complete with formal charges*.

# 5

# Chemical Equations and Quantitative Relations

OBJECTIVES

(a) You should be able to demonstrate your knowledge of the following terms by defining them, describing them, or giving specific examples of them:

>Avogadro's number [5.1]
>Born-Haber cycle [5.10]
>calorie [5.7]
>calorimeter [5.7]
>combustion [5.7]
>empirical (also called simplest) formula [5.4]
>endothermic [5.7]
>enthalpy [5.7]
>enthalpy of combustion [5.8]
>enthalpy of formation [5.8]

enthalpy of reaction [5.7]
exothermic [5.7]
heat capacity [5.7]
joule [5.7]
lattice energy [5.10]
law of conservation of mass [5.5]
law of constant composition [5.3]
law of Hess [5.8]
limiting reagent [5.6]
mole [5.1]
molecular formula [5.4]
standard state [5.8]

(b) You should be able to perform calculations to obtain
numbers of moles and molecules, percent composition,
empirical formulas, and molecular formulas.

(c) You should be able to balance chemical equations and
use them in stoichiometric calculations.

(d) You should be able to calculate heats of reaction
(enthalpies of reaction) and heat capacities.

(e) You should be able to use the law of Hess and to
construct Born-Haber cycles for thermochemical calculations.

EXERCISES

I. Chemical calculations should always be performed
with the units of measurement included with all numbers. Work the following to gain expertise in using
proper units, i.e., labels:

1. Approximately 200 million tires are discarded
annually. If the average tire weighs 30 pounds,
how many pounds of tires are discarded in a year?

2. How many kilograms of tires are discarded per
year? See problem 1 of this section.

3. If 25% of the rubber is recoverable for use as
blacktop material, how many grams of tires per
year could be recycled? See problem 1 of this
section.

II. Work the following problems. Recall that a factor
is meaningless without corresponding units, the label.

1. A diamond weighs 2.42 carats. Knowing that a diamond is pure carbon, calculate the number of moles of carbon in the diamond. One carat is 0.200 g.

2. How many atoms are in the diamond of problem 1 of this section?

3. Most iron is obtained from $Fe_2O_3$. What is the percentage of iron (by weight) in $Fe_2O_3$?

4. Gibbsite, $Al_2O_3 \cdot 3H_2O$, is a naturally occurring material from which aluminum is produced.[1] What is the percentage of aluminum (by weight) in gibbsite?

5. What is the percentage of copper (by weight) in malachite, $CuCO_3 \cdot Cu(OH)_2$?[1]

6. Chemical analyses are often performed to determine the percent composition of a pure material, and from the information the empirical formula can be determined. An organic material is analyzed and found to contain 75.92% C, 17.71% N, and 6.37% H. Determine the empirical formula of the compound.

7. Tritopine, an alkaloid isolated from opium, has been found to be 74.0% C, 7.90% H, 14.0% O, and 4.10% N by weight. What is the empirical formula? The molecular weight is known to be 682 g/mol. What is the molecular formula?

8. Commercially iron is obtained from the reduction of hematite, $Fe_2O_3$:

$$Fe_2O_3 + CO \rightarrow Fe + CO_2$$

How many grams of iron can be obtained from 5.24 g $Fe_2O_3$?

9. How many grams of iron can be obtained by the reaction of 2.78 g CO with excess $Fe_2O_3$? (See problem 8 of this section.)

---

1. Many minerals and other materials are combinations of two or more compounds in specific ratios. Gibbsite is a combination of aluminum oxide, $Al_2O_3$, and water in a 1:3 ratio.

10. The carbon monoxide needed for producing iron by the reaction given in problem 8 of this section is obtained by the reaction of oxygen with carbon, which is in the form of coke:

$$C + O_2 \rightarrow CO$$

How much iron can be obtained from 1.47 g C?

11. Refer to problem 8 of this section and answer the following:
    a. How much iron can be obtained from 4.02 g $Fe_2O_3$ and 1.78 g CO?
    b. What are the masses of all compounds after the reaction is complete?

12. How much iron can be obtained by reducing a short ton (1 short ton = 2000 lb) of iron ore that contains 84.6% hematite, $Fe_2O_3$? (See problem 8 of this section.)

13. What is the minimum amount of carbon monoxide required for the complete reduction of all hematite in a short ton of ore that is 84.6% hematite? (See problem 8 of this section.)

14. Ethyl alcohol, $C_2H_5OH$, is a product of the fermentation of sugars. In the presence of the enzyme invertase, sucrose undergoes conversion into invert sugar.[2]

$$C_{12}H_{22}O_{11} + H_2O \xrightarrow{\text{invertase}} 2C_6H_{12}O_6$$

The invert sugar is converted into ethyl alcohol and carbon dioxide by the enzyme zymase. Usually only half the sucrose is converted into alcohol, a 50% yield.

$$C_6H_{12}O_6 \xrightarrow{\text{zymase}} 2C_2H_5OH + 2CO_2$$

How much ethyl alcohol, $C_2H_5OH$, can be obtained from 171 g sucrose, $C_{12}H_{22}O_{11}$?

---

2. A catalyst such as an enzyme is not written in the balanced equation of a reaction; rather, it is written over the reaction arrow of the equation. Zymase increases the rate of conversion of sucrose to alcohol such that the conversion proceeds at a reasonable rate.

15. A 0.7234 g sample of hemoglobin is isolated and analyzed for iron. If the hemoglobin is found to contain 0.342% Fe and it is assumed that four ferrous ions, $Fe^{2+}$, are incorporated in each hemoglobin molecule, calculate the molecular weight of hemoglobin.

16. What is the mass of one atom of gold?

17. A dye called bromosulfaphthalein, BSP, is excreted almost entirely by the liver, and the presence of the dye has been used since 1925 in a general liver function test. The dosage of BSP is proportional to body weight. If the dose of BSP should be 5.0 mg per kg body weight and a BSP solution is available that is 5.0% BSP by weight, what weight of the BSP solution is needed to test the liver function of a person weighing 110 lb?

18. A solution contains 54.7% sulfuric acid by weight. What weight of solution will contain 124 g $H_2SO_4$?

19. An ore known to contain hematite, $Fe_2O_3$, as the only iron containing species is analyzed for iron. It is found to contain 27.42% Fe by weight. What is the percentage of hematite (by weight) in the ore?

III. Use the data in the following table when needed to solve the following thermochemical problems. For alternate molecules write the equation of the reaction yielding the indicated $\Delta H_f$ in the space provided.

Enthalpies of formation at $25^\circ$C

| MOLECULE | $\Delta H_f$ (kcal) | $\Delta H_f$ (kJ) | EQUATION |
|---|---|---|---|
| CO(g) | −26.4 | −100.4 | |
| $CO_2$(g) | −94.1 | −393.7 | |
| $Fe_2O_3$(s) | −196.5 | −822.2 | |
| HCl(g) | −22.1 | −92.5 | |
| $H_2O$(g) | −57.8 | −241.8 | |
| $H_2O$(l) | −68.3 | −285.8 | |

| | | |
|---|---|---|
| $H_2SO_4(l)$ | -193.9 | -811.3 |
| $I_2(g)$ | +14.9 | +62.3 |
| $NH_3(l)$ | -16.0 | -66.9 |
| $NO(g)$ | +21.6 | +90.4 |
| $NO_2(g)$ | +12.6 | +52.7 |
| $O_3(g)$ | +34.0 | +142.3 |
| $PCl_3(g)$ | -73.2 | -306.3 |
| $PCl_5(g)$ | -95.4 | -399.2 |
| $POCl_3(g)$ | -141.5 | -592.0 |
| $SO_2(g)$ | -71.0 | -297.0 |
| $SO_3(g)$ | -94.4 | -395.0 |

1. What is the enthalpy change at 25°C for the following reaction:

$$SO_3(g) \; + \; H_2O(l) \longrightarrow H_2SO_4(l)$$

2. What is the enthalpy change at 25°C for the following reaction:

$$PCl_5(g) \; + \; H_2O(g) \longrightarrow POCl_3(g) \; + \; 2HCl(g)$$

3. The enthalpy of combustion of pentane, $C_5H_{12}(g)$, to $CO_2(g)$ and $H_2O(l)$ is -845.2 kcal/mol [or -3526.3 kJ/mol]. What is the enthalpy of formation of pentane at 25°C?

4. A 12.45 g sample of $P_4O_{10}(s)$ is reacted with a stoichiometrically equivalent quantity of water in a vessel placed in a calorimeter containing 950.0 g of water. The temperature of the calorimeter and its contents increases from 22.815°C to 26.885°C. If the calorimeter has a heat absorbing capacity equal to that of 165.0 g $H_2O$, what is the enthalpy change for the following reaction?

$$P_4O_{10}(s) \; + \; 6H_2O(l) \longrightarrow 4H_3PO_4(aq)$$

Assume the specific heat of water to be 1.000 cal/g°C.

5. Calculate the lattice energy of potassium chloride from the following data.  For potassium the ionization energy is 99 kcal/mol, and the enthalpy of sublimation is 21 kcal/mol.  For chlorine the dissociation energy is 58 kcalmol, and the electron affinity is -88 kcal/mol.  The enthalpy of formation of KCl is -104 kcal/mol.

ANSWERS TO
EXERCISES

I. Factor-label method

1. $6.0 \times 10^9$ lb
   [5.2]

You normally use the factor-label method in solving problems, but you may not have thought of the process in precisely this way:

$$? \text{ lb} = 200 \times 10^6 \text{ tires} \left(\frac{30 \text{ lb}}{1 \text{ tire}}\right)$$

$$= 6000 \times 10^6 \text{ lb}$$

$$= 6.0 \times 10^9 \text{ lb}$$

The unwanted labels cancel.  If you do not have the proper label, the answer cannot possibly be correct.

2. $2.7 \times 10^9$ kg
   [5.2]

$$? \text{ kg} = 6.0 \times 10^9 \text{ lb} \left(\frac{1.00 \text{ kg}}{2.20 \text{ lb}}\right)$$

$$= 2.7 \times 10^9 \text{ kg}$$

Problems 1 and 2 of this section are not fundamentally different.  A number with its label is multiplied by a ratio that has equivalent quantities in numerator and denominator. The mathematical operation is set up such that the unwanted label cancels and the desired label remains.

3. $6.8 \times 10^{11}$ g recycled
   [5.2]

$$? \text{ g recycled} = 2.7 \times 10^9 \text{ kg} \left(\frac{1000 \text{ g}}{1.00 \text{ kg}}\right)\left(\frac{25 \text{ recycled}}{100}\right)$$

The calculation, if performed to this point, yields $2.7 \times 10^{12}$ g.

$$= 6.8 \times 10^{11} \text{ g recycled}$$

The answer can be obtained directly from the initial data by using a logic chain and avoiding intermediate calculations:

$$? \text{ g recycled} = 200 \times 10^6 \text{ tires} \left(\frac{30 \text{ lb}}{1 \text{ tire}}\right) \left(\frac{1000 \text{ g}}{2.20 \text{ lb}}\right) \left(\frac{25 \text{ recycled}}{100}\right)$$

$$= 6.8 \times 10^{11} \text{ g recycled}$$

The preceding is a faster and easier method of performing calculations, but there is a danger. *Do not forget that this method is the same as doing the problem in steps.* Note the units remaining after each multiplication in this problem. Observe that the procedure is the same as the stepwise analysis, except that intermediate answers are not recorded. Note the progressive change in units.

$$? \text{ g recycled} = 200 \times 10^6 \text{ tires} \left(\frac{30 \text{ lb}}{\text{tire}}\right) \left(\frac{1 \text{ kg}}{2.20 \text{ lb}}\right) \left(\frac{1000 \text{ g}}{1 \text{ kg}}\right) \left(\frac{25 \text{ recycled}}{100}\right)$$

$$\text{lb} \rightarrow \text{kg} \rightarrow \text{g} \rightarrow \text{g recycled}$$

## II. Chemical calculations

1. $4.03 \times 10^{-2}$ mol C [5.1]

$$? \text{ mol C} = 2.42 \text{ carat C} \left(\frac{0.200 \text{ g C}}{1 \text{ carat C}}\right) \left(\frac{1 \text{ mol C}}{12.01 \text{ g C}}\right)$$

The mass of carbon has been calculated.

$$= 0.0403 \text{ mol C}$$
$$= 4.03 \times 10^{-2} \text{ mol C}$$

2. $2.43 \times 10^{22}$ atoms C [5.1]

$$? \text{ atoms C} = 4.03 \times 10^{-2} \text{ mol C} \left(\frac{6.022 \times 10^{23} \text{ atoms C}}{1 \text{ mol C}}\right)$$

$$= 2.43 \times 10^{22} \text{ atoms C}$$

The logic chain, sometimes called the sausage approach, could also be used with the original data:

$$? \text{ atoms C} = 2.42 \text{ carat C} \left(\frac{0.200 \text{ g C}}{1 \text{ carat C}}\right) \left(\frac{1 \text{ mol C}}{12.01 \text{ g C}}\right) \left(\frac{6.022 \times 10^{23} \text{ atoms C}}{1 \text{ mol C}}\right)$$

$$= 2.43 \times 10^{22} \text{ atoms C}$$

3. $69.9434\%$ Fe [5.3]

The formula indicates the ratio of moles. From this the ratio of masses must be calculated to determine percent composition.

In one mole of $Fe_2O_3$ there would be how many grams Fe and how many grams O?

$$? \text{ g Fe/1 mol } Fe_2O_3 = \left(\frac{55.847 \text{ g Fe}}{1 \text{ mol Fe}}\right)\left(\frac{2 \text{ mol Fe}}{1 \text{ mol } Fe_2O_3}\right)$$

$$= 111.694 \text{ g Fe/1 mol } Fe_2O_3$$

$$? \text{ g O/1 mol } Fe_2O_3 = \left(\frac{15.9994 \text{ g O}}{1 \text{ mol O}}\right)\left(\frac{3 \text{ mol O}}{1 \text{ mol } Fe_2O_3}\right)$$

$$= 47.9982 \text{ g O/1 mol } Fe_2O_3$$

The mass of one mole of $Fe_2O_3$ is the sum of the masses of the component parts, 159.692 g $Fe_2O_3$ per 1 mol $Fe_2O_3$.

Percent composition is the mass of a component divided by the mass of the whole times 100. Therefore,

$$\% \text{ Fe in } Fe_2O_3 = \left(\frac{111.694 \text{ g Fe}}{159.692 \text{ g } Fe_2O_3}\right) 100$$

$$= 69.9434\% \text{ Fe}$$

4. 34.59019% Al   First we calculate the mass of one mole of gibb-
   [5.3]          site:

$$? \text{ g Al/1 mol } Al_2O_3 \cdot 3H_2O = \left(\frac{26.98154 \text{ g Al}}{1 \text{ mol Al}}\right)\left(\frac{2 \text{ mol Al}}{1 \text{ mol } Al_2O_3 \cdot 3H_2O}\right)$$

$$= 53.96308 \text{ g Al/1 mol } Al_2O_3 \cdot 2H_2O$$

$$? \text{ g O/1 mol } Al_2O_3 \cdot 3H_2O = \left(\frac{15.9994 \text{ g O}}{1 \text{ mol O}}\right)\left(\frac{6 \text{ mol O}}{1 \text{ mol } Al_2O_3 \cdot 3H_2O}\right)$$

$$= 95.9964 \text{ g O/1 mol } Al_2O_3 \cdot 3H_2O$$

$$? \text{ g H/1 mol } Al_2O_3 \cdot 3H_2O = \left(\frac{1.0079 \text{ g H}}{1 \text{ mol H}}\right)\left(\frac{6 \text{ mol H}}{1 \text{ mol } Al_2O_3 \cdot 3H_2O}\right)$$

$$= 6.0474 \text{ g H/1 mol } Al_2O_3 \cdot 3H_2O$$

mass of one mole of gibbsite = 53.96308 g + 6.0474 g + 95.9964 g = 156.0069 g

Then we calculate the percentage of aluminum:

$$\% \text{ Al} = \left(\frac{\text{mass Al}}{\text{mass Al}_2\text{O}_3 \cdot 3\text{H}_2\text{O}}\right) 100$$

$$= \left(\frac{53.96308 \text{ g}}{156.0069 \text{ g}}\right) 100$$

$$= 34.59019 \% \text{ Al}$$

5. 57.4775% Cu
[5.3]

First we calculate the mass of a mole of malachite:

$$? \text{ g CuCO}_3 \cdot \text{Cu(OH)}_2 = 2 \text{ mol Cu}\left(\frac{63.546 \text{ g Cu}}{1 \text{ mol Cu}}\right) + 5 \text{ mol O}\left(\frac{15.9994 \text{ g O}}{1 \text{ mol O}}\right) +$$

$$1 \text{ mol C}\left(\frac{12.011 \text{ g C}}{1 \text{ mol C}}\right) + 2 \text{ mol H}\left(\frac{1.0079 \text{ g H}}{1 \text{ mol H}}\right)$$

$$= 221.11 \text{ g}$$

Then we calculate the percentage of copper:

$$\% \text{ Cu} = \left(\frac{\text{mass Cu}}{\text{mass CuCO}_3 \cdot \text{Cu(OH)}_2}\right) 100$$

$$= \left(\frac{127.092 \text{ g}}{221.116 \text{ g}}\right) 100$$

$$= 57.4775\% \text{ Cu}$$

6. $C_5H_5N$
[5.4]

We must determine the ratio of moles of atoms in the molecule. We assume for convenience that there is one hundred grams of material. From the percentages given in the problem, there would be 75.92 g C, 17.71 g N, and 6.37 g H in the sample.

$$? \text{ mol C} = 75.92 \text{ g C}\left(\frac{1 \text{ mol C}}{12.011 \text{ g C}}\right)$$

$$= 6.321 \text{ mol C}$$

$$? \text{ mol N} = 17.71 \text{ g N}\left(\frac{1 \text{ mol N}}{14.007 \text{ g N}}\right)$$

$$= 1.264 \text{ mol N}$$

$$? \text{ mol H} = 6.37 \text{ g H}\left(\frac{1 \text{ mol H}}{1.008 \text{ g H}}\right)$$

$$= 6.319 \text{ mol H}$$

The compound has the formula $C_{6.321}N_{1.264}H_{6.3119}$, but this is not an acceptable representation since atoms combine in whole number ratios. To reduce any formula to the nearest whole number ratio, we divide by the smallest number. Thus, the empirical formula is

$$\frac{C_{6.321}N_{1.264}H_{6.319}}{1.264 \quad 1.264 \quad 1.264} = C_5NH_5$$

For this compound the empirical formula is the same as the molecular formula, and the molecule is pyridine, which has the structure

7. empirical formula: $C_{21}H_{27}O_3N$; molecular formula: $C_{42}H_{54}O_6N_2$

We assume for convenience that we have a 100 g sample, and we determine the mole ratios:

$$? \text{ mol C} = 74.0 \text{ g C} \left(\frac{1 \text{ mol C}}{12.01 \text{ g C}}\right)$$

$$= 6.16 \text{ mol C}$$

$$? \text{ mol H} = 7.90 \text{ g H} \left(\frac{1 \text{ mol H}}{1.008 \text{ g H}}\right)$$

$$= 7.84 \text{ mol H}$$

$$? \text{ mol O} = 14.0 \text{ g O} \left(\frac{1 \text{ mol O}}{16.00 \text{ g O}}\right)$$

$$= 0.875 \text{ mol O}$$

$$? \text{ mol N} = 4.10 \text{ g N} \left(\frac{1 \text{ mol N}}{14.01 \text{ g N}}\right)$$

$$= 0.293 \text{ mol N}$$

The nearest whole number ratio for this compound is $C_{21}H_{27}O_3N$, which has a molecular weight of 341 g/mol. The molecular weight is found by determining what multiple the weight of the empirical formula is of the actual molecular weight:

$$\frac{\text{actual molecular weight}}{\text{molecular weight of empirical formula}} = ?$$

$$\frac{682 \text{ g/mol}}{341 \text{ g/mol}} = 2$$

Thus, the molecular formula is twice the empirical formula, and the molecular formula is

$$C_{42}H_{54}O_6N_2$$

8. 3.67 g Fe
   [5.5]

First we make sure the chemical equation is balanced. The equation as written in the problem is not balanced. The balanced equation

$$Fe_2O_3 + 3CO \longrightarrow 2Fe + 3CO_2$$

states that 1 mol $Fe_2O_3$ and 3 mol CO react to give 2 mol Fe and 3 mol $CO_2$. The molecular weights of all compounds can be determined from atomic weights. The molecular weights are: $Fe_2O_3$, 159.692 g/mol; CO, 28.010 g/mol; Fe, 55.847 g/mol; and $CO_2$, 44.010 g/mol. From this information it is obvious that the chemical equation also says that 159.692 g $Fe_2O_3$ and (3 x 28.010) g CO react to yield (2 x 55.847) g Fe and (3 x 44.010) g $CO_2$.

$$Fe_2O_3 + 3CO \longrightarrow 2Fe + 3CO_2$$

$$1 \text{ mol } Fe_2O_3 + 3 \text{ mol } CO \longrightarrow 2 \text{ mol } Fe + 3 \text{ mol } CO_2$$

$$159.692 \text{ g } Fe_2O_3 + (3 \times 28.010) \text{ g } CO \longrightarrow (2 \times 55.847) \text{ g } Fe + (3 \times 28.010) \text{ g } CO_2$$

$$159.692 \text{ g } Fe_2O_3 + 84.030 \text{ g } CO \longrightarrow 111.694 \text{ g } Fe + 84.030 \text{ g } CO_2$$

Remember this type of relationship; such relationships are necessary for solving problems involving chemical reactions. The logic chain can be used to solve the problem:

$$? \text{ g Fe} = 5.24 \text{ g } Fe_2O_3 \left( \frac{(2 \times 55.85) \text{ g Fe}}{159.7 \text{ g } Fe_2O_3} \right)$$

$$= 3.67 \text{ g Fe}$$

As long as we keep track of the units, we can use other logic chains:

$$? \text{ g Fe} = 5.24 \text{ g Fe}_2O_3 \left(\frac{1 \text{ mol Fe}_2O_3}{159.7 \text{ g Fe}_2O_3}\right) \left(\frac{2 \text{ mol Fe}}{1 \text{ mol Fe}_2O_3}\right) \left(\frac{55.85 \text{ g Fe}}{1 \text{ mol Fe}}\right)$$

<table>
<tr><td>mol Fe$_2$O$_3$<br>calculated<br>from molec.<br>wt. of Fe$_2$O$_3$</td><td>→</td><td>mol Fe<br>calculated<br>knowing that<br>2 mol Fe<br>are obtained<br>from 1 mol<br>Fe$_2$O$_3$</td><td>→</td><td>g Fe<br>calculated<br>using<br>atomic wt.<br>of Fe</td></tr>
</table>

$$= 3.67 \text{ g Fe}$$

A logic chain will always yield the correct answer
if we make sure all unwanted units cancel and all
ratios by which we multiply the original data are
equal to 1.

9. 3.70 g Fe

Since all the CO reacts and some Fe$_2$O$_3$ is left over,
the original amount of CO limits the amount of Fe
that can be produced:

$$? \text{ g Fe} = 2.78 \text{ g CO} \left(\frac{(2 \times 55.85) \text{ g Fe}}{(3 \times 28.01) \text{ g CO}}\right)$$

$$= 3.70 \text{ g Fe}$$

10. 4.56 g Fe

First we make sure all equations are balanced:

$$2C + O_2 \rightarrow 2CO$$

2 mol C                2 mol CO
(2 x 12.011) g C    (2 x 28.010) g CO

Using the information in the preceding equation and
the one for Fe$_2$O$_3$ reduction (see problem 8 of this
section), one can write many logic chains. Some of
the following are more efficient than others, but
your concern should be to find a method that is log-
ical to you:

$$? \text{ g Fe} = 1.47 \text{ g C} \left(\frac{(2 \times 28.01) \text{ g CO}}{(2 \times 12.01) \text{ g C}}\right) \left(\frac{(2 \times 55.85) \text{ g Fe}}{(3 \times 28.01) \text{ g CO}}\right)$$

g CO                →       g Fe obtained from
calculated                  stoichiometry

$$= 4.56 \text{ g Fe}$$

or

$$? \text{ g Fe} = 1.47 \text{ g C} \left(\frac{1 \text{ mol C}}{12.01 \text{ g C}}\right)\left(\frac{2 \text{ mol Fe}}{3 \text{ mol C}}\right)\left(\frac{55.85 \text{ g Fe}}{1 \text{ mol Fe}}\right)$$

| initial number of mol C | → | mol Fe obtained from reaction stoichiometry | → | g Fe obtained from molecular weight of Fe |

$$= 4.56 \text{ g Fe}$$

or

$$? \text{ g Fe} = 1.47 \text{ g C} \left(\frac{1 \text{ mol C}}{12.01 \text{ g C}}\right)\left(\frac{1 \text{ mol CO}}{1 \text{ mol C}}\right)\left(\frac{2 \text{ mol Fe}}{3 \text{ mol CO}}\right)\left(\frac{55.85 \text{ g}}{1 \text{ mol Fe}}\right)$$

| initial mol C | → | mol CO from reaction stoichiometry | → | mol Fe from reaction stoichiometry | → | g Fe from molecular weight of Fe |

$$= 4.56 \text{ g Fe}$$

11. a.  2.37 g Fe
    [5.5]

In this problem the quantities of two reagents are specified. Most probably the reagents are not mixed in the exact reaction ratio. Therefore, one of them will not react completely and the other will. In order to work the problem, we must identify the reagent that reacts completely, since it limits the amount of product that can be obtained:

Method 1:

If we have an electronic calculator, we can solve the problem very quickly by assuming first one and then the other reagent is limiting. If $Fe_2O_3$ were limiting,

$$? \text{ g Fe} = 4.02 \text{ g Fe}_2O_3 \left(\frac{(2 \times 55.85) \text{ g Fe}}{(1 \times 159.69) \text{ g Fe}_2O_3}\right)$$

$$= 2.81 \text{ g Fe}$$

If CO were limiting,

$$? \text{ g Fe} = 1.78 \text{ g CO} \left(\frac{(2 \times 55.85) \text{ g Fe}}{(3 \times 28.01) \text{ g CO}}\right)$$

$$= 2.37 \text{ g Fe}$$

It is obvious that only 2.37 g Fe could be produced. Since all the CO is consumed to produce this amount of iron, the remaining $Fe_2O_3$ cannot react. The reactants that would yield the least amount of product is the limiting reagent.

Method 2:

First we calculate the number of moles of each reactant:

$$? \text{ mol } Fe_2O_3 = 4.02 \text{ g } Fe_2O_3 \left( \frac{1 \text{ mol } Fe_2O_3}{159.69 \text{ g } Fe_2O_3} \right)$$

$$= 0.0252 \text{ mol } Fe_2O_3$$

$$? \text{ mol } CO = 1.78 \text{ g } CO \left( \frac{1 \text{ mol } CO}{28.01 \text{ g } CO} \right)$$

$$= 0.0635 \text{ mol } CO$$

The chemical equation says that $Fe_2O_3$ and CO react in a mole ratio of 1 to 3. Therefore, three times the number of moles of $Fe_2O_3$ must be available as CO. In this case there should be (3 x 0.0252) mol, or 0.0756 mol, of CO available to completely react with the $Fe_2O_3$. This much CO is not available; therefore, all the $Fe_2O_3$ cannot react and CO is the limiting reagent. We use the limiting reagent to continue the problem:

$$? \text{ g } Fe = 0.0635 \text{ mol } CO \left( \frac{(2 \times 55.85) \text{ g } Fe}{3 \text{ mol } CO} \right) = 2.36 \text{ g } Fe$$

Occasionally, the answer obtained by one method may differ from that obtained by another method by one unit in the last significant digit. In this problem the difference in the values for the weight of Fe results because of the rounding off to 0.0635 mol CO.

b. 2.37 g Fe,
   0.00 g CO,
   0.64 g $Fe_2O_3$,
   2.79 g $CO_2$,
   [5.5]

After the reaction is complete, 2.37 g Fe is formed and 0.00 g CO remains. The mass of $CO_2$ and the mass of residual $Fe_2O_3$ need to be calculated:

$$? \text{ g } CO_2 = 0.0635 \text{ mol } CO \left( \frac{(3 \times 44.01) \text{ g } CO_2}{3 \text{ mol } CO} \right) = 2.79 \text{ g } CO_2$$

The mass of $Fe_2O_3$ remaining can be calculated in several ways. Two possible ways are given:

Method 1:

? g $Fe_2O_3$ remaining = initial g $Fe_2O_3$ - reacted g $Fe_2O_3$

$$= 4.02 \text{ g } Fe_2O_3 - 0.0635 \text{ mol CO} \left( \frac{(1 \times 159.69) \text{ g } Fe_2O_3}{3 \text{ mol CO}} \right)$$

$$= 4.02 \text{ g } Fe_2O_3 - 3.38 \text{ g } Fe_2O_3$$

$$= 0.64 \text{ g } Fe_2O_3$$

Method 2:

Since mass is conserved in a chemical reaction, the mass of $Fe_2O_3$ remaining is the total mass of all initial materials, 4.02 g $Fe_2O_3$ + 1.78 g CO = 5.80 g initial material, minus the mass of all material that reacted, i.e., the products, 2.37 g Fe + 2.79 g $CO_2$ = 5.16 g reacted material.  Thus,

$$\begin{array}{r} 5.80 \text{ g initial material} \\ -5.16 \text{ g reacted material} \\ \hline 0.64 \text{ g unreacted material, } Fe_2O_3 \end{array}$$

12. 592 lb Fe
    [5.5]

This problem can be solved with a logic chain:

? lb Fe = 2000 lb ore $\left( \dfrac{84.6 \text{ lb } Fe_2O_3}{100 \text{ lb ore}} \right) \left| \left( \dfrac{(2 \times 55.85) \text{ lb Fe}}{(1 \times 159.69) \text{ lb } Fe_2O_3} \right) \right|$ = 592 lb Fe

| calculate | calculate |
|---|---|
| amount of | mass of Fe |
| $Fe_2O_3$ from  → → → | from reaction |
| % composition | stoichiometry |

13. $8.90 \times 10^2$
    lb CO
    [5.5]

This problem can be solved with a logic chain:

? lb CO = 2000 lb ore $\left( \dfrac{84.6 \text{ lb } Fe_2O_3}{100 \text{ lb ore}} \right) \left( \dfrac{(3 \times 28.01) \text{ lb CO}}{(1 \times 159.69 \text{ lb}) \ Fe_2O_3} \right)$

$$= 8.90 \times 10^2 \text{ lb CO}$$

14. 46.1 g $C_2H_5OH$
    [5.5]

$$? \text{ g } C_2H_5OH = 171 \text{ g } C_{12}H_{22}O_{11} \left(\frac{1 \text{ mol } C_{12}H_{22}O_{11}}{342 \text{ g } C_{12}H_{22}O_{11}}\right)\left(\frac{1 \text{ mol } C_{12}H_{22}O_{11} \quad \text{reacting}}{2 \text{ mol } C_{12}H_{22}O_{11} \quad \text{present}}\right)$$

$$\left(\frac{2 \text{ mol } C_6H_{12}O_6}{1 \text{ mol } C_{12}H_{22}O_{11}}\right)\left(\frac{2 \text{ mol } C_2H_5OH}{1 \text{ mol } C_6H_{12}O_6}\right)\left(\frac{46.1 \text{ g } C_2H_5OH}{1 \text{ mol } C_2H_5OH}\right)$$

$$= 46.1 \text{ g } C_2H_5OH$$

15.  $6.53 \times 10^4$ g/mol
     [5.1, 5.3]           A logic chain allows us to solve the problem very
                          quickly if we recognize that 4 mol iron, or (4 x
                          55.84) g Fe, are in 1 mol Hb:

$$? \text{ mol wt Hb} = \left(\frac{1.000 \text{ g Hb}}{0.00342 \text{ g Fe}}\right)\left(\frac{(4 \times 55.85)\text{g Fe}}{1 \text{ mol Hb}}\right)$$

                                        from the
                                        % composition

$$= 6.53 \times 10^4 \text{ g/mol}$$

                          The molecular weight of Hb is known to be
                          64,456 g/mol.

16.  $3.271 \times 10^{-22}$     $? \text{ g Au/atom Au} = \left(\dfrac{196.9665 \text{ g Au}}{1 \text{ mol Au}}\right)\left(\dfrac{1 \text{ mol Au}}{6.022 \times 10^{23} \text{ atoms Au}}\right)$
     g/atom
     [5.1]
                                $= 3.271 \times 10^{-22} \text{ g/atom}$

17.  5.0 g soln.
     [5.2]

$$? \text{ g soln.} = \left(\frac{5.0 \text{ mg BSP}}{1 \text{ kg}}\right)\left(\frac{1 \text{ kg}}{2.2 \text{ lb}}\right)\left(110 \text{ lb}\right)\left(\frac{100 \text{ mg soln.}}{5.0 \text{ mg BSP}}\right)\left(\frac{1 \text{ g soln.}}{1000 \text{ mg soln.}}\right)$$

                   wt BSP/lb wt BSP            wt soln            wt soln.

          = 5.0 g soln.

18.  227 g soln        $? \text{ g soln.} = 124 \text{ g } H_2SO_4 \left(\dfrac{100 \text{ g soln.}}{54.7 \text{ g } H_2SO_4}\right) = 227 \text{ g soln.}$
     [5.2]

19.  39.20% $Fe_2O_3$     In the logic chain we make sure all unwanted units
     [5.3]            cancel:

$$? \text{ \% } Fe_2O_3 = 27.42\% \text{ Fe} \left(\frac{(1 \times 159.69) \text{ g } Fe_2O_3}{(2 \times 55.847) \text{ g Fe}}\right)$$

$$= 39.20\% \text{ } Fe_2O_3$$

or

$$? \% Fe_2O_3 = \left(\frac{27.42 \text{ g Fe}}{100.00 \text{ g Fe}_2O_3}\right) \left(\frac{(1 \times 159.69) \text{ g Fe}_2O_3}{(2 \times 55.847) \text{ g Fe}}\right)$$

$$= \frac{0.3920 \text{ g Fe}_2O_3}{1.000 \text{ g ore}}$$

$$= \frac{39.20 \text{ g Fe}_2O_3}{100.00 \text{ g ore}}$$

$$= 39.20\% \text{ Fe}_2O_3$$

III. Thermochemical calculations

Equations [5.7]

$$C(\text{graphite}) + \tfrac{1}{2}O_2(g) \rightarrow CO(g)$$

$$2Fe(s) + 3/2\ O_2(g) \rightarrow Fe_2O_3(s)$$

$$H_2(g) + \tfrac{1}{2}O_2(g) \rightarrow H_2O(g)$$

$$H_2(g) + S(s) + 2O_2(g) \rightarrow H_2SO_4(l)$$

$$\tfrac{1}{2}N_2(g) + 3/2\ H_2(g) \rightarrow NH_3(l)$$

$$\tfrac{1}{2}N_2(g) + O_2(g) \rightarrow NO_2(g)$$

$$P(s) + 3/2\ Cl_2(g) \rightarrow PCl_3(g)$$

$$P(s) + \tfrac{1}{2}O_2(g) + 3/2\ Cl_2(g) \rightarrow POCl_3(g)$$

$$S(s) + 3/2\ O_2(g) \rightarrow SO_3(g)$$

1. $\Delta H = -31.2$ kcal [or $-130.5$ kJ] [5.8]

We sum any available equations including the enthalpies such that the final equation and the enthalpy for that equation is obtained. In the final equation $SO_3(g)$ is to the left of the reaction arrow, so an equation with known enthalpy change is needed that contains $SO_3(g)$ on the left. From the table we choose the equation and standard enthalpy of formation of $SO_3(g)$:

$$S(s) + 3/2\ O_2(g) \rightarrow SO_3(g) \qquad \Delta H = -94.4 \text{ kcal}$$
$$[\text{or } -395.0 \text{ kJ}]$$

We invert the equation so that $SO_3(g)$ is on the left. Inverting the equation changes the sign of the value of $\Delta H$.

$$SO_3(g) \rightarrow S(s) + 3/2\ O_2(g) \qquad \Delta H = +94.4\ \text{kcal} \qquad (1)$$
$$[\text{or } +395.0\ \text{kJ}]$$

In the final equation $H_2O(l)$ is also needed on the left. From the table we choose the equation and standard enthalpy of formation of $H_2O(l)$, invert the equation, and change the sign of $\Delta H$:

$$H_2O(l) \rightarrow H_2(g) + 1/2\ O_2(g) \qquad \Delta H = +68.3\ \text{kcal} \qquad (2)$$
$$[\text{or } +285.8\ \text{kJ}]$$

Sulfuric acid is needed on the right in the final equation, so we choose the equation and standard enthalpy of formation of $H_2SO_4(l)$:

$$H_2(g) + S(s) + 2\ O_2(g) \rightarrow H_2SO_4(l) \qquad (3)$$
$$\Delta H = -193.9\ \text{kcal}$$
$$[\text{or } -811.3\ \text{kJ}]$$

We sum equations (1), (2), and (3), including the $\Delta H$ values:

$$SO_3(g) \rightarrow S(s) + 3/2\ O_2(g) \qquad \Delta H = +94.4\ \text{kcal}$$
$$[\text{or } +395.0\ \text{kJ}]$$

$$H_2O(l) \rightarrow H_2(g) + 1/2\ O_2(g) \qquad \Delta H = +68.3\ \text{kcal}$$
$$[\text{or } +285.8\ \text{kJ}]$$

$$H_2(g) + S(s) + 2O_2(g) \rightarrow H_2SO_4(l) \qquad \Delta H = -193.9\ \text{kcal}$$
$$[\text{or } -811.3\ \text{kJ}]$$

$$\overline{SO_3(g) + H_2O(l) \rightarrow H_2SO_4(l)} \qquad \Delta H = -31.2\ \text{kcal}$$
$$[\text{or } -130.5\ \text{kJ}]$$

To find the enthalpy change of a reaction we need

only find appropriate thermochemical data for reactions that, when mathematically combined, give the desired reaction.

*For any reaction, $\Delta H$ equals the sum of the enthalpies of formation of the products minus the sum of the enthalpies of formation of the reactants. Each $\Delta H_f$ must be multiplied by the number of moles that appear before the corresponding molecule in the balanced chemical equation.*

In this problem

$$\Delta H = \Delta H_f \text{ (products)} - \Delta H_f \text{ (reactants)}$$
$$= (-193.9 \text{ kcal}) - (-94.4 \text{ kcal} - 68.3 \text{ kcal})$$
$$= -31.2 \text{ kcal} \quad \text{[or } 130.5 \text{ kJ]}$$

2.  $\Delta H = -32.5$ kcal
        [or $-136.0$ kJ]
    [5.8]

|  | $\Delta H$ (kcal) |
|---|---|
| $P(s) + 3/2 \, Cl_2(g) + \frac{1}{2}O_2(g) \rightarrow POCl_3(g)$ | $-141.5$ |
| $H_2(g) + Cl_2(g) \rightarrow 2HCl(g)$ | $2(-22.1)$ |
| $PCl_5(g) \rightarrow P(s) + 5/2 \, Cl_2(g)$ | $+95.4$ |
| $H_2O(g) \rightarrow H_2(g) + \frac{1}{2}O_2(g)$ | $+57.8$ |

$PCl_5(g) \quad + H_2O(g) \rightarrow POCl_3(g) + 2HCl(g) \quad -32.5$

3.  $\Delta H = -35.1$ kcal
        [or $146.8$ kJ]
    [5.8]

First we balance the equation:

$$C_5H_{12}(g) + 8O_2(g) \rightarrow 5CO_2(g) + 6H_2O(l)$$

$$\Delta H = -845.2 \text{ kcal}$$

From the law of Hess we know

$$\Delta H = \Delta H_f \text{ (products)} - \Delta H_f \text{ (reactants)}$$

$$\Delta H_{\text{combustion}} = 5\Delta H_f(CO_2) + 6\Delta H_f(H_2O) - \Delta H_f(C_5H_{12}) - 8\Delta H_f(O_2)$$

We rearrange the preceding equation to find the enthalpy of formation of pentane:

$$\Delta H_f(C_5H_{12}) = 5\Delta H_f(CO_2) + 6\Delta H_f(H_2O) - 8\Delta H_f(O_2) - \Delta H_{combustion}$$

Then we substitute values for all known enthalpies:

$$\Delta H_f(C_5H_{12}) = 5(-94.1 \text{ kcal}) + 6(-68.3 \text{ kcal}) - 8(0) + 845.2 \text{ kcal}$$

$$= -35.1 \text{ kcal}$$

4. $\Delta H = -103.5$ kcal/mol [or $-433.0$ kJ/mol] [5.7]

We calculate the change in temperature:

$$\Delta T = 26.885°C - 22.815°C = +4.070°C$$

Thus, heat is evolved by the reaction:

$$? \text{ cal} = (950.0 \text{ g } H_2O + 165.0 \text{ g } H_2O)(4.070°C)(1.000 \text{ cal/g }°C)$$

$$= 4538 \text{ cal}$$

The preceding amount of heat is liberated by the reaction of 12.45 g of $P_4O_{10}$. The enthalpy change per mole of $P_4O_{10}$ can be determined as follows:

$$? \text{ kcal/mol} = \left(\frac{4538 \text{ cal}}{12.45 \text{ g } P_4O_{10}}\right)\left(\frac{283.9 \text{ g } P_4O_{10}}{1 \text{ mol } P_4O_{10}}\right)\left(\frac{1 \text{ kcal}}{1000 \text{ cal}}\right)$$

$$= 103.5 \text{ kcal/mol}$$

Therefore,

$$\Delta H = -103.5 \text{ kcal/mol} \quad [\text{or } -433.0 \text{ kJ/mol}]$$

The sign is negative since heat is *released*.

5. $\Delta H = -165$ kcal [or $690$ kJ] [5.9]

The lattice energy is the energy associated with the process.

$$K^+(g) + Cl^-(g) \rightarrow KCl(s) \qquad \Delta H_{l.e.} = ?$$

A Born-Haber cycle can be used to determine the value of $\Delta H_{l.e.}$ of KCl. From the law of Hess we know that

$$\Delta H_{l.e.} = \Delta H_f(KCl) - \Delta H_f(K^+) - \Delta H_f(Cl^-)$$

The value of $\Delta H_f(KCl)$ is given, but we must determine $\Delta H_f(K^+)$ and $\Delta H_f(Cl^-)$ from a series of reactions, since $\Delta H_f$ pertains to the formation of a species from elements in standard states.

We add the ionization energy of potassium, $\Delta H_{i.p.} = 99$ kcal, which pertains to the equation

$$K(g) \rightarrow K^+(g) + e^-$$

and the enthalpy of sublimation, $\Delta H_{subl.} = 21$ kcal, which pertains to the equation

$$K(s) \rightarrow K(g)$$

to obtain a $\Delta H_f$ of $K^+(g)$, $\Delta H_f(K^+) = 120$ kcal, which pertains to the equation

$$K(s) \rightarrow K^+(g) + e^-$$

Similarly, we add the electron affinity of chlorine, $\Delta H_{e.a.} = -88$ kcal, which pertains to the equation

$$Cl(g) + e^- \rightarrow Cl^-(g)$$

and half the dissociation energy of $Cl_2$, $\frac{1}{2}\Delta H_{diss.} = 29$ kcal, which pertains to the equation

$$\tfrac{1}{2}Cl_2(g) \rightarrow Cl(g)$$

to obtain $\Delta H_f(Cl^-)$, $\Delta H_f(Cl^-) = -59$ kcal, which pertains to the equation

$$\tfrac{1}{2} Cl_2(g) + e^- \rightarrow Cl^-(g)$$

To obtain $\Delta H_{l.e.}$ we add the following equations and enthalpies:

| | |
|---|---|
| $K(s) + \tfrac{1}{2}Cl_2(g) \rightarrow KCl(s)$ | $\Delta H_f(KCl) = -104$ kcal |
| $K^+(g) + e^- \rightarrow K(s)$ | $-\Delta H_f(K^+) = -120$ kcal |
| $Cl^-(g) \rightarrow \tfrac{1}{2}Cl_2(g) + e^-$ | $-\Delta H_f(Cl^-) = +59$ kcal |
| $K^+(g) + Cl^-(g) \qquad KCl(s)$ | $\Delta H_{l.e.}(KCl) = -165$ kcal |

Review the calculation of the lattice energy of sodium chloride in Section 5.9 of your text.

SELF-TEST

Complete the test in 20 minutes:

1. Calculate the number of atoms in a 3.20 g sample of copper.

2. In which of the following compounds is the mass percentage of calcium greatest?

   a. $Ca_3(PO_4)_2$     c. $CaI_2$

   b. $CaO$           d. $CaCO_3$

3. Calculate the enthalpy of formation of $CS_2(l)$ from the following values:

enthalpy of combustion of $CS_2(l)$ to $CO_2(g)$ and $SO_2(g)$: -257.1 kcal/mol [-1075.7 kJ/mol]

enthalpy of combustion of $C(s)$ to $CO_2(g)$: -94.1 kcal/mol [or -393.7 kJ/mol]

enthalpy of combustion of $S(s)$ to $SO_2(g)$: -71.0 kcal/mol [or -297.1 kJ/mol]

4. A certain organic chemical is shown by analyses to contain by weight 43.90% carbon, 3.05% hydrogen, 9.76% oxygen, and 43.29% chlorine. What is the empirical formula of the compound?

5. The compound $S_4N_3Cl$ can be formed by the reaction

   $3\ S_4N_4 + 2\ S_2Cl_2 \rightarrow 4\ S_4N_3Cl$

How many grams of $S_4N_3Cl$ can be formed with 5.00 g $S_4N_4$ and 3.00 g $S_2Cl_2$?

6. Calculate the enthalpy change of the reaction

   $BaO(s) + CO_2(g) \rightarrow BaCO_3(s)$

from the following standard enthalpies of formation at 298°K: $BaO(s)$, -133.5 kcal/mol [or -558.6 kJ/mol]; $CO_2(g)$, -94.05 kcal/mol [or 393.50 kJ/mol]; and $BaCO_3(s)$, -290.8 kcal/mol [or 1216.7 kJ/mol].

# 6
# Gases

OBJECTIVES     (a) You should be able to demonstrate your knowledge
of the following terms by defining them, describing
them, or giving specific examples of them:

        absolute zero [6.4]
        atmospheric pressure [6.2]
        Avogadro's principle [6.7]
        barometer [6.2]
        Boyle's Law [6.3]
        Celsius temperature scale [6.4]
        Charles' Law [6.4]
        critical pressure [6.14]
        critical temperature [6.14]
        Dalton's law of partial pressure [6.10]

density [6.5]
effusion [6.12]
gas [6.1, 6.6]
Gay-Lussac's law of combining volumes [6.8]
Graham's law of effusion [6.12]
ideal gas [6.4, 6.5]
Kelvin (also called absolute) temperature scale [6.4]
kinetic energy [6.6]
manometer [6.2]
pressure [6.2]
standard temperature and pressure, abbreviated STP [6.4]
torr [6.2]
vapor pressure [6.10]
van der Waals equation [6.13]

(b) If all but one of the variables (pressure, $P$, volume, $V$, number of moles, $n$, and temperature, $T$) of the ideal gas law are given, you should be able to calculate the unknown variable.

(c) If a gas under a defined set of conditions is changed to a new set of conditions, you should be able to use Boyle's and Charles' laws in calculations to determine the complete new set of conditions.

(d) You should understand vapor pressure and partial pressures and be able to perform calculations involving them.

(e) You should be able to calculate relative rates of gaseous effusion.

(f) You should understand Gay-Lussac's law of combining volumes and be able to perform calculations involving chemical reactions of gases.

EXERCISES

I. Answer each of the following with *true* or *false*. If a statement is false, correct it.

_____

1. One atmosphere equals 760 torr.

_____

2. Boyle's law states that the volume of a gas varies inversely with the pressure under which it is measured.

_____

_____

_____

_____

_____

_____

_____

_____

_____

_____

3. Charles' law states the the volume of a gas varies directly with temperature in $^{\circ}C$.

4. STP stands for standard temperature, $273^{\circ}K$, and pressure, 1 atmosphere.

5. The value of $R$ is 0.08206 liter atm $^{\circ}K^{-1}$ mol$^{-1}$.

6. $PV = nRT$

7. Temperatures in $^{\circ}C$ are changed to the Kelvin scale by adding 100:  $T = t + 100$.

8. A mole of ideal gas contains 6.022 x $10^{23}$ molecules and occupies 22.414 liters at STP.

9. In a mixture of gases the total pressure is the sum of the partial pressures of the components.

10. According to Graham's law of effusion, gases with larger molecular weights effuse through small openings more rapidly than gases with smaller molecular weights.

11. The van der Waals equation accounts for the fact that real gas molecules have no volume and exert no attractive forces.

12. Below the critical temperature it is impossible to liquefy a gas regardless of pressure.

II. Complete the following statements with one of the following:

increases
decreases
remains the same

1. If the temperature of a gas is increased and the pressure on the system is unchanged, the volume of the gas _____.

2. If a gas is enclosed in a rigid container and the container is heated, the pressure exerted by the gas _____.

3. A vessel contains 2.5 mol of oxygen.  If an additional 2.5 mol of oxygen is added to the vessel, the pressure _____.

4. A gas is allowed to expand from 1 liter to 22.4 liters.  The number of moles of gas _____.

5. As a gas is heated, the average kinetic energy of the molecules _____.

6. As a gas is compressed without changing the temperature, the average kinetic energy of the molecules _____.

7. As a gas is compressed at constant temperature, the average mean free path of a molecule _____.

8. As the volume of a gas is increased at constant temperature, the pressure _____.

9. A cylinder contains oxygen at a pressure of 1500 pounds per square inch, and some gas is released from it.  The pressure _____. The number of moles of gas in the cylinder _____.  If there is no temperature change, the average kinetic energy of the molecules in the cylinder _____ and the mean free path _____.

10. An evacuated one liter flask is opened at sea level.  If there is no temperature change, the number of moles of gas in the flask _____. The flask is sealed and carried to a mountain top.  During transport the pressure of the contained gas _____.  The flask is again opened.  The number of moles of gas in the flask _____ and the pressure _____.

11. A weather balloon is released from a station in Texas.  As the baloon rises, its size _____ due to decreased atmospheric pressure.

12. A diver carries a tank that could supply a 30-minute oxygen supply at sea level.  The effective oxygen supply _____ as the diver explores at a depth of 50 feet if temperature is constant.

13. The combustion of hydrogen is represented by

$$2H_2(g) + O_2(g) \rightarrow 2H_2O(g)$$

If the temperature and volume do not change, the pressure _____ as the reaction proceeds in a closed container. If an appreciable temperature increase occurs during the reaction, the final pressure _____ due to the temperature change.

14. A gas in a 1 liter container is heated from $0^\circ C$ to $100^\circ C$. Simultaneously the volume of the container increases to 2 liters. The pressure _____ due to the temperature change and _____ due to the volume change.

15. Octane is burned at constant pressure in an expandable vessel:

$$2C_8H_{18}(g) + 25O_2(g) \rightarrow 16CO_2(g) + 18H_2O(g)$$

As the reaction proceeds, the number of moles of gas _____ and the volume of gas _____.

16. A naturally occuring mixture of $^{35}Cl$ and $^{37}Cl$ is enclosed in a cylinder. As a small leak allows gas to escape very slowly, a piston maintains the trapped sample at STP. The number of molecules in the cylinder _____. The density of the trapped gas _____. The molecular weight of the trapped gas _____.

17. If a gas sample is isothermally expanded, the pressure _____.

18. If a gas is heated isobarically, its volume _____.

19. A mole of helium at STP is enclosed in a rigid container. Argon is gradually added to the container. The pressure _____. The number of moles of helium _____. The partial pressure of helium _____. The partial pressure of argon _____.

20. A rigid one liter vessel contains equimolar concentrations of He and Ar. As the temperature

is increased, the partial pressure of He, $p_{He}$, _____ and the partial pressure of Ar, $p_{Ar}$, _____. The total pressure _____.

21. Gas is gradually escaping from a container. The number of moles of gas, $n$, _____. The volume and pressure are maintained constant. The temperature of the gas _____.

III.  Work the following problems:

1. A vessel containing argon at 0.10 atm is heated from $0^oC$ to $100^oC$. If the volume does not change, what is the final pressure?

2. A 75 ml sample of gas is heated at constant pressure from $0^oC$ to $33^oC$. What is the volume at $33^oC$?

3. What volume will an ideal gas occupy at absolute zero?

4. A gas in a 1.00 liter container is allowed to expand to a volume of 5.00 liters. If the initial pressure is 748 torr, what is the final pressure if the temperature does not change?

5. A 50 ml gas sample is cooled from $100^oC$ to $12^oC$ while the pressure is held constant. What is the final volume?

6. What volume will 16 g of oxygen gas occupy at STP?

7. What volume in ml will 0.500 mol of an ideal gas occupy at STP?

8. What is the density, in g/ml, of oxygen at STP?

9. How many moles of gas at 1.00 atm and $25^oC$ are contained in a 5.00 liter vessel?

10. A balloon containing 1.00 liter of He at STP is purchased in an air-conditioned store, in which the temperature is $25^oC$, and carried outside where it heats to $45^oC$. If there is no pressure change, what is the final volume?

11. If the pressure on the balloon described in problem 10 of this section does change from 745 torr to 757 torr during the temperature change, what is the final volume?

12. A gas sample is heated from $-10^{\circ}C$ to $87^{\circ}C$, and the volume is increased from 1.00 liter to 3.30 liters. If the initial pressure is 0.750 atm, what is the final pressure?

13. A 1.0 liter flask contains 3.5 mol of gas at $27^{\circ}C$. What is the pressure of the gas?

14. An evacuated 1.00 liter flask weighs 104.35 g. A quantity of nitrogen gas is added to the flask. After the nitrogen is added, the flask and contents weigh 105.68 g, and the pressure exerted by the gas is 2.00 atm. What is the temperature inside the flask?

15. When magnesium metal is burned in air, solid MgO forms:

$$2Mg(s) + O_2(g) \rightarrow 2MgO(s)$$

What volume of oxygen at STP is needed to react with 0.500 g Mg?

16. The complete combustion of octane yields carbon dioxide and water:

$$2C_8H_{18}(g) + 25O_2(g) \rightarrow 16CO_2(g) + 18H_2O(g)$$

What volume of gas is produced from the complete combustion of 0.670 g octane if the temperature is $400^{\circ}C$ and the pressure is 1.2 atm?

17. Hydrogen gas can be generated by the action of hydrochloric acid on magnesium:

$$Mg(s) + 2\ HCl(aq) \rightarrow Mg^{+2}(aq) + 2\ Cl^-(aq) + H_2(g)$$

If all the hydrogen from the reaction of 2.00 g Mg is collected over water at 1.00 atm and $25^{\circ}C$, what volume will the dried gas occupy at STP?

18. Traces of oxygen can be removed from streams of gas by reaction with heated copper:

    $$2Cu(s) + O_2(g) \rightarrow 2CuO(s)$$

    If the original copper weighs 0.3750 g and the oxidized copper weighs 0.3800 g, what volume of $O_2$ at STP was removed from the gas stream?

19. In a normal breath 2.00 liters of gas can be inhaled. How many moles of gas at STP are in this volume? If 21% (by volume) of the inhaled gas molecules is oxygen, how many grams of oxygen are in this volume?

20. The atmospheric pressure on Mt. Everest is 0.330 atm with a temperature of -10°C. How many grams of oxygen are inhaled in a 2.00 liter breath?

21. What is the total % decrease in oxygen intake per breath when a person goes from sea level to the mountain top? (See problems 19 and 20 of this section.)

22. If pure oxygen is inhaled from an oxygen cylinder on a mountain top at 0.330 atm and -10°C, how many grams of oxygen would be inhaled per 2.00 liter breath?

23. One mole of hemoglobin, which has a weight of 66,280 g, can bind four moles of oxygen gas. What volume of $O_2$ at STP can 100 ml of blood with a hemoglobin concentration of 160 g/liter carry?

24. Sulfur dioxide concentrations in the atmosphere can easily reach 0.20 ppm, which is 0.20 ml $SO_2$ per $10^6$ ml air. How many molecules of $SO_2$ at STP are inhaled in a 2.00 liter breath?

25. About $1.5 \times 10^8$ metric tons (1 metric ton = 1000 kg) of CO are released into the atmosphere each year. What is the volume of this quantity of CO at STP?

26. At STP 0.66 mol of $H_2$ and 0.33 mol of $O_2$ are mixed in an expandable container. After ignition the product of the reaction reaches a temperature of 1300°C with a pressure of 800 torr. What is the volume of the reaction product?

$$2H_2(g) + O_2(g) \rightarrow 2H_2O(g)$$

27. At STP 0.66 mol of $H_2$ and 0.66 mol of $O_2$ are mixed in an expandable container. After ignition the remaining gases reach a temperature of 1300°C and a pressure of 800 torr. What is the total volume of the remaining gases?

28. Atmospheric air is a gaseous mixture of water vapor, oxygen, carbon dioxide, nitrogen, and traces of other species. What is the mole fraction of nitrogen gas at STP if $p_{O2}$ = 159 torr, $p_{CO2}$ = 0.23 torr, and $p_{H2O}$ = 23.8 torr?

29. A gas, X, diffuses 3.1 times faster than fluorine gas. What is the molecular weight of the gas X?

30. A storage tank at JFK space center can contain 900,000 gallons, which is approximately 3.4 million liters, of liquid hydrogen, which has a density of 0.070 g/cc at -253°C. What volume in liters would this hydrogen occupy at STP if it were evaporated?

31. W. W. Ruby estimated the total mass of atmospheric oxygen to be $15 \times 10^{20}$ g. If all this oxygen were at STP, what volume in liters would it occupy?

32. Uranium 235 and uranium 238 are separated by the effusion difference of the hexafluorides $^{235}UF_6$ and $^{238}UF_6$. What is the ratio of the effusion rates of the hexafluorides?

33. What is the density of phosgene gas, $COCl_2$, at STP?

34. A 100 ml flask contains 0.162 g of an unknown

gas at 760 torr and 100°C.  What is the molecular weight of the gas?

35. Acetylene is formed by the reaction of water and calcium carbide:

$$CaC_2(s) + 2H_2O(l) \rightarrow C_2H_2(g) + Ca(OH)_2(s)$$

What volume of acetylene can be produced at STP from 4.00 g $CaC_2$ and 2.75 g water?

36. Acetylene is burned to form water and carbon dioxide:

$$2C_2H_2(g) + 5O_2(g) \rightarrow 2H_2O(g) + 4CO_2(g)$$

Answer the following:
a. What volume of $CO_2$ at STP is formed from 3.00 g $C_2H_2$?
b. What volume of $CO_2$ at STP is formed from 2.00 liters $C_2H_2$ at STP?

37. The van der Waals constants for ammonia are given in Table 6.3 of your text.  Use this data to determine the pressure at which one mole of $NH_3$ will occupy 22.4 liters at 25°C.

38. Use the data in Table 6.3 of your text to determine which of the gases listed has the strongest intermolecular interactions and which has the largest molecular volume.

39. One (1.0) ml of liquid water at 25°C, which has density of 1.00 g/ml, is placed in an evacuated 10 liter vessel, which is also at 25°C.  What is the pressure inside the vessel?  What is the mass of water in the gaseous state?

40. One (1.00) ml of liquid water at 25°C is added to dry helium that is contained in a 10.0 liter cylinder at 25°C and 100 atm.  What is the partial pressure of water vapor in the vessel?  What is the partial pressure of He?  What is the total pressure?

41. At what temperature will all the water in problem 39 be volatilized and exert a pressure of 1.0 atm?  The density of water at 25°C is 1.0 g/ml.

42. One (1.0) liter of a gas collected over water at STP weighs 1.135 g before drying. What is the molecular weight of the unknown gas?

43. A gas mixture is known to contain only helium and nitrogen. What are the partial pressures of each gas if the density of the mixture is 0.475 g/liter at STP?

44. What would be the total volume of the atmosphere at STP if the atmospheric concentration of $O_2$ were 21% by volume? (See problem 31 of this section.)

45. Would the actual volume of the earth's atmosphere be larger of smaller than that estimated in problem 44 of this section? Why?

46. In the atmosphere upon irradiation with ultraviolet light, oxygen is converted to oxone, $O_3$:

$$3O_2 \xrightarrow{h\nu} 2O_3$$

The symbol $h\nu$ above the arrow in the equation for the reaction indicates the quantum of radiation necessary to cause the reaction to proceed. This same reaction can be carried out under controlled laboratory conditions. If 1.47 liters of $O_2$ in a sealed container at 1.01 atm is irradiated until 5.24% of the oxygen reacts, what is the final pressure inside the flask? Assume that the temperature also rises from 25°C to 54°C during the reaction.

ANSWERS TO
EXERCISES

I. Properties of gases

1. True [6.2]

2. True [6.3]

3. False [6.4]        Volume varies directly with temperature in °K.

4. True [6.4]

5. True [6.5]        In SI units $R$ is 8.314 J K$^{-1}$ mol$^{-1}$.

6. True [6.5]

7. False [6.4]          $T = t + 273$

8. True [5.1, 6.7]

9. True [6.10]

10. False [6.12]        A molecule with a smaller molecular weight diffuses
                        more quickly.

11. False [6.13]        In the van der Waals equation $a$ corrects for inter-
                        molecular interactions and $b$ corrects for molecular
                        volume.

12. False [6.14]        Above the critical temperature liquefaction cannot
                        occur.


## II. Behavior of gases

1. increases [6.4]

2. increases [6.4]

3. increases [6.5]

4. remains the
   same [6.5]

5. increases [6.6]

6. remains the
   same [6.6]

7. decreases [6.6]

8. decreases [6.3]

9. decreases [6.5]      Gas escapes.
   decreases [6.5]
   remains the
   same [6.6]
   increases [6.6]      Molecules are less tightly packed.

10. increases [6.5]     Gas enters the flask until the pressure inside the
                        tank equals the pressure outside.
    remains the         The volume, number of moles, and temperature remain
    same [6.5]          constant.

    decreases [6.5]        Gas escapes.
    decreases [6.5]


11. increases [6.2]

12. decreases [6.5]        More gas will remain in the tank at the higher
                           pressures encountered beneath sea level.

13. decreases [6.8]        Three moles of reactants produce only two moles of
                           product.

    increases

14. increases [6.5]
    decreases [6.5]

15. increases [6.5]
    increases [6.5]

16. decreases [6.5]
    increases [6.12]       The $^{35}$Cl isotope effuses more quickly than $^{37}$Cl;
                           therefore, the remaining gas is enriched in the
                           heavier isotope.
    increases [2.9]        Atomic weights are weighted averages of isotopic
                           masses.

17. decreases [6.3]        The word *isothermal* means constant temperature.

18. increases [6.4]        The word *isobaric* means constant pressure.

19. increases [6.10]
    remains the
    same [6.5]
    remains the
    same [6.10]
    increases [6.5]

20. increases
    [6.5, 6.10]
    increases
    [6.5, 6.10]
    increases [6.6]

21. decreases [6.5]
    increases [6.5]

III.   Gas law calculations

1. 0.14 atm
   [1.3, 6.4]

An increase in temperature results in an increase in pressure.  In the gas laws temperatures must be expressed in $^\circ K$, whereas pressures may be expressed in any pressure units.(See Example 6.3 of your text.) Thus,

$$? \text{ atm} = 0.10 \text{ atm} \left( \begin{array}{c} \text{temperature} \\ \text{correction} \\ \text{factor} \end{array} \right)$$

Notice that the ratio of temperatures must be larger than 1 to reflect an increase in pressure:

$$? \text{ atm} = 0.10 \text{ atm} \left( \frac{373^\circ K}{273^\circ K} \right) = 0.14 \text{ atm}$$

Note that only two significant figures are reported since 0.10 atm has two significant figures.  You may find it convenient to tabulate information for use in Boyle's and Charles' laws in the following way.  Tabulations such as this should be helpful if you have a tendency to invert correction factors.

|   | INITIAL CONDITIONS | FINAL CONDITIONS |
|---|---|---|
| $T$ | $(0 + 273)^\circ K$ | $(100 + 273)^\circ K$ |
| $P$ | 0.10 atm | ? |
| $V$ | constant | constant |

Since the temperature increases and the volume is constant, the pressure must increase; therefore, the temperature correction factor must be larger than 1.

$$? \text{ atm} = 0.10 \text{ atm} \left( \frac{373 \ ^\circ K}{273 \ ^\circ K} \right) = 0.14 \text{ atm}$$

2. 84 ml [6.4]

An increase in temperature is accompanied by an increase in volume.  In the gas laws volume may be expressed in any volume units, whereas temperature must be expressed in $^\circ K$.  Thus, we find

$$? \text{ ml} = 75 \text{ ml} \left( \frac{306 \ ^\circ K}{273 \ ^\circ K} \right) = 84 \text{ ml}$$

3. 0 [6.5]

An ideal gas has no molecular volume and compresses to zero volume at absolute zero.

4. 150 torr
   [6.3]

An increase in volume is accompanied by a decrease in pressure. (See Example 6.2 of your text.)   Thus, we find

$$? \text{ atm} = 748 \text{ torr} \left(\frac{1.00 \text{ liter}}{5.00 \text{ liters}}\right) = 150 \text{ torr}$$

5. 38 ml [6.4]

A decrease in temperature is accompanied by a decrease in volume.   Thus, we find

$$? \text{ ml} = 50 \text{ ml} \left(\frac{285°K}{373°K}\right) = 38 \text{ ml}$$

According to Charles' law, the volume of an ideal gas varies directly with absolute temperature. (See Example 6.3 of your text.)

6. 11 liters $O_2$

   [5.1,  5.3,  6.5,  6.7]

First we calculate the number of moles of oxygen gas.  Oxygen is diatomic, i.e., it exists as $O_2$ molecules, and the molecular weight is 32.0.

$$? \text{ mol } O_2 = \frac{16 \text{ g } O_2}{32.0 \text{ g } O_2/1 \text{ mol } O_2} = 0.500 \text{ mol } O_2$$

Each mole of ideal gas occupies 22.4 liters at STP, i.e., 273°K and 1 atm.   Therefore,

$$? \text{ liter } O_2 = 0.500 \text{ mol } O_2 \left(\frac{22.4 \text{ liters } O_2}{1 \text{ mol } O_2}\right)$$

$$= 11 \text{ liters } O_2$$

We can also directly use the ideal gas law:

$$PV = nRT$$

$$V = \frac{nRT}{P}$$

$$= \frac{(0.500 \text{ mol})(0.0821 \text{ liter atm } °K^{-1} \text{ mol}^{-1})(273°K)}{1 \text{ atm}}$$

$$= 11 \text{ liters}$$

7. $1.12 \times 10^4$ ml
   [6.7]

A mole of an ideal gas occupies 22.4 liters at STP, and each liter contains 1000 ml.   Therefore,

$$? \text{ ml} = 0.500 \text{ mol} \left(\frac{22.4 \text{ liters}}{1 \text{ mol}}\right)\left(\frac{1000 \text{ ml}}{1 \text{ liter}}\right)$$

$$= 1.12 \times 10^4 \text{ ml}$$

Note that all units cancel except ml.  It is imperative to include proper units in all calculations.

8. $1.43 \times 10^{-3}$ g/ml
   [6.7]

At STP a mole of oxygen occupies 22.4 liters.   Each mole weighs 32.0 g.   Therefore,

$$? \text{ g/ml} = \left(\frac{1 \text{ mol O}_2}{22.4 \text{ liters O}_2}\right)\left(\frac{32.0 \text{ g O}_2}{1 \text{ mol O}_2}\right)\left(\frac{1 \text{ liter O}_2}{1000 \text{ ml O}_2}\right)$$

$$1.43 \times 10^{-3} \text{ g/ml}$$

9. 0.204 mol
   [6.5]

We use the ideal gas law:

$$PV = nRT$$

$$n = \frac{(1.00 \text{ atm}) (5.00 \text{ liters})}{(0.08206 \text{ liter atm }^\circ\text{K}^{-1}\text{mol}^{-1}) (298^\circ\text{K})}$$

$$= 0.204 \text{ mol}$$

If $0.08206$ liter atm $^\circ\text{K}^{-1}\text{mol}^{-1}$ is the value used for $R$, volume must be expressed in liters, temperature in $^\circ\text{K}$, and pressure in atmospheres.

10. 1.07 liters
    [6.4]

According to Charles' law, an increase in temperature is accompanied by an increase in volume.   Therefore,

$$? \text{ liters} = \left(1.00 \text{ liter}\right)\left(\frac{318^\circ\text{K}}{298^\circ\text{K}}\right) = 1.07 \text{ liter}$$

11. 1.05 liters
    [6.4]

Charles' law predicts that an increase in temperature is accompanied by an increase in volume, and Boyle's law predicts that an increase in pressure is accompanied by a decrease in volume.   Combining laws, we find

$$? \text{ liters} = 1.00 \text{ liters} \left(\frac{318^\circ\text{K}}{298^\circ\text{K}}\right) \left(\frac{745 \text{ torr}}{757 \text{ torr}}\right)$$

$$= 1.05 \text{ liters}$$

Note that the temperature correction factor must be larger than one to reflect an increase in volume. The pressure correction factor must be smaller than one to account for a decrease in volume due to an increase in pressure.   The pressure change does not have a significant effect in this problem; however, this is not the normal situation.

12. 0.311 atm
    [6.3, 6.4, 6.5]

An increase in temperature causes an increase in pressure, and an increase in volume causes a decrease in pressure.   Therefore,

$$? \text{ atm} = 0.750 \text{ atm} \cdot \left(\frac{360^{\circ}\text{K}}{263^{\circ}\text{K}}\right) \left(\frac{1.00 \text{ liter}}{3.30 \text{ liters}}\right)$$

$$= 0.311 \text{ atm}$$

If you did not solve this problem correctly the first time, tabulate the data as shown in problem 1 of this section:

|   | INITIAL CONDITIONS | FINAL CONDITIONS |
|---|---|---|
| $T$ | $(-10 + 273)^{\circ}\text{K}$ | $(87 + 273)^{\circ}\text{K}$ |
| $P$ | 0.750 atm | ? |
| $V$ | 1.00 liter | 3.30 liters |

Use the preceding logic to solve the problem.

13. 86 atm [6.5]

We use the ideal gas law:

$$PV = nRT$$

$$P = \frac{nRT}{V}$$

$$= \frac{(3.5 \text{ mol})(0.0821 \text{ liter atm }^{\circ}\text{K}^{-1} \text{ mol}^{-1})(300^{\circ}\text{K})}{1.0 \text{ liter}}$$

$$= 86 \text{ atm}$$

If we wish the pressure to be expressed in torr, we convert atmosphere to torr:

$$? \text{ torr} = 86 \text{ atm} \left(\frac{760 \text{ torr}}{1 \text{ atm}}\right) = 6.5 \times 10^{4} \text{ torr}$$

14. $513^{\circ}$K, or $240^{\circ}$C

If a problem concerns an ideal gas and temperature, pressure, and volume are not changed, we can use $PV = nRT$. In this problem we wish to calculate temperature:

$$T = \frac{PV}{nR}$$

Since $P$, $V$, and $R$ are known, we need to compute the value of $n$:

$$? \text{ mol N} = \frac{(105.68 - 104.35) \text{ g N}_2}{(28.02 \text{ g N}_2/1 \text{ mol N}_2)} = 0.04747 \text{ mol N}_2$$

Substituting values into ideal gas law, we find

$$T = \frac{(2.00 \text{ atm})(1.00 \text{ liter})}{(0.04747 \text{ mol})(0.08206 \text{ liter atm } {}^{\circ}K^{-1} \text{ mol}^{-1})}$$

$$= 513 {}^{\circ}K$$

and

$$T = (513 - 273)(1 {}^{\circ}C) = 240 {}^{\circ}C$$

15.  0.230 liter $O_2$
     [6.9]

The balanced equation shows that 1 mol $O_2$, which occupies 22.4 liters at STP, combines with 2 mol Mg, (2 x 24.31)g Mg. This relationship can be used to compute the volume of $O_2$ gas:

$$? \text{ liters } O_2 = 0.500 \text{ g Mg} \left( \frac{22.4 \text{ liters } O_2}{(2 \times 24.31) \text{ g Mg}} \right)$$

$$= 0.230 \text{ liter } O_2$$

See Example 6.12 of your text.

16.  4.6 liters gas
     [6.8]

We compute the volume of octane at STP (see Example 6.13 of your text):

$$? \text{ liters } C_8H_{18} = \left( \frac{0.679 \text{ g } C_8H_{18}}{(114.2 \text{ g } C_8H_{18}/1 \text{ mol } C_8H_{18})} \right) \left( \frac{22.4 \text{ liters } C_8H_{18}}{1 \text{ mol } C_8H_{18}} \right)$$

$$= 0.132 \text{ liter}$$

The balanced equation shows that 16/2, or 8, times the volume of octane is the volume of $CO_2$ produced, and 18/2, or 9, times the volume of octane is the volume of water vapor produced. Therefore, the total volume of gas produced at STP would be

$$V_{total} = V_{CO_2} + V_{H_2O}$$

$$= (8 \times 0.132) \text{liter} + (9 \times 0.132) \text{liter}$$

$$= 2.24 \text{ liters}$$

Changing from STP to the actual conditions of 400°C and 1.20 atm, we find

$$? \text{ liters} = 2.24 \text{ liters} \left( \frac{(673 {}^{\circ}K)(1.0 \text{ atm})}{(273 {}^{\circ}K)(1.2 \text{ atm})} \right)$$

$$= 4.6 \text{ liters gas}$$

17. 1.84 liters $H_2$

    [6.8]

Since the water vapor is removed before the volume is measured, its effect need not be considered. Therefore,

$$? \text{ liters } H_2 = 2.00 \text{ Mg}\left(\frac{22.4 \text{ liters } H_2}{24.30 \text{ g Mg}}\right)$$

$$= 1.84 \text{ liters } H_2$$

18. 0.00350 liter,
    or 3.50 ml, $O_2$

    [6.9]

We calculate the mass of oxygen:

$$? \text{ g} = 0.3800 \text{ g} - 0.3750 \text{ g} = 0.0050 \text{ g}$$

We calculate the volume occupied by this quantity of oxygen at STP

$$? \text{ liters } O_2 = \left(\frac{0.00150 \text{ g } O_2}{32.0 \text{ g } O_2/1 \text{ mol } O_2}\right)\left(\frac{22.4 \text{ liters } O_2}{1 \text{ mol } O_2}\right)$$

$$= 0.00350 \text{ liter}$$

or

$$? \text{ ml } O_2 = 0.00350 \text{ liter } (1000 \text{ ml/1 liter}) = 3.50 \text{ ml}$$

19. 0.60 g $O_2$
    [6.5, 6.7]

First we calculate the number of moles of gas at STP:

$$PV = nRT$$

$$n = \frac{PV}{RT}$$

$$= \frac{(1.00 \text{ atm})(2.00 \text{ liters})}{(0.08206 \text{ liter atm } {}^\circ K^{-1} \text{ mol}^{-1})(273 {}^\circ K)}$$

$$= 8.92 \times 10^{-2} \text{ mol}$$

Alternatively, since 1 mol of ideal gas at STP occupies 22.4 liters,

$$? \text{ mol gas} = 2.00 \text{ liters}\left(\frac{1 \text{ mol gas}}{22.4 \text{ liters gas}}\right)$$

$$= 8.93 \times 10^{-2} \text{ mol gas}$$

Twenty-one percent of this quantity is $O_2$. Therefore,

$$? \text{ g } O_2 = 8.93 \times 10^{-2} \text{ mol gas}\left(\frac{0.21 \text{ mol } O_2}{1 \text{ mol gas}}\right)\left(\frac{32.0 \text{ g } O_2}{1 \text{ mol } O_2}\right)$$

$$= 0.60 \text{ g } O_2$$

20.  0.20 g $O_2$
     [6.5]

On the top of Mt. Everest,

$$n = \frac{(0.330 \text{ atm})(2.00 \text{ liters})}{(0.0821 \text{ liter atm } {}^\circ K^{-1} \text{ mol}^{-1})(263 {}^\circ K)}$$

$$= 3.06 \times 10^{-2} \text{ mol}$$

and

$$?g\ O_2 = 3.06 \times 10^{-2} \text{ mol gas} \left(\frac{0.21 \text{ mol } O_2}{1 \text{ mol gas}}\right)\left(\frac{32.0 \text{ g } O_2}{1 \text{ mol } O_2}\right)$$

$$= 0.20 \text{ g } O_2$$

21.  67%

The total % *decrease* is calculated by dividing the difference of the two values by the higher value and then multiplying by 100:

$$?\ \% \text{ decrease} = \left(\frac{(0.60 - 0.20) \text{ g } O_2}{0.60 \text{ g } O_2}\right) \quad 100 = 67\%$$

If the problem had asked for the total % *increase* in oxygen intake per breath when a person goes from the mountain top to sea level, 0.20 g $O_2$ would be in the denominator.  In percentage calculations the correct value or reference value is always in the denominator.

22.  0.979 g $O_2$

We use the ideal gas law to calculate $n$:

$$PV = nRT$$

$$n = \frac{PV}{RT}$$

$$= \frac{(0.330 \text{ atm})(2.00 \text{ liters})}{(0.08206 \text{ liter atm } {}^\circ K^{-1} \text{ mol}^{-1})(263 {}^\circ K)}$$

$$= 3.06 \times 10^{-2} \text{ mol}$$

Then we calculate the grams of $O_2$:

$$?\ g\ O_2 = 3.06 \times 10^{-2} \text{ mol } O_2 \ (32.0 \text{ g } O_2/1 \text{ mol } O_2)$$

$$= 0.979 \text{ g } O_2$$

Note that more oxygen is inhaled per 2.00 liter breath when a person breathes from a cylinder of oxygen on Mt. Everest than when he or she breathes normally at sea level.

23. 21.6 ml $O_2$
    [6.8]

This problem describes a balanced chemical equation. Note that one mole of hemoglobin combines with four moles of $O_2$. Therefore, if we let Hb stand for hemoglobin, the equation is

$$Hb + 4O2 \rightarrow Hb \cdot 4O2$$

in which $Hb \cdot 4O_2$ is the product. The problem can be solved simply by determining the number of grams of Hb available:

$$? \text{ liters } O_2 = 16.0 \text{ g Hb}\left(\frac{(4 \times 22.4) \text{ liters } O_2}{66,280 \text{ g Hb}}\right)$$

$$= 0.0216 \text{ liter } O_2$$

or

$$? \text{ ml } O_2 = 0.0216 \text{ liter } O_2\left(\frac{1000 \text{ ml } O_2}{1 \text{ liter } O_2}\right)$$

$$= 21.6 \text{ ml } O_2$$

24. $1.1 \times 10^{16}$
    molecules $SO_2$

First we calculate the volume of $SO_2$:

$$? \text{ liters } SO_2 = 2.00 \text{ liters air}\left(\frac{0.20 \text{ liter } SO_2}{1 \times 10^6 \text{ liters air}}\right)$$

$$= 4.00 \times 10^{-7} \text{ liter } SO_2$$

Then we calculate the number of moles of $SO_2$ from the volume of $SO_2$, and finally we calculate the number of molecules of $SO_2$ from the number of moles of $SO_2$. Combining these two calculations, we find

$$? \text{ molecules } SO_2 = 4.00 \times 10^{-7} \text{ liters } SO_2\left(\frac{6.02 \times 10^{23} \text{ molecules } SO_2}{22.4 \text{ liters } SO_2}\right)$$

$$= 1.1 \times 10^{16} \text{ molecules } SO_2$$

Can you calculate the mass of $SO_2$?

25. $1.2 \times 10^{14}$
    liters CO

We calculate the volume at STP:

$$? \text{ liters CO} = 1.5 \times 10^8 \text{ metric tons CO} \left(\frac{10^6 g}{1 \text{ metric ton}}\right)\left(\frac{22.4 \text{ liters CO}}{28.0 \text{ g CO}}\right)$$

$$= 1.2 \times 10^{14} \text{ liters CO}$$

The factors used are

$$10^3 \text{ kg} \cong 1 \text{ metric ton}$$

$$10^3 \text{ g} = 1 \text{ kg}$$

$$22.4 \text{ liters CO} = 1 \text{ mol CO} = 28.0 \text{ g CO}$$

See Section 5.2 of your text.

26. 81 liters $H_2O$      The oxygen and hydrogen are mixed in the proper molar ratio for complete reaction, 2 mol $H_2$ to 1 mol $O_2$. Therefore, either the number of moles of $O_2$ or $H_2$ can be used to begin the calculation. First we calculate the volume of water vapor at STP:

$$? \text{ liters } H_2O = 0.33 \text{ mol } O_2\left(\frac{(2 \times 22.4) \text{ liters } H_2O}{1 \text{ mol } O_2}\right) = 15 \text{ liters } H_2O$$

or

$$? \text{ liters } H_2O = 0.66 \text{ mol } H_2\left(\frac{(2 \times 22.4) \text{ liters } H_2O}{2 \text{ mol } H_2}\right) = 15 \text{ liters } H_2O$$

Then we adjust the volume to the actual temperature and pressure conditions. An increase in temperature is accompanied by an increase in volume, and an increase in pressure is accompanied by a decrease in volume. Thus,

$$? \text{ liters } H_2O = 15 \text{ liters } H_2O\left(\frac{1573°K}{273°K}\right)\left(\frac{760 \text{ torr}}{800 \text{ torr}}\right)$$

$$= 81 \text{ liters } H_2O$$

27. $1.2 \times 10^2$ liters gas      It takes only 0.33 mol $O_2$ to completely react with all the hydrogen. The excess $O_2$, 0.33 mol $O_2$, will remain in the container. The volume occupied by $H_2O(g)$ is 81 liters (see problem 26 of this section). The volume occupied by the excess $O_2$ is

$$V = \frac{(0.33 \text{ mol})(0.0821 \text{ liter atm } °K^{-1} \text{ mol}^{-1})(1573°K)}{800 \text{ torr}\left(\frac{1 \text{ atm}}{760 \text{ torr}}\right)} = 40 \text{ liters}$$

The total volume of all gases remaining after ignition is

$$V_{total} = 81 \text{ liters } H_2O + 40 \text{ liters } O_2$$

$$= 1.2 \times 10^2 \text{ liters gas}$$

Alternatively, we could reason that 0.66 mol $H_2$ reacts to form 0.66 mol $H_2O$ and 0.33 mol $O_2$ remains unreacted.  A total of 0.99 mol gas, 0.66 mol $H_2O$ and 0.33 mol $O_2$, remains after the reaction. Therefore,

$$V = \frac{(0.99 \text{ mol}) (0.0821 \text{ liter atm } {}^{\circ}K^{-1}mol^{-1}) \quad (1573{}^{\circ}K)}{800 \text{ torr}\left(\dfrac{1 \text{ atm}}{760 \text{ torr}}\right)}$$

$$= 1.2 \times 10^2 \text{ liters}$$

28. 0.759

The partial pressure of a component gas such as $N_2$ is directly proportional to the mole fraction of that component gas and the total pressure:

$$p_{N_2} = x_{N_2} \, p_{total}$$

The total pressure is the sum of the partial pressures of the components:

$$p_{total} = p_{N_2} + p_{O_2} + p_{CO_2} + p_{H_2O}$$

Therefore,

$$p_{N_2} = p_{total} - p_{O_2} - p_{CO_2} - p_{H_2O}$$

$$= 760 \text{ torr} - 159 \text{ torr} - 0.23 \text{ torr} - 23.8 \text{ torr}$$

$$= 577 \text{ torr}$$

and

$$x_{N_2} = \frac{p_{N_2}}{p_{total}} = \frac{577 \text{ torr}}{760 \text{ torr}} = 0.759$$

29. 4.0   [6.12]          Graham's law of effusion states

$$\frac{r_A}{r_B} = \sqrt{\frac{M_B}{M_A}}$$

Therefore,

$$\frac{3.1}{1} = \sqrt{\frac{38.0}{M_A}}$$

$$M_A = 4.0$$

30. 2.6 x $10^9$ liters    We calculate the number of moles of liquid hydrogen:
    $H_2$ [5.2]

$$? \text{ mol } H_2 = 3.4 \times 10^6 \text{ liters } H_2 \left(\frac{10^3 \text{ cm}^3 \text{ } H_2}{1 \text{ liter } H_2}\right)\left(\frac{0.070 \text{ g } H_2}{1 \text{ cm}^3 \text{ } H_2}\right)\left(\frac{1 \text{ mol } H_2}{2.02 \text{ g } H_2}\right)$$

$$= 1.18 \times 10^8 \text{ mol } H_2$$

Then we can calculate the volume of gas at STP:

$$? \text{ liters } H_2 = 1.18 \times 10^8 \text{ mol } H_2 \left(\frac{22.4 \text{ liters } H_2}{1 \text{ mol } H_2}\right)$$

$$= 2.6 \times 10^9 \text{ liters } H_2$$

As the gas at STP, $H_2$ occupies approximately 1000 times the volume of the liquid.

31. 1.0 x $10^{21}$ liters   At STP each mol $O_2$, i.e. 32.0 g $O_2$, occupies 22.4
    $O_2$ [6.7]             liters. Therefore,

$$? \text{ liters } O_2 = 15 \times 10^{20} \text{ g } O_2 \left(\frac{22.4 \text{ liters } O_2}{32.0 \text{ g } O_2}\right)$$

$$= 1.0 \times 10^{21} \text{ liters } O_2$$

32. 0.996          According to Graham's law,
    [5.3, 6.12]

$$\frac{r_A}{r_B} = \sqrt{\frac{M_B}{M_A}}$$

The molecular weights of the two gases must be calculated:

molecular weight of $^{235}UF_6$ = 235 + 6(19) = 349

molecular weight of $^{238}UF_6$ = 238 + 6(19) = 352

Substituting the molecular weights into the equation of Graham's law, we find

$$\frac{r_{238_{UF_6}}}{r_{235_{UF_6}}} = \frac{349}{352} = \sqrt{0.992} = 0.996$$

33.  4.42 g/liter

Since the molecular weight of phosgene is 99.0 g/mol, at STP 99.0 occupies 22.4 liters. Therefore,

$$\text{density} = \frac{\text{weight}}{\text{volume}}$$

$$= \frac{99.0 \text{ g}}{22.4 \text{ liters}}$$

$$= 4.42 \text{ g/liter}$$

34. 49.6 mol
   [5.2, 6.5]

The ideal gas law can be used.  Since $n$, the number of moles, is the weight of material, $g$, divided by the molecular weight, $M$, we can write

$$PV = nRT$$

$$PV = \left(\frac{g}{M}\right) RT$$

$$M = \frac{gRT}{PV}$$

$$= \frac{(0.162 \text{ g}) (0.08206 \text{ liter atm } {}^\circ K^{-1} \text{ mol}^{-1}) (373 {}^\circ K)}{(1 \text{ atm}) (0.100 \text{ liter})}$$

$$= 49.6 \text{ g/mol}$$

The preceding method is preferred.  Any problem in which the temperature, pressure, and volume are not changed can be solved directly with the ideal gas law; problems that involve a change in any of these can be solved easily with a tabular method such as that suggested in problem 1 of this section.

The molecular weight of an ideal gas is the mass of gas that would occupy 22.4 liters at STP. Thus, we could solve the problem in the following way:

First, we calculate the volume of 0.162 g of gas at STP:

$$? \text{ ml gas} = 100.0 \text{ ml gas} \left(\frac{273^\circ\text{K}}{273^\circ\text{K}}\right)$$

$$= 73.2 \text{ ml gas, or } 0.0732 \text{ liter gas}$$

Then we calculate the density of the gas at STP:

$$\text{density} = \frac{\text{weight}}{\text{volume}} = \left(\frac{0.162 \text{ g}}{0.0732 \text{ liter}}\right) = 2.21 \text{ g/liter}$$

Finally we calculate the molecular weight:

$$M = 2.21 \text{ g/liter} \left(\frac{22.4 \text{ liters}}{1 \text{ mol}}\right) = 49.5 \text{ g/mol}$$

We can also use another method:

$$? \text{ g/mol} = \left(\frac{0.162\text{g}}{100 \text{ ml}}\right)\left(\frac{373 \text{ }^\circ\text{K}}{273 \text{ }^\circ\text{K}}\right)\left(\frac{1000 \text{ ml}}{1 \text{ liter}}\right)\left(\frac{22.4 \text{ liters}}{1 \text{ mol}}\right)$$

$$= 49.6 \text{ g/mol}$$

Be careful with this method. Volume is in the denominator so the temperature correction factor must be inverted.

35. 1.40 liters
     $C_2H_2$
     [5.6, 6.9]

First we determine which reagent limits the reaction by calculating the number of moles of each:

$$? \text{ mol CaC}_2 = 4.00 \text{ g CaC}_2 \left(\frac{1 \text{ mol CaC}_2}{64.10 \text{ g CaC}_2}\right)$$

$$= 0.0624 \text{ mol CaC}_2$$

$$? \text{ mol H}_2\text{O} = 2.75 \text{ g H}_2\text{O} \left(\frac{1 \text{ mol H}_2\text{O}}{18.02 \text{ g H}_2\text{O}}\right)$$

$$= 0.153 \text{ mol H}_2\text{O}$$

Excess water is available since only 0.125 mol $H_2O$ is needed. The limiting reagent is $CaC_2$.

Therefore,

$$? \text{ liters } C_2H_2 = 0.0624 \text{ mol CaC}_2 \left( \frac{22.4 \text{ liters } C_2H_2}{1 \text{ mol CaC}_2} \right)$$

$$= 1.40 \text{ liters } C_2H_2$$

36. a. 5.17 liters
       $CO_2$ [6.9]

We find

$$? \text{ liters } CO_2 = 3.00 \text{ g } C_2H_2 \left( \frac{(4 \times 22.4) \text{ liters } CO_2}{2(26.02) \text{ g } C_2H_2} \right)$$

$$= 5.17 \text{ liters } CO_2$$

   b. 4.00 liters
      $CO_2$ [6.9]

You should be able to do part (b) of this problem by inspection. Gay-Lussac's law states that in a chemical reaction the ratio of moles is the same as the ratio of volumes measured at the same temperature and pressure.

37. 1.085 atm
    [6.13]

The van der Waals equation is

$$\left( P + \frac{n^2 a}{V^2} \right) \left( V - nb \right) = nRT$$

Rearranging the equation, we find

$$P = \frac{nRT - \left( \frac{n^2 a}{V} \right) + \left( \frac{n^3 ab}{V^2} \right)}{V - nb}$$

Solving for $P$, we find

$$P = 1.085 \text{ atm}$$

From the ideal gas law an answer of 1.092 atm is obtained.

38. $Cl_2$ [6.13]

The value of $a$ is largest for chlorine gas; thus, the intermolecular attraction of chlorine gas molecules is greater than that of the other gas molecules. Notice that the value of $a$ is smallest for helium, the molecules of which show little intermolecular interaction. The value of $a$ is large for molecules with a large dipole and a nonbonding electron pair, such as ammonia molecules.

Chlorine gas is also the largest of the gases listed in Table 6.3 of your text; it has the largest value of $b$, a constant which is directly proportional to molecular radius.

39. 0.23 g $H_2O$
    [6.5, 6.10]

Table 6.2 of your text gives the vapor pressure of water at $25^{\circ}C$:

$$P_{H_2O} = 0.0313 \text{ atm}$$

Even if excess water were added to the vessel, the vapor pressure of water would still be 0.0313 atm. The mass of gaseous water can be calculated from the ideal gas law:

$$PV = nRT = (g/m) \text{ } RT$$

$$g = \frac{mPV}{RT}$$

$$g = \frac{(18.02 \text{ g/mol})(0.0313 \text{ atm})(10.0 \text{ liter})}{(0.0821 \text{ liter atm } ^{\circ}K^{-1} \text{ mol}^{-1})(298^{\circ}K) \text{ })}$$

$$= 0.23 \text{ g } H_2O$$

From the density, 1.00 g/ml, we find that only 0.23 ml $H_2O$ evaporated:

$$? \text{ ml } H_2O = 0.23 \text{ g } H_2O \left( \frac{1.00 \text{ ml } H_2O}{1.00 \text{ g } H_2O} \right) = 0.23 \text{ ml } H_2O$$

40. 100 atm [6.10]

The partial pressure of water vapor is 0.0313 atm, and the partial pressure of He is 100 atm. The total pressure is the sum of the two partial pressure:

$$P = P_{H_2O} + P_{He}$$

$$= 0.0313 \text{ atm} + 100 \text{ atm}$$

$$= 100 \text{ atm}$$

41. $2.2 \times 10^2 \text{ } ^{\circ}K$
    [5.2, 6.5, 6.14]

First we calculate the number of moles in 1.0 ml $H_2O(1)$:

$$? \text{ mol } H_2O = 1.0 \text{ ml } H_2O \left(\frac{1.0 \text{g } H_2O}{1 \text{ ml } H_2O}\right) \left(\frac{1 \text{ mol } H_2O}{18.0 \text{ g } H_2O}\right)$$

$$= 0.0556 \text{ mol } H_2O$$

Then we calculate the temperature from the ideal gas law:

$$T = \frac{PV}{nR} = \frac{(1.0 \text{ atm})(10.0 \text{ liter})}{(0.0556 \text{ mol})(0.0821 \text{ liter atm } ^\circ K^{-1} \text{ mol}^{-1})}$$

$$= 2.2 \times 10^2 \, ^\circ K$$

This is above the critical temperature of water, $647.2 \, ^\circ K$, so no water can exist in liquid form.

42. 25.3 g/mol
    [6.5]

Since the gas was weighed before drying, the mass of water vapor must be subtracted to obtain the mass of the unknown gas. We use the ideal gas law to find the mass of the water vapor:

$$? \text{ g } H_2O = \frac{MPV}{RT}$$

$$= \frac{(18.0 \text{ g/mol})(0.0060 \text{ atm})(1.00 \text{ liter})}{(0.0821 \text{ liter atm } ^\circ K^{-1} \text{ mol}^{-1})(273 ^\circ K)}$$

$$= 0.0048 \text{ g}$$

The mass of unknown gas is

1.135g - 0.005g = 1.130g

The molecular weight of the unknown gas is

$$? \text{ g/mol} = \left(\frac{1.130 \text{ g}}{1.00 \text{ liter}}\right)\left(\frac{22.4 \text{ liters}}{1 \text{ mol}}\right)$$

$$= 25.3 \text{ g/mol}$$

43. $p_{He}$ = 549 torr

    and

    $p_{N_2}$ = 210 torr
    [2.9]

A mole of the gas mixture would weigh

(0.475 g/liter)(22.4 liter/mol) = 10.64 g/mol

If we let $x$ equal the mole fraction of He and $1-x$ equal the mole fraction of $N_2$,

$$x(4.00 \text{ g/mol}) + (1-x)(28.0 \text{ g/mol}) = 10.64 \text{ g/mol}$$

$$x = 0.723$$

$$1-x = 0.277$$

The partial pressure of helium is the total pressure times the mole fraction of helium:

$$p_{He} = X_{He} \, p_{total}$$

$$= (0.723)(760 \text{ torr})$$

$$= 549 \text{ torr, or } 0.723 \text{ atm}$$

Similarly for nitrogen:

$$p_{N_2} = X_{N_2} \, p_{total}$$

$$= (0.277)(760 \text{ torr})$$

$$= 210 \text{ torr, or } 0.277 \text{ atm}$$

44. $4.8 \times 10^{21}$ liters gas

We find

$$? \text{ liters gas} = 1.0 \times 10^{21} \text{ liters O}_2 \left( \frac{1 \text{ liter gas}}{0.21 \text{ liter O}_2} \right)$$

$$= 4.8 \times 10^{21} \text{ liters gas}$$

45. larger

The volume of the atmosphere would be larger. Atmospheric pressure decreases with height above the Earth's surface; thus, the average pressure cannot be 1 atmosphere but a lower value. An average temperature of $273^\circ$K is also low, but the effect due to temperature is smaller than that due to pressure.

46. 1.09 atm

Since this problem contains so many variables, it is best to tabulate the information:

|   | INITIAL CONDITIONS | FINAL CONDITIONS |
|---|---|---|
| $P$ | 1.01 atm | ? |
| $V$ | 1.47 liters | 1.47 liters |
| $T$ | $(25 + 273)\,^{\circ}K$ | $(54 + 273)\,^{\circ}K$ |
| $n$ | $?\ n_{O_2}$ | $?\left(n_{O_2} + n_{O_3}\right)$ |

We use the ideal gas law, $PV = nRT$, to determine the initial number of moles of $O_2$:

$$n = \frac{PV}{RT} = 6.07 \times 10^{-2} \text{ mol } O_2$$

We then calculate the number of moles of $O_2$ that react (5.24% of the original amount):

$$n_{O_2} \text{ reacting} = (0.0524)\,(6.07 \times 10^{-2} \text{ mol } O_2)$$

$$= 3.18 \times 10^{-3} \text{ mol } O_2$$

and the number of moles of $O_2$ remaining:

$$n_{O_2} \text{ remaining} = (6.07 \times 10^{-2} \text{ mol}) - (3.18 \times 10^{-3} \text{ mol})$$

$$= 5.75 \times 10^{-2} \text{ mol } O_2$$

We can use the number of moles of $O_2$ reacting and the balanced equation to determine the number of moles of $O_3$ formed:

$$n_{O_3} = 3.18 \times 10^{-3} \text{ mol } O_2 \left(\frac{2 \text{ mol } O_3}{3 \text{ mol } O_2}\right)$$

$$= 2.12 \times 10^{-3} \text{ mol } O_3$$

The total amount of gas remaining after the reaction is

$$n_{total} = n_{O_2} + n_{O_3}$$

$$= (5.75 \times 10^{-2} \text{ mol } O_2) + (2.12 \times 10^{-3} \text{ mol } O_3)$$

$$= 5.96 \times 10^{-2} \text{ mol gas}$$

From $PV = nRT$ we find the final pressure:

$$PV = nRT = \frac{(5.96 \times 10^{-2} \text{ mol})(0.08206 \text{ liter atm mol}^{-1} {}^{\circ}K^{-1})(327^{\circ}K)}{1.47 \text{ liters}}$$

$$= 1.09 \text{ atm}$$

**SELF-TEST**

Complete the test in 50 minutes:

I.  Work the following problems:

1.  A bulb filled with a gas to a pressure of 760 mm weighs 116.3124 g.  When the bulb is heated to 88°C at a pressure of 760 mm, 100 ml of gas is expelled and the bulb and remaining gas weigh 116.2584 g.  What is the molecular weight of the gas?

2.  A 1.168 g sample of an oxide, $XO_2$, reacts with exactly 500 ml of $H_2$ gas measured at STP.  What is the molecular weight of $XO_2$?

$$XO_2(s) + 2H_2(g) \rightarrow X(s) + 2H_2O(g)$$

3.  A 100 ml sample of gas is collected over water at temperature $T_1$, and the wet gas is found to exert a pressure of 750 mm at that temperature.  The same sample of gas is found to occupy 97.1 ml at $T_1$ when dry and under a pressure of 760 mm.  Calculate the vapor pressure of water at $T_1$ from these data.

4.  In a mixture of $C_2H_6$ and $O_2$ confined in a 1.00 liter container, the partial pressure of $C_2H_6$ is 160 mm and the partial pressure of $O_2$ is 560 mm.  The mixture is ignited and reacts according to the equation:

$$2C_2H_6(g) + 7O_2(g) \rightarrow 4CO_2(g) + 6H_2O(g)$$

The temperature is constant throughout the experiment and is high enough so that the $H_2O$ is gaseous.   What is the total pressure of the final mixture?

5. Calculate the density of a gas in g/liter at STP if a given volume of the gas effuses through an apparatus in 5.00 minutes and the same volume of oxygen at the same temperature and pressure effuses through the apparatus in 6.30 minutes.

6. Assume that 10 liters of $NH_3(g)$ and 10 liters of $O_2(g)$ are mixed and react according to the equation:

$$4NH_3(g) + 5O_2(g) \rightarrow 4NO(g) + 6H_2O(g)$$

If the conditions under which the gases are measured are constant and such that all materials are gaseous, list the volumes of all materials at the conclusion of the reaction.

7. Calculate the molecular weight of a gas that has a density of 1.59 g/liter at $50^\circ C$ and a pressure of 730 mm.

II.   Complete the statement or answer the question:

1. Of the two gases, $H_2$ (molecular weight = 2) and $C_5H_{12}$ (molecular weight = 72), under the same conditions of temperature and pressure, _____ would effuse _____ times more rapidly through a given orifice.

2. If 1.00 liter samples of $H_2(g)$ and $C_5H_{12}(g)$ are considered, both at STP, which sample has the larger number of molecules? _____
The molecules of which sample have the larger average kinetic energy? _____

3. The partial pressure of oxygen in a flask containing 32 g of $O_2$ and 32 g of $H_2$ is _____ of the total pressure.   (Atomic weights: O = 16.0, H = 1.0)

4.  The behavior of real gases deviates from that
    described by the ideal gas law because the ideal
    gas law fails to take into account _____
    and _____.

III.  Using the coordinates given, sketch the approximate
      shape of the curve for an ideal gas:

a.  pressure vs. volume

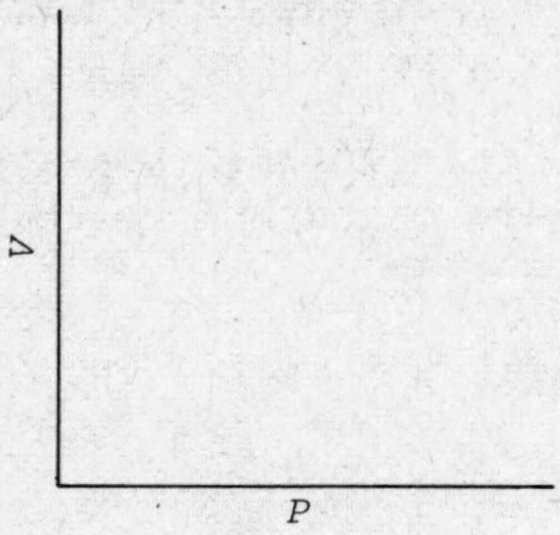

c.  absolute temper-
    ature vs. volume

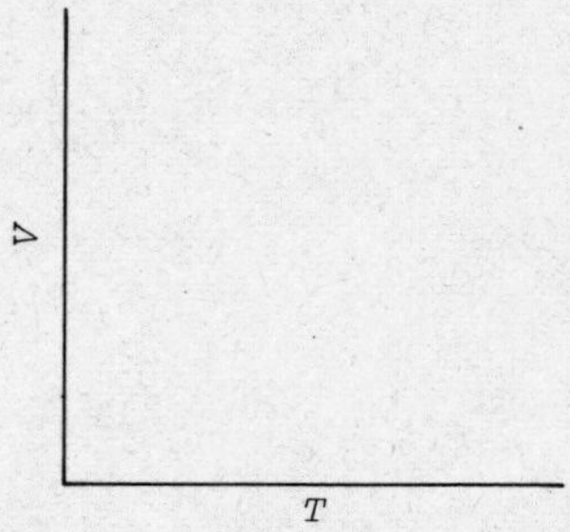

b.  pressure vs. the
    product of pres-
    sure and volume

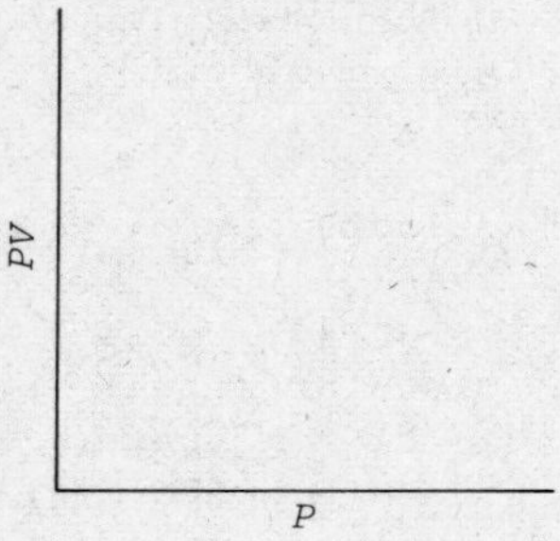

d.  energy distribu-
    tion of molecules

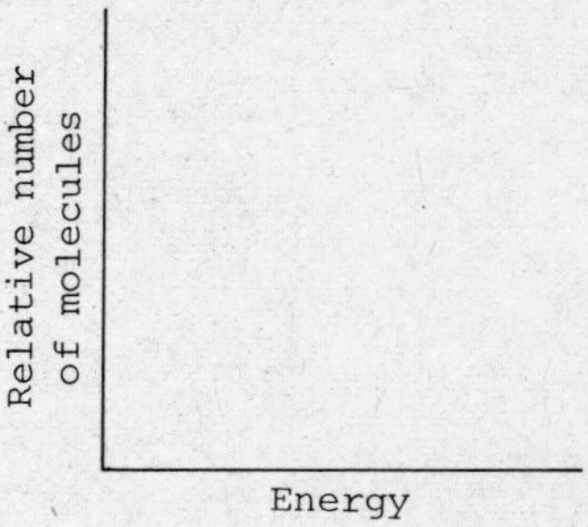

# 7
# Liquids and Solids

OBJECTIVES      (a) You should be able to demonstrate your knowledge of
the following terms by defining them, describing
them, or giving specific examples of them:

       Bragg equation [7.10]
       Clausius-Clapeyron equation [7.5]
       crystal lattice structures [7.9, 7.11, 7.12]
       dislocations [7.14]
       entropy of vaporization [7.5]
       equilibrium [7.3]
       evaporation [7.2]
       lattice energy [7.13]
       liquid [7.1]
       melting point [7.6]

metastable [7.8]
molar heat of crystallization [7.6]
molar heat of fusion [7.6]
molar heat of sublimation [7.8]
molar heat of vaporization [7.5]
normal boiling point [7.4]
normal freezing point [7.6]
phase diagram [7.8]
point defects [7.14]
semiconductor [7.14]
solid [7.12]
supercooling [7.6]
surface tension [7.1]
triple point [7.8]
Trouton's rule [7.5]
unit cell [7.9]
vapor pressure [7.3, 7.7]
viscosity [7.1]

(b) You should be able to determine the heat of vaporization, $\Delta H_v$, from the vapor pressure of a liquid at
two temperatures and rearrange the Clausius-Clapeyron
equation to determine either vapor pressure or
temperature if given the other variables.

(c) You should be able to interpret phase diagrams.

(d) You should be familiar with the various types of
crystal lattices and be able to calculate densities
from structural data and vice versa.

(e) You should be able to use the Bragg equation to
determine the distance between diffraction planes.

EXERCISES     I.  Answer each of the following with an entry from the
list on the right:

1. At STP water can exist at          a. decrease(s)
   100°C as either gas or _____.

2. The phase in which molecular       b. increase(s)
   motion is most restricted is
   the _____ phase.
                                        c. remains the
3. As a liquid is heated, the             same
   vapor pressure _____.

4. _____ is a property of
   liquids that describes resistance to flow.

5. In the _____ state matter
   assumes the shape of and completely fills the container
   into which it is placed.

6. A drop of liquid has a spher-
   ical shape due to the property
   of _____ .

7. As pressure is decreased,
   boiling points _____ .

8. As the temperature of a
   liquid increases, the rate
   of evaporation _____ .

9. _____ is the condition in
   which the rates of two opposite
   tendencies are equal.

10. The pressure of vapor in equi-
    librium with a liquid at a
    given temperature is called
    _____ .

11. At the _____ the vapor
    pressure of a liquid equals
    one atmosphere pressure.

12. The equation

    $$\frac{\Delta H_v}{T_b} = 21 \text{ cal/}^{\circ}\text{K mol}$$

    is a statement of _____ .

13. The _____ is the tempera-
    ture at which liquid and solid
    are in equilibrium under one
    atmosphere pressure.

14. _____ is the process in
    which a solid passes directly
    to the vapor phase.

15. An _____ semiconductor
    material contains an excess
    number of electrons.

d. gas

e. liquid

f. solid

g. surface
   tension

h. viscosity

i. Hess's law

j. Trouton's
   rule

k. melting point

l. normal boiling
   point

m. sublimation

n. equilibrium

o. vapor pressure

p. *n*-type

q. *p*-type

r. simple

s. body-centered

16. The _____ cubic crystal is     t. face-centered
    the cubic crystal that has
    the most empty space.

17. Each atom has six nearest
    neighbors in the _____
    cubic crystal.

18. There are two atoms per unit
    cell in the _____ cubic
    crystal.

19. The _____ cubic crystal is
    the most densely packed
    cubic crystal.

20. The diagram

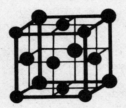

    is a representation of the
    _____ cubic crystal.

II.  Sulfur can exist in several crystalline forms.  The
     phase diagram in Figure 7.1 is that for monoclinic

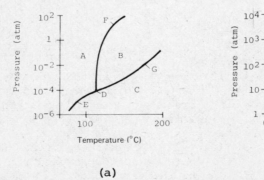

(a)                              (b)

Figure 7.1  Phase diagrams for (a) monoclinic sulfur and (b) carbon

sulfur.  The unit cell of monoclinic sulfur does
not have any equal edge dimensions and one of the
angles is not 90°:

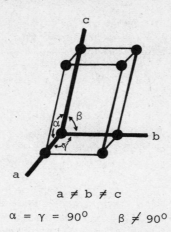

$$a \neq b \neq c$$

$$\alpha = \gamma = 90° \qquad \beta \neq 90°$$

Solid carbon can exist as diamond and as graphite.
Diamond can only be formed at pressures above $10^4$
atmosphere; after it is formed it cannot easily be
destroyed at atmospheric pressures even though
graphite is the preferred structure.

Use the information in the phase diagrams of Figure
7.1 and answer each of the following questions with
an entry from the list on the right:

1.  In the phase diagram for sulfur
    area $A$ represents the _____
    phase.

2.  In the phase diagram for sulfur
    area $B$ represents the _____
    phase.

3.  In the phase diagram for sulfur
    area $C$ represents the _____
    phase.

4.  In the phase diagram for sulfur
    point $D$ represents the _____

5.  In the phase diagram for sulfur
    line $E$ represents the _____
    equilibrium.

a. sulfur

b. carbon

c. sulfur and
   carbon

d. gas-liquid

e. gas-solid

f. liquid-
   solid

g. liquid-
   gas

h. solid-gas

6. In the phase diagram for sulfur line *F* represents the _____ equilibrium.

7. In the phase diagram for sulfur line *G* represents the _____ equilibrium.

8. Which element has the lower melting point: sulfur or carbon?

   _____

9. When sulfur is heated from $100^{\circ}C$ to $150^{\circ}C$ under 1 atmosphere pressure, a _____ phase change occurs.

10. When carbon is heated from $100^{\circ}C$ to $150^{\circ}C$ under 1 atmosphere pressure, a _____ phase change occurs.

11. As the pressure exerted on a sample of sulfur at $100^{\circ}C$ is reduced from 1 atmosphere to $10^{-6}$ atmosphere, a _____ phase change occurs.

12. As the pressure exerted on a sulfur sample maintained at $150^{\circ}C$ is reduced from 1 atmosphere to $10^{-6}$ atmosphere, a _____ phase change occurs.

13. Which carbon phase exists at $3000^{\circ}C$ and 1 atmosphere pressure? _____

14. Which element never exists as a liquid at 1 atmosphere pressure: carbon or sulfur? _____

15. Which is more dense: liquid or solid sulfur? _____

i. solid-liquid

j. none

k. gas

l. liquid

m. solid

n. triple point

III. Identify the phase(s) present in each region of the following curve and discuss the energy changes occurring as a solid is heated from point 1 to point 6:

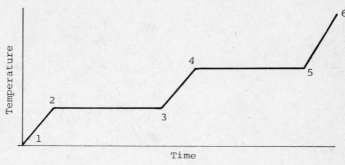

1. Line segment 1-2: _____

2. Line segment 2-3: _____

3. Line segment 3-4: _____

4. Line segment 4-5: _____

5. Line segment 5-6: _____

IV. Answer each of the following:

1. Argon crystallizes in a face-centered cubic structure at $-189^{\circ}C$. If the density of the solid is 1.7 $g/cm^3$, calculate the edge length of a unit cell.

2. An oxide of zirconium forms a face-centered cubic lattice of zirconium ions, and the tetrahedral holes are occupied by oxide ions. What is the formula of the zirconium oxide?

3. Crystal studies have shown that a certain sulfide of manganese forms a face-centered cubic lattice with sulfide ions filling the octahedral holes. What is the formula of the compound?

4. In the mineral spinel one-eighth of the tetrahedral holes are occupied by $Mg^{2+}$ ions and one-half of the octahedral holes are occupied by

$Al^{3+}$ ions. Oxide ions occupy the corners and face-centers of the unit cell. What is the formula of spinel?

5. The mineral perovskite has a crystalline structure in which the oxide ions occupy the face-centers and the larger cations, $Ca^{2+}$, occupy the corners of the unit cell. The smaller cations, $Ti^{4+}$, occupy the octahedral holes formed exclusively by the oxide ions. What is the formula of perovskite?

6. If both form the same type of lattice, which would have the larger electrostatic potential energy per mole: MgO or $CuSO_4$?

7. Which of the following liquids should exert the smaller vapor pressure at $25^{\circ}C$: $CCl_4$ or $H_2O$?

8. Which is the largest for a particular substance: $\Delta H_{vaporization}$, $\Delta H_{fusion}$, or $\Delta H_{sublimation}$?

9. Which of the following ionic compounds should have the highest melting temperature: $CaSO_4$, CaS, MgO, or $NH_4NO_3$?

10. Which of the following liquids should have the smaller heat of vaporization: $NH_3$ or $BCl_3$?

11. The heat of vaporization of $S_2Cl_2$ at the normal boiling point is 63.9 cal/g. Use Trouton's rule to predict the normal boiling point of this compound.

12. In the diffraction of a gold crystal with X-rays of a wavelength equal to 1.54Å, a first-order reflection is shown at an angle of $22^{\circ}10'$. What is the distance between the diffracted planes?

13. Aluminum crystallizes in a face-centered cubic unit cell with the length of an edge equal to 4.05Å. Assume the atoms are hard spheres and each face-centered atom touches the four corner atoms of its face. Calculate the radius of a hard-sphere atom.

# ANSWERS TO EXERCISES

1. liquid [7.4]
2. solid [7.7]
3. increases [7.3]
4. viscosity [7.1]
5. gas [6.1]
6. surface ten-
   sion [7.1]
7. decreases [7.4]
8. increases [7.2]
9. equilibrium
   [7.3]
10. vapor pressure
    [7.3]
11. normal boiling
    point [7.4]
12. Trouton's rule
    [7.5]
13. melting point
    [7.6]
14. sublimation
    [7.8]
15. *n*-type [7.14]
16. simple [7.9]
17. simple [7.9]
18. body-centered
    [7.9]
19. face-centered
    [7.11]
20. face-centered
    [7.9]

I. Physical states of matter

It is important to remember the terms that apply to phase changes.  The following diagram may be of help:

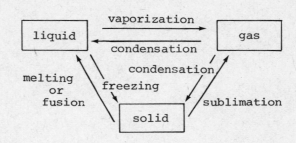

II. Phase diagrams for monoclinic sulfur and carbon.

See Section 7.8 of your text for a discussion of phase diagrams.

1. solid
2. liquid
3. gas
4. triple point

5. gas-solid or
   solid-gas
6. liquid-solid or
   solid-liquid
7. gas-liquid or
   liquid-gas

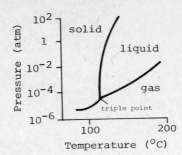

8. sulfur

Carbon does not form a liquid phase at 1 atmosphere; it sublimes when heated to $3652^{\circ}$C.

9. solid-liquid

Sulfur

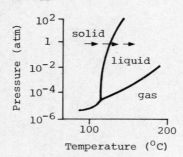

10. none

Carbon

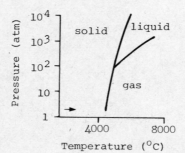

11. solid-gas

**Sulfur**

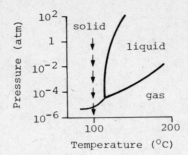

12. liquid-gas

**Sulfur**

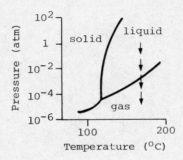

13. solid

**Carbon**

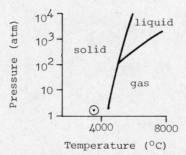

14. carbon

Carbon

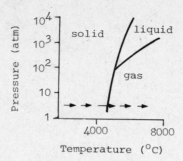

15. solid

As pressure is increased, the material becomes more dense and the solid form is produced:

Sulfur

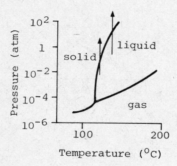

III.  Phase and energy changes

See Sections 7.4 and 7.6 of your text.

1. Line segment
   1-2:

In this region the solid phase is present.  The temperature and kinetic energy of the molecules are increasing.

2. Line segment
   2-3:

Solid and liquid are in equilibrium at the melting point.  The temperature and the kinetic energy are constant.  The potential energy increases since the system is still being heated.

3. Line segment
   3-4

In this region the liquid is present.  The temperature and the kinetic energy are increasing.

4. Line segment
   4-5

Liquid and vapor are in equilibrium at the boiling point.  The temperature and the kinetic energy are constant.  The potential energy is increasing.

5. Line segment
   5-6

In this region the gas phase is present.  The temperature and kinetic energy are increasing.

## IV. Molecular interactions in solids and liquids

1. 5.3 x 10$^{-8}$ cm,
   or 5.3 Å
   [7.9]

Argon forms a face-centered cubic crystal, and there are four atoms in a unit cell.  First we calculate the mass of the four atoms:

$$? \text{ g Ar} = 4 \text{ Ar atoms} \left(\frac{1 \text{ mol Ar}}{6.022 \times 10^{23} \text{ atoms Ar}}\right)\left(\frac{39.948 \text{ g Ar}}{1 \text{ mol Ar}}\right)$$

$$= 2.653 \times 10^{-22} \text{ g Ar}$$

Then we use the density to calculate the volume occupied by the unit cell:

$$? \text{ cm}^3 = 2.653 \times 10^{-22} \text{ g Ar}\left(\frac{1 \text{ cm}^3}{1.7 \text{ g Ar}}\right)$$

$$= 1.6 \times 10^{22} \text{ cm}^3$$

Finally we calculate the edge of the unit cell:

$$? \text{ cm} = \sqrt[3]{\text{cm}^3} = \sqrt[3]{1.6 \times 10^{-22} \text{ cm}^3} = \sqrt[3]{160 \times 10^{-24} \text{ cm}^3}$$

$$= 5.4 \times 10^{-8} \text{ cm}$$

2. ZrO$_2$
   [7.12]

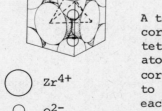

Zr$^{4+}$

O$^{2-}$

The zirconium ions are in a face-centered cubic lattice, and there are four zirconium ions per unit cell.  There are eight tetrahedral holes; thus, there are eight oxide ions.  The simplest formula is ZrO$_2$.

A tetrahedral hole is located directly behind a corner in a face-centered cubic unit cell.  The tetrahedron is formed with a corner atom and the atoms in the centers of the faces that form the corner.  A tetrahedral hole is shown in the diagram to the left.  There are eight tetrahedral holes in each cell.

3. MnS
   [7.12]

The manganese ions form a face-centered cubic lattice, and there are four manganese ions per unit
cell.  There are four octahedral holes; thus,
there are four sulfide ions.  The simplest formula
is MnS.

One-quarter of an octahedral hole is shown in the
following diagram of the unit cell.  There are twelve
such quarter holes in a unit cell.  One complete
octahedral hole is in the center of the unit cell.

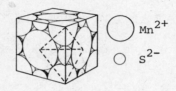

4. MgAl$_2$O$_4$
   [7.12]

The Mg$^{2+}$ ions occupy one-eighth of the tetrahedral
holes.  Since there are 8 tetrahedral holes, there
is (1/8)(8), or 1, Mg$^{2+}$ ion.

There is one octahedral hole in the center and one
on each edge.  Each edge is shared by 4 unit cells,
so there are ¼(12), or 3, edge octahedral holes per
unit cell.  The total number of octahedral holes
per unit cell is therefore 4.  The Al$^{3+}$ ions occupy
half the octahedral holes; thus, there are ½(4),
or 2, Al$^{3+}$ ions in a cell.

Oxide ions occupy the corners and face centers.
There are 8 corners, each shared by 8 unit cells;
thus, there is (1/8)(8), or 1, oxide ion per unit
cell.  There are 6 faces, each shared by 2 unit
cells; thus, there are ½(6), or 3, oxide ions per
unit cell.

The formula of spinel is therefore MgAl$_2$O$_3$.

5. CaTiO$_3$
   [7.12]

There are 8 corner Ca$^{2+}$ ions, each shared by 8
unit cells; thus, there is (1/8)(8), or 1, Ca$^{2+}$
ion per unit cell.

There are 6 face-center positions containing an oxide ion, each shared by another unit cell; thus, there are ½(6), or 3, oxide ions per unit cell.

The $Ti^{4+}$ ion is in the center of the unit cell, in the one octahedral hole formed exclusively by oxide ions.  It is not shared with any other unit cell, so there is one $Ti^{4+}$ ion per unit cell.

The formula of perovskite is therefore $CaTiO_3$.

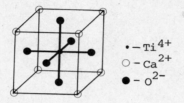

$\bullet - Ti^{4+}$
$\circ - Ca^{2+}$
$\bullet - O^{2-}$

6. MgO
   [7.13]

The equation for calculating potential energy is

$$PE = -\left(\frac{(Z_c)(Z_a)e^2}{r}\right)NA$$

For both MgO and $CuSO_4$,

$z_a = 2$

$z_c = 2$

and $e^2$ and $N$ are constants.

The Madelung constant, $A$, is the same for both compounds since they have the same crystalline structure.  Since $r$ is the only variable and since it is in the denominator, the numerical value of $PE$ would be greater for whichever compound has the smaller value of $r$, the ionic radius.  From the positions in the periodic table, we would expect $Mg^{2+}$ to be slightly smaller than $Cu^{2+}$.  Also, we would expect $O^{2-}$ to be much smaller than $SO_4^{2-}$.  Thus, since

$$r_{MgO} < r_{CuSO_4}$$

we would predict

$$|PE_{MgO}| > |PE_{CuSO_4}|$$

7. $H_2O$
   [7.3]

Liquids with strong intermolecular attractive forces have low vapor pressures. Carbon tetrachloride is a nonpolar, tetrahedral molecule and water is a polar, bent molecule:

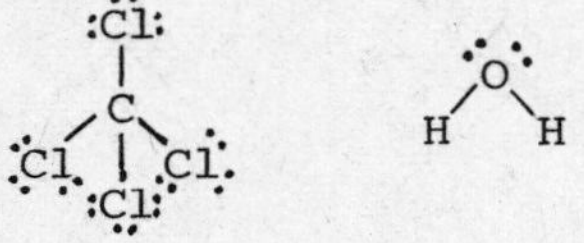

The only attractive forces that exist between $CCl_4$ molecules are relatively weak van der Waals forces; however, there are rather strong dipole-dipole attractive forces between $H_2O$ molecules. Thus, the vapor pressure of water is lower than that of carbon tetrachloride.

8. $\Delta H_{sublimation}$
   [5.9, 7.5, 7.6]

Sublimation is the process in which a solid goes directly to the vapor state; it requires more energy than a phase change from solid to liquid or liquid to vapor. At any given temperature

$$\Delta H_{sublimation} = \Delta H_{fusion} + \Delta H_{vaporization}$$

We obtain the values of the heats of fusion and vaporization for water at $0^{\circ}C$ from thermodynamic tables and use them with the law of Hess to calculate the heat of sublimation of water at $0^{\circ}C$ (see Section 5.8 of your text):

$H_2O(s) \xrightarrow{0^{\circ}C} H_2O(l)$    $\Delta H_f = 1.43\ kcal/mol$

$H_2O(l) \xrightarrow{0^{\circ}C} H_2O(g)$    $\Delta H_v = 47.50\ kcal/mol$

$H_2O(s) \xrightarrow{0^{\circ}C} H_2O(g)$    $\Delta H_{subl.} = 48.93\ kcal/mol$

9. MgO
   [7.13]

The substance with the highest melting point is that with the strongest intermolecular attractive forces, i.e., higher temperatures are required to break the stronger lattice. The greater the value of the electrostatic potential energy, the more difficult it is to break the lattice:

$$PE = - \frac{(Z_c)(Z_a)e^2}{r} NA$$

In the equation $e^2$ and $N$ are constants, and the values of $A$ for the series of compounds given in this problem are similar (see Table 7.4 of your text). The compounds have cesium chloride, $A = 1.763$, sodium chloride, $A = 1.748$, or zinc blende, $A = 1.638$, structures. None could have fluorite or rutile. Of the compounds in the series, $NH_4NO_3$ is the only one with ionic charges of 1+ and 1-; this fact eliminates $NH_4NO_3$ as a possibility because all others have $Z_c = Z_a = 2$. To decide among the remaining three substances, we consider $r$. Since $r$ is in the denominator, the smallest value of $r$ will give the largest value of potential energy, $PE$. The $Mg^{2+}$ ion is smaller than $Ca^{2+}$, and $O^{2-}$ is smaller than either $S^{2-}$ or $SO_4^-$; thus, MgO has the smallest value of $r$ and the largest value of $PE$.

10. $BCl_3$
    [7.5]

In general, the lower the heat of vaporization, the weaker the intermolecular forces of attraction. Boron trichloride, $BCl_3$, is a nonpolar molecule with attractive forces weaker than those of the polar ammonia molecule, $NH_3$.

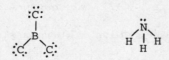

11. 140°C
    [7.5]

Trouton's rule states

$$\frac{\Delta H_v}{T_b} = 21 \text{ cal/}^{\circ}\text{K mol}$$

To find the normal boiling point, $T_b$, we rearrange and solve the preceding equation:

$$T_b = \frac{\Delta H_V}{21 \text{ cal/}^\circ\text{K mol}}$$

The molar heat of vaporization, $\Delta H_V$, should be in units of cal/mol:

$$\Delta H_V = 63.9 \text{ cal/g} \left(\frac{135.0 \text{ g } S_2Cl_2}{1 \text{ mol } S_2Cl_2}\right)$$

$$= 8.63 \times 10^3 \text{ cal/mol}$$

Finally we substitute the value of $\Delta H_V$ into Trouton's equation and determine $T_b$:

$$T_b = \frac{8.63 \times 10^3 \text{ cal/mol}}{21 \text{ cal/}^\circ\text{K mol}}$$

$$= 410^\circ\text{K}$$

Changing $T_b$ from Kelvin to centigrade, we find

$$T_b = 140^\circ\text{C}$$

12. 2.04Å
    [7.10]

We substitute values into the Bragg equation:

$$n\lambda = 2d \sin\phi$$

$$1(1.54\text{Å}) = 2(d)(0.377)$$

To determine $\sin 22^\circ 10'$, we look up the value in a table or use a slide rule or calculator. The value, 0.377, was obtained on a slide rule.

We then solve the Bragg equation:

$$d = \frac{1.54\text{Å}}{2(0.377)} = 2.04\text{Å}$$

13. 1.43Å
    [7.9 ]

This is a problem of geometry. Envision the unit cell:

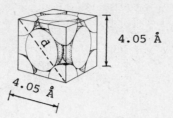

4.05 Å

4.05 Å

Using the Pythagorean theorem, we calculate the length of the hypotenuse, $d$:

$$d = \sqrt{2}\,(4.05 \text{ Å})\ \ =\ \ 5.73 \text{ Å}$$

The length $d$ corresponds to 4 radii; thus, $r = d/4$ = 1.43 Å.  The value of $r$ is slightly larger than expected (See Figure 3.1 of your text).

SELF-TEST        I.  Complete the test in 20 minutes:

   1. Complete the phase diagram of $H_2O$.  Label the
      triple point, the normal boiling point, the normal
      freezing point, and the solid, liquid, and vapor
      phases.  Clearly show the approximate slopes of all
      lines.

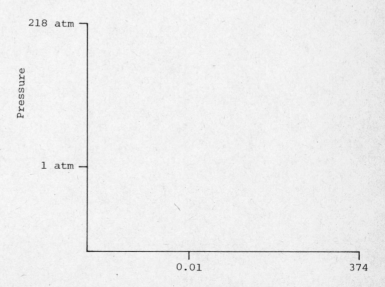

Temperature ($^{\circ}$C)

2. How many atoms are in the unit cell of
   a. a simple cubic lattice
   b. a body-centered cubic lattice

3. At $328^\circ K$ the vapor pressure of a liquid is
   0.823 atm. If the heat of vaporization is 7.07
   Kcal/mol, what is the normal boiling point of
   the liquid?  Do not use Trouton's rule.

4. What is the atomic weight of an element that
   crystallizes as a face-centered cubic crystal,
   the density of which is 8.94 $g/cm^3$?  The length
   of the diagonal through the center of the cube
   is 6.10 Å.

# 8
# Oxygen and Hydrogen

OBJECTIVES

(a) You should be able to demonstrate your knowledge of the following terms by defining them, describing them, or giving specific examples of them:

    acidic oxides [8.3]
    alkaline oxides [8.3]
    allotropy [8.4]
    amphoterism [8.3]
    displacement [8.7]
    disproportionation [8.5]
    equivalent weight [8.6]
    hydride ion [8.8]
    hydrogen bond [8.9]
    hydronium ion [8.8]

oxidation [8.5]
oxide ion [8.2]
oxidizing agent [8.5]
ozone [8.4]
peroxide ion [8.2]
reducing agent [8.5]
reduction [8.5]
superoxide ion [8.2]

(b) You should be familiar with the chemical and physical properties of oxygen and hydrogen.

(c) You should be familiar with simple reactions of oxygen and/or hydrogen with elements and compounds. Your instructor will emphasize the important reactions.

(d) You should be able to balance oxidation-reduction equations by using the oxidation-number method and/or the ion-electron method.

(e) You should be able to determine equivalent weights and use them in calculations involving chemical reactions.

EXERCISES

I. In this section you will find questions that involve concepts and methods presented in Chapters 1 to 8 of your text. These questions are designed to help you meet the objectives of Chapter 8 and to serve as a brief review. Answer each of the following:

1. Is the following statement *true* or *false*? The density of hydrogen gas is lower than that of any other gas.

2. Hydrogen gas can be prepared commercially by each of the following methods. Balance the equations.

a. $C(s) + H_2O(g) \xrightarrow{1000^\circ C} CO(g) + H_2(g)$

b. $H_2O(l) \xrightarrow{electrolysis} H_2(g) + O_2(g)$

c. $Fe(s) + H_2O(g) \xrightarrow{400^\circ C} Fe_3O_4(s) + H_2(g)$

d. $Zn(s) + H^+(aq) \longrightarrow Zn^{2+}(aq) + H_2(g)$

3. Determine the oxidation number of each atom in problem 2 of this section.

4. Identify the reducing agent in each reaction in problem 2 of this section.

5. Which compound would you expect to have a higher normal boiling point: $O_2$ or $O_3$?

6. Draw a Lewis structure of the peroxide ion, $O_2^{2-}$, and the superoxide ion, $O_2^-$.

7. Write the molecular-orbital configuration of $O_2^{2-}$.

8. What is the bond order of the superoxide anion, $O_2^-$?

9. Which type of magnetic behavior would you expect for a solid containing the superoxide anion?

10. How many protons are in the nucleus of an oxygen atom?

11. Write the quantum numbers of each electron of an oxygen atom.

12. Which type of crystalline solid would oxygen form?

13. Which of the following molecules should contain the least polar bond: $OF_2$, $H_2O_2$, or $SO_2$?

14. Yellow phosphorus is very toxic and is used in some rat poisons. Since elemental phosphorus is normally not a constituent of biological material, the following nonquantitative screening test can be used to detect it:

Elemental phosphorus, $P_4$, is converted in the presence of water to hypophosphorous acid, $H_3PO_2$, and phosphine, $PH_3$ (reaction 1). The hypophosphorous acid converts silver ions to metallic silver (reaction 2). The phosphine reacts with the silver ions of $AgNO_3$ to form silver phosphide, $Ag_3P$ (reaction 3). Phosphine and silver phosphide each give a brown stain to filter paper impregnated with $AgNO_3$.

Do the following:

a. Write a balanced equation for each reaction of the test.

b. Identify the oxidizing and reducing agents in each redox reaction of the test.

c. Determine the equivalent weight of each phosphorus-containing compound in the redox reaction(s) of the test.

15. Balance each of the following equations. All reactions occur in acid solution.

   a. $MnO_4^- + H_2C_2O_4 \rightarrow Mn^{2+} + CO_2$

   b. $Sn^{2+} + HgCl_2 \rightarrow Hg_2Cl_2 + Sn^{4+} + Cl^-$

   c. $MnO_4^- + Mo^{3+} \rightarrow MoO_4^{2-} + Mn^{2+}$

   d. $Ce^{4+} + H_3AsO_3 \rightarrow Ce^{3+} + H_3AsO_4$

16. Balance each of the following equations. All reactions occur in alkaline solution.

   a. $IO_4^- + I^- \rightarrow IO_3^- + I_2$

   b. $CO(NH_2)_2 + OBr^- \rightarrow CO_2 + N_2 + Br^-$

   c. $Al \rightarrow AlOH_4^- + H_2$


II. Oxygen

On Earth oxygen composes an estimated 46.5% of the crust, 89% of the water, 23.2% of the air, and possibly as much as 50% of the minerals (all percentages are by weight). Living organisms on Earth contain approximately one oxygen atom for every three other atoms. The reactions of oxygen are numerous and of great importance. We will consider only a few examples of such reactions. Answer the following:

1. Oxygen in the atmosphere reacts in the presence of ultraviolet light to produce ozone, $O_3$. The ozone absorbs most of the ultraviolet light in

the upper atmosphere and protects Earth from the damaging rays. Write a balanced equation for the production of ozone and give the oxidation number of oxygen in ozone.

2. Oxygen is consumed in plant and animal respiration and carbon dioxide and water are produced. For example,

$$C_6H_{12}O_6(aq) + O_2(g) \rightarrow CO_2(g) + H_2O(l)$$
glucose

$$C_{12}H_{22}O_{11}(aq) + 12O_2(g) \rightarrow 12CO_2(g) + 11H_2O(l)$$
maltose

The $CO_2$ that is produced enters the carbon dioxide cycle and the $H_2O$ enters the water cycle. Carbon dioxide dissolves in the oceans where it can react with $Ca^{2+}$ to form carbonate deposits:

$$Ca^{2+}(aq) + H_2O(l) + CO_2(g) \rightarrow CaCO_3(s) + 2H^+(aq)$$

Photosynthesis by the phytoplankton in the sea returns oxygen to the atmosphere:

$$6CO_2(g) + 6H_2O(l) \rightarrow 6O_2(g) + C_6H_{12}O_6(aq)$$

Which of the preceding processes would irreversibly reduce the amount of $O_2$ in the atmosphere under ordinary conditions?

3. Rocks exposed to the atmosphere can be oxidized by oxygen; for example, iron(II) oxide is oxidized to iron(III) oxide. Write the equation for the atmospheric oxidation of iron(II) oxide.

4. Oxygen oxidizes many compounds and elements. The fuels we burn consume oxygen and produce $CO_2$. Write a balanced equation for the complete oxidation of coal (assume it is only carbon). Also write a balanced equation for the complete oxidation of $C_8H_{18}$, octane.

5. Oxygen oxidizes the volcanic gases CO, $SO_2$, and $H_2$. Write a balanced equation for the complete oxidation of each of the volcanic gases.

## III. Hydrogen

Hydrogen composes 63% of the total number of atoms in living organisms on Earth and 91% of the number of atoms in our universe.  The ability of certain molecules to form hydrogen bonds is important.  Without hydrogen bonding life would be different; for example, water would boil at a much lower temperature, and DNA would not be a double helix.

1. The mass of the Earth's atmosphere is approximately $5 \times 10^5$ short tons (1 short ton = 2000 lb).  If 0.5% of the mass is $H_2$, how many short tons of $H_2$ are in the atmosphere?

2. The Earth contains approximately $1.5 \times 10^{24}$ g of free water.  How many short tons of hydrogen combined with oxygen to form the Earth's free water?

## ANSWERS TO EXERCISES

### I. Review

1. True
   [6.7]

If we consider hydrogen to be an ideal gas, one mole of that gas would occupy 22.4 liters.  Since 1 mol of $H_2$ is 2 g $H_2$, there would be 2 g of $H_2$ in 22.4 liters.  Similarly, the density of helium would be 4 g/22.4 liters.  All other gases are heaver than hydrogen or helium and would have a larger density.

2., 3., 4.
   [3.10, 8.5]

Oxidation numbers are given above the balanced equations and reducing agents are identified by the letter $R$ below the equation:

a. $\overset{0}{C}(s) + \overset{1+\ 2-}{H_2O}(g) \xrightarrow{1000^\circ C} \overset{2+2-}{CO}(g) + \overset{0}{H_2}(g)$
      $R$

b. $2\overset{1+\ 2-}{H_2O}(l) \xrightarrow{electrolysis} 2\overset{0}{H_2}(g) + \overset{0}{O_2}(g)$
      $R$

c. $3\overset{0}{Fe}(s) + 4\overset{1+2-}{H_2O} \xrightarrow{400^\circ C} \overset{8/3+2-}{Fe_3O_4}(s) + 4\overset{0}{H_2}(g)$
      $R$

d. $\overset{0}{Zn}(s) + 2\overset{1+}{H^+}(aq) \longrightarrow \overset{2+}{Zn^{2+}}(aq) + \overset{0}{H_2}(g)$
      $R$

5. $O_3$
   [7.4]

The molecular weight of $O_3$ is greater than that of $O_2$, and the attractive forces of $O_3$, which has a dipole moment, are stronger than those of $O_2$, which has no dipole moment.

6. $$\left[ :\ddot{O} : \ddot{O}: \right]^{2-}$$

$$\left[ :\ddot{O} : \ddot{O}\cdot \right]^{-}$$

   [3.6]

7. [4.4]

The molecular-orbital configuration of $O_2^{2-}$ is

8. $1\frac{1}{2}$
   [4.4, 8.2]

$(\sigma 1s)^2 (\sigma *1s)^2 (\sigma 2s)^2 (\sigma *2s)^2 (\sigma 2p)^2 (\pi 2p)^4 (\pi *2p)^4$.

9. paramagnetic
   [2.16]

There is one unpaired electron in $O_2^{-}$.

10. 8 protons
    [2.4, 2.7]

11. [2.15]

| Electron | $n$ | $l$ | $m$ | $s$ |
|----------|-----|-----|-----|-----|
| 1 | 1 | 0 | 0 | $+\frac{1}{2}$ |
| 2 | 1 | 0 | 0 | $-\frac{1}{2}$ |
| 3 | 2 | 0 | 0 | $+\frac{1}{2}$ |
| 4 | 2 | 0 | 0 | $-\frac{1}{2}$ |
| 5 | 2 | 1 | +1 | $+\frac{1}{2}$ |
| 6 | 2 | 1 | 0 | $+\frac{1}{2}$ |
| 7 | 2 | 1 | -1 | $+\frac{1}{2}$ |
| 8 | 2 | 1 | +1 | $-\frac{1}{2}$ |

12. [4.8]

Oxygen would form a nonpolar molecular solid with van der Waals forces holding the molecules together.

13. $H_2O_2$
    [3.9]

In peroxide the O—O bond is nonpolar.

14. [8.5, 8.6]
    a.

Reaction (1)

$$P_4 + 6H_2O \rightarrow 3H_3PO_2 + PH_3$$

is a disproportionation reaction:

$$3\left[P_4 + 8H_2O \quad \rightarrow \quad 4H_3PO_2 + 4H^+ + 4e^-\right]$$

$$P_4 + 12H^+ + 12e^- \quad \rightarrow \quad 4PH_3$$

$$\overline{4P_4 + 24H_2O \quad \rightarrow \quad 12H_3PO_2 + 4PH_3}$$

or

$$P_4 + 6H_2O \rightarrow 3H_2PO_2 + PH_3$$

Reaction (2)

$$H_3PO_2 + 2H_2O + 4AgNO_3 \rightarrow 4HNO_3 + H_3PO_4 + 4Ag$$

is a redox reaction:

$$H_3PO_2 + 2H_2O \quad \rightarrow \quad H_3PO_4 + 4H^+ + 4e^-$$

$$4\left[e^- + Ag^+ + NO_3^- \rightarrow Ag + NO_3^-\right]$$

$$\overline{H_3PO_2 + 2H_2O + 4Ag^+ + 4NO_3^- \rightarrow 4H^+ + H_3PO_4 + 4Ag + 4NO_3^-}$$

or

$$H_3PO_2 + 2H_2O + 4AgNO_3 \rightarrow 4HNO_3 + H_3PO_4 + 4Ag$$

Reaction (3)

$$PH_3 + 3AgNO_3 \rightarrow 3HNO_3 + Ag_3P$$

is not a redox reaction.

When you balance any equation by the ion-electron
method, you can check your answer by answering
each of the following questions:

a. Is the same number of atoms of each element on
   both sides of the equation?

b. Is the total charge the same on each side of the
   equation?

c. Do particular ions or molecules appear only once
   in the equation?

d. Are the coefficients in the equation the lowest
   whole numbers?

If the answer to each question is *yes*, you have balanced the equation correctly.

b.  Reaction (1) is a disproportionation reaction since $P_4$ is both the oxidizing and reducing agent and undergoes oxidation and reduction.

In reaction (2) $H_3PO_2$ is oxidized and is therefore the reducing agent. The silver ion, $Ag^+$, is reduced and is therefore the oxidizing agent.

c.  In reaction (1) $P_4$ is converted to $H_3PO_2$:

$$\overset{0}{P_4} + 8H_2O \rightarrow 4\overset{1+}{H_3PO_2} + 4H^+ + 4e^-$$

Each atom of phosphorus undergoes an increase in oxidation number of 1, from 0 to 1+. The total increase for 4 P atoms is 4, and the equivalent weight of $P_4$ is 4(30.974)/4, or 30.974.

In the conversion of $P_4$ to $PH_3$

$$\overset{0}{P} + 12H^+ + 12e^- \rightarrow 4\overset{3-}{PH_3}$$

each atom of phosphorus undergoes a decrease in oxidation number of 3, from 0 to 3-. The total decrease for 4 P atoms is 12, and the equivalent weight of $P_4$ is 4(30.974)/12, or 10.32.

In reaction (2)

$$H_2PO_2 + 2H_2O + 4AgNO_3 \rightarrow 4HNO_3 + H_3PO_4 + 4Ag$$

each atom of phosphorus undergoes an increase in oxidation number of 4, from 1+ to 5+. Thus, the equivalent weight of $H_3PO_2$ is 65.82/4, or 16.46.

15. [8.5]  a. $2MnO_4^- + 5H_2C_2O_4 + 6H^+ \rightarrow 2Mn^{2+} + 10CO_2 + 8H_2O$

b. $Sn^{2+} + 2HgCl_2 \rightarrow Hg_2Cl_2 + Sn^{4+} + 2Cl^-$

c. $8H_2O + 3MnO_4^- + 5Mo^{3+} \rightarrow 5MoO_4^{2-} + 3Mn^{2+} + 16H^+$

d. $2Ce^{4+} + H_3AsO_3 + H_2O \rightarrow 2Ce^{3+} + H_3AsO_4 + 2H^+$

16. [8.5]  a. $IO_4^- + 2I^- + H_2O \rightarrow IO_3^- + I_2 + 2OH^-$

b. $CO(NH_2)_2 + 3OBr^- \rightarrow CO_2 + N_2 + 3Br^- + 3H_2O$

c. $2OH^- + 2Al + 6H_2O \rightarrow 2Al(OH)_4^- + 3H_2$

## II. Oxygen

1. The equation for the conversion of oxygen into ozone is

$$3\overset{0}{O_2} \overset{h\nu}{\rightarrow} 3\overset{0}{O_3}$$

The oxidation number of an oxygen atom in $O_3$ is 0.

2. The process described by the equation

$$Ca^{2+}(aq) + H_2O + CO_2(g) \rightarrow CaCO_3(s) + 2H^+(g)$$

ties up oxygen as a solid carbonate deposit. Respiration consumes oxygen, but the oxygen can be released by photosynthesis.

3. The equation for the atmospheric oxidation of FeO is

$$4FeO(s) + O_2(g) \rightarrow 2Fe_2O_3(s)$$

4. The equations for the oxidations of C(s) and $C_8H_{18}( )$ are

$$C(s) + O_2(g) \rightarrow CO_2(g)$$

$$2C_8H_{18}(l) + 25O_2(g) \rightarrow 16CO_2(g) + 18H_2O(g)$$

5. The equations for the oxidations of the volcanic gases are

$$2CO(g) + O_2(g) \rightarrow 2CO_2(g)$$

$$2SO_2(g) + O_2(g) \rightarrow 2SO_3(g)$$

$$2H_2(g) + O_2(g) \rightarrow 2H_2O(g)$$

## III. Hydrogen

1. $2.5 \times 10^{13}$ short tons $H_2$

We calculate the number of short tons of $H_2$ in the atmosphere of Earth:

? short tons $H_2$ = 5 x $10^{15}$ short tons atmosphere $\left(\dfrac{5 \text{ short tons } H_2}{1000 \text{ short tons atmosphere}}\right)$

$$= 2.5 \times 10^{13} \text{ short tons } H_2$$

2. 1.8 x $10^{17}$ short tons $H_2$

We calculate the number of short tons of $H_2$ as follows:

? short tons $H_2$ = 1.5 x $10^{24}$ g $H_2O$ $\left(\dfrac{1 \text{ mol } H_2O}{18.0 \text{ g } H_2O}\right)\left(\dfrac{2.02 \text{ g } H_2}{1 \text{ mol } H_2O}\right)\left(\dfrac{1 \text{ lb } H_2}{4.54 \text{ g } H_2}\right)$

$$\left(\dfrac{1 \text{ short ton } H_2}{2000 \text{ lb } H_2}\right)$$

$$= 1.8 \times 10^{17} \text{ short tons } H_2$$

SELF-TEST

Complete the test in 30 minutes:

1. Balance the following equations:

   a. $ClO_2 \rightarrow ClO_2^- + ClO_3^-$          (basic soln.)

   b. $IO_4^- + H_2AsO_3^- \rightarrow IO_3^- + H_2AsO_4^-$     (acid soln)

   c. $Sb + NO_3^- \rightarrow Sb_4O_6 + NO$          (acid soln.)

   d. $O_3 + I^- \rightarrow H_2O + I_2$          (acid soln.)

   e. $NO_3^- + H_2S \rightarrow NO + S$          (acid soln.)

   f. $Cr_2O_7^{2-} + Cl^- \rightarrow Cr^{3+} + Cl_2$          (acid soln.)

2. Determine the equivalent weight (to three significant figures) of each reactant in the equations in problem 1 of the self-test.

3. Complete the following:

   a. $Mg(s) + H_2O(g) \rightarrow$

   b. $H_2(g) + Cl_2(g) \rightarrow$

   c. $Na(s) + H_2O(l) \rightarrow$

# 9
# Solutions

OBJECTIVES

(a) You should be able to demonstrate your knowledge of the following terms by defining them, describing them, or giving specific examples of them:

activity [9.13]
activity coefficient [9.13 ]
azeotrope [9.11]
colligative properties [9.10]
Debye-Hückel theory [9.13]
electrolyte [9.12]
end point [9.7]
entropy [9.4]
fractional distillation [9.11]
Gibbs free energy, $G$ [9.4]

heat of solution [9.4]
Henry's law [9.5]
hydration [9.3]
ideal solution [9.8]
indicator [9.7]
Le Chatelier's principle [9.5]
ligands [9.3]
molality, $m$ [9.6]
molarity, $M$ [9.6]
mole fraction, $X$ [9.6]
normality, $N$ [9.6]
osmosis [9.10]
osmotic pressure [9.10]
Raoult's law [9.8]
saturation [9.1]
simple distillation [9.11]
solute [9.1]
solvent [9.1]
standard solution [9.7]
supersaturation [9.1]
titration [9.7]
van't Hoff factor [9.13]
weight percentage [9.6]

(b) You should be able to determine the molarity, $M$, the molality, $m$, and the normality, $N$, of a solution. You should also be able to determine the weight percentage of solute in a solution and the mole fraction, $X$, of a component of a solution.

(c) You should be able to work problems involving titrations and dilutions.

(d) You should be able to determine the vapor pressure of a component of an ideal solution.

(e) You should be able to determine molecular weights from data on freezing point depression and boiling point elevation.

EXERCISES

I. Work the following problems to gain experience in using various expressions of concentration. Remember the equations for molarity, $M$, normality, $N$, molality, $m$, and mole fraction, $X$:

$$M = \frac{\text{number of moles of solute}}{\text{number of liters of solution}}$$

$$N = \frac{\text{number of gram equivalent weights of solute}}{\text{number of liters of solution}}$$

$$m = \frac{\text{number of moles of solute}}{\text{number of kilograms of solvent}}$$

$$X = \frac{\text{number of moles of a component}}{\text{total number of moles of all components}}$$

1. A total of 4.00 g of $AgNO_3$ is dissolved in a small quantity of water and diluted to 1.00 liter. What is the molarity of the final $AgNO_3$ solution?

2. The silver nitrate solution of problem 1 of this section is to be used for titrating the chloride in natural ground water:

$$Ag^+(aq) + Cl^-(aq) \longrightarrow AgCl(s)$$

What is the normality of the silver nitrate solution?

3. A solution is $0.625M$ $KMnO_4$. What is the normality of the solution if the permanganate is to be used in the reaction

$$MnO_4^- + Fe^{2+} \longrightarrow Mn^{2+} + Fe^{3+}$$

4. How many grams of $K_2Cr_2O_7$ are needed to prepare 1.00 liter of $0.100M$ $K_2Cr_2O_7$?

5. A beverage contains 0.174 g of ethanol, $C_2H_5OH$, and 4.72 g of water, $H_2O$. What is the mole fraction of ethanol?

6. What is the molality of ethanol in the solution described in problem 5 of this section?

7. How many grams of $K_2Cr_2O_7$ are needed to prepare 60 ml of a $0.15M$ $K_2Cr_2O_7$ solution?


II. Often it is necessary to prepare dilute solutions from concentrated solutions by dilution. Picture the following dilution:

(a) A specified amount of solute from container A

is transferred to container B, which is empty:

(b) The solute in container B is diluted by the
addition of solvent from container C:

The same amount of solute exists in container B
*before* and *after* dilution, only the amount of
solvent is changed.  If we designate the condition before dilution by 1 and that after
dilution by 2 and express the amount of solute
transferred in moles,

$$\text{moles}_1 = \text{moles}_2$$

and

$$M_1 V_1 = M_2 V_2$$

in which $M$ is molarity and $V$ is volume.

If we express the amount of solute transferred
in equivalents,

$$\text{equivalents}_1 = \text{equivalents}_2$$

and

$$N_1 V_1 = N_2 V_2$$

in which $N$ is normality and $V$ is volume.

We can also express the amount of solute transferred in grams:

$$\text{grams}_1 = \text{grams}_2$$

and

$$(\text{g/100 ml solvent})_1 \, (V_1) = (\text{g/100 ml solvent})_2 \, (V_2)$$

Work the following problems:

1. What volume of a 0.45$M$ $KMnO_4$ solution is needed to prepare 1.0 liter of 0.10$M$ $KMnO_4$?

2. What volume of a 30.5% HCl solution, which has a density of 1.12 g/ml, is needed to prepare 500 ml of 0.200$M$ HCl?

3. What volume of a standard riboflavin solution that has a density of 1.00 mg/ml is necessary to prepare 100 ml of a solution that has a density of 0.100 mg/ml?

4. What is the molarity of an acetic acid solution if 25 ml of this solution is diluted to 100 ml to form a 0.75$M$ solution?

5. How many milliliters of 0.065$M$ riboflavin, vitamin $B_2$, are necessary to prepare 1.0 liter of riboflavin solution that contains a solute concentration of 0.10 mg/ml? The molecular weight of riboflavin is 376.

III. Work the following problems. The problems concern material presented in Chapter 9 and will serve as a review.

1. In which of the following liquids should the gas HF be most soluble:  $CF_4$, $Br_2$, or $H_2O$?

2. Which of the following expressions of concentration is (are) independent of temperature: molarity, molality, or normality?

3. Which of the following solutions would have the lowest freezing point:  1.00$m$ $AlCl_3$, 1.00$m$ $CaCl_2$, or 1.00$m$ NaCl?

4. The solubility curve for $CoSO_4$ is shown in Figure 9.1 of the study guide. Refer to the curve and determine whether the solution process of $CoSO_4$ in $H_2O$ is endothermic or exothermic.

5. Is the absolute value of the lattice energy of $CoSO_4$ greater than that of the hydration energy? (See Figure 9.1 of the study guide.)

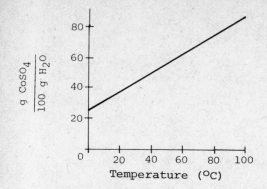

Figure 9.1   Solubility curve for CoSO$_4$

6. The aluminum compound Al$_2$(SO$_4$)$_3$ is a strong electrolyte. What is the vapor pressure above a solution prepared with 0.100 mol of Al$_2$(SO$_4$)$_3$ and 9.00 mol of H$_2$O at 30°C if the vapor pressure of pure H$_2$O at 30°C is 0.0395 atm?

7. Using Toepfer's reagent indicator, one can determine the concentration of free HCl present in gastric juice by titration with NaOH; normal concentrations are between 0 and 0.040$N$. If 5.0 ml of gastric juice requires 1.0 ml of 0.10$N$ NaOH to reach the salmon pink end point, determine whether a patient's level of free HCl is normal.

8. The heating of 400 ml of BaCl$_2$ solution, which has a density of 1.09 g/ml, produces 43.7 g of dry, solid residue. What is the molarity of the original BaCl$_2$ solution? What is the molality?

9. Is Henry's law valid for all solutions?

10. If NaCl costs $56.00 per ton and CaCl$_2$ costs $88.00 per ton, which compound would be more economical to prevent icy roads?

11. According to Raoult's law, the lowering of vapor pressure of the solvent in dilute solutions of nonvolatile solutes is proportional to what?

12. Does the entropy of a natural process increase

or decrease?

13. Which of the following would be expected to have the greater hydration energy:  $BeCl_2$ or $BaCl_2$?

14. At $25^{\circ}C$ the osmotic pressure of 100 ml of β-lactoglobulin solution containing 1.49 g of that protein was found to be 0.0100 atm. Calculate the molecular weight of the protein.

15. The vapor pressure of pure water at $25^{\circ}C$ is $3.12 \times 10^{-2}$ atm.  What is the vapor pressure of a 100 g sample of water in which 27.4 g of sucrose, $C_{12}H_{22}O_{11}$, is dissolved?

16. The amount of chloride in a blood sample can be determined by titrating a prepared sample of blood with mercuric nitrate:

$$2\ Cl^- + Hg^{2+} \rightarrow HgCl_2$$

The end point is observed when the diphenyl-carbazone indicator changes color.  Very small amounts of blood are used and microburets are used in the titration.  If it takes 2.42 ml of $0.0100N$ mercuric nitrate solution to titrate the chloride in a 0.200 ml blood sample, what is the chloride concentration in milliequivalents per liter?

17. Bacteria responsible for gangrene, *Clostridium welchii*, are anaerobic organisms, i.e., their growth is inhibited by molecular oxygen.  Why are patients with gangrene sometimes placed in a high pressure, oxygen-rich chamber (a hyperbaric chamber)?

18. Water boils at $95^{\circ}C$ in Denver, Colorado.  Why? Estimate how much NaCl should be added to 1000 g $H_2O$ to raise the boiling point of water in Denver to $100^{\circ}C$.

ANSWERS TO EXERCISES

I. Expressions of concentration

Refer to Section 9.6 of your text if you have any questions on this material.

1. 0.0235$M$

We calculate the molarity:

$$? \ M \ AgNO_3 = \frac{\text{number of moles of } AgNO_3}{\text{number of liters of } AgNO_3 \text{ soln.}}$$

$$= \frac{(4.00 \text{ g } AgNO_3)\left(\dfrac{1 \text{ mol } AgNO_3}{169.9 \text{ g } AgNO_3}\right)}{1 \text{ liter } AgNO_3 \text{ soln.}}$$

$$= 0.0235M \ AgNO_3$$

2. 0.0235$N$

For the reaction

$$Ag^+(aq) + Cl^-(aq) \rightarrow AgCl(s)$$

the equivalent weight of $AgNO_3$ is equal to the molecular weight of $AgNO_3$. Therefore,

$$N = M = 0.0235$$

3. 3.12$N$

For the reaction

$$5e^- + 8H^+ + Mn\overset{7+}{O_4^-} \rightarrow Mn^{2+} + 4H_2O$$

the equivalent weight of $KMnO_4$ is 1/5 of the molecular weight of $KMnO_4$. Thus, the normality is 5 times the molarity since 1 mol $KMnO_4$ is 5 equivalents:

$$? \ N \ KMnO_4 = \left(\frac{0.625 \text{ mol } KMnO_4}{1 \text{ liter } KMnO_4 \text{ soln.}}\right)\left(\frac{5 \text{ equivalents } KMnO_4}{1 \text{ mol } KMnO_4}\right)$$

$$= \left(\frac{3.12 \text{ equivalents } KMnO_4}{1 \text{ liter } KMnO_4 \text{ soln.}}\right)$$

$$= 3.12N \ KMnO_4$$

4. 29.4 g

We calculate the number of grams of $K_2Cr_2O_7$ as follows:

$$MK_2Cr_2O_7 = \frac{\text{number of moles of } K_2Cr_2O_7}{\text{number of liters of } K_2Cr_2O_7 \text{ soln.}}$$

$$0.100M \ K_2Cr_2O_7 = \frac{? \text{ g } K_2Cr_2O_7 \left(\dfrac{1 \text{ mol } K_2Cr_2O_7}{294.2 \text{ g } K_2Cr_2O_7}\right)}{1 \text{ liter } K_2Cr_2O_7 \text{ soln.}}$$

$$? \text{ g } K_2Cr_2O_7 = \frac{\left(0.100M \text{ } K_2Cr_2O_7\right)\left(1 \text{ liter } K_2Cr_2O_7 \text{ soln.}\right)\left(294.2 \text{ g } K_2Cr_2O_7\right)}{1 \text{ mol } K_2Cr_2O_7}$$

$$= 29.4 \text{ g } K_2Cr_2O_7$$

We could also use a logic chain:

$$? \text{ g } K_2Cr_2O_7 = \left(\frac{0.100 \text{ mol } K_2Cr_2O_7}{1 \text{ liter } K_2Cr_2O_7 \text{ soln.}}\right)\left(\frac{294.2 \text{ g } K_2Cr_2O_7}{1 \text{ mol } K_2Cr_2O_7}\right)\left(1 \text{ liter } K_2Cr_2O_7 \text{ soln.}\right)$$

$$= 29.4 \text{ g } K_2Cr_2O_7$$

5. 0.0142    We calculate the mole fraction of ethanol as follows:

$$? \text{ mol } C_2H_5OH = 0.174 \text{ g } C_2H_5OH \left(\frac{1 \text{ mol } C_2H_5OH}{46.08 \text{ g } C_2H_5OH}\right)$$

$$= 0.003776 \text{ mol } C_2H_5OH$$

$$? \text{ mol } H_2O = 4.72 \text{ g } H_2O \left(\frac{1 \text{ mol } H_2O}{18.02 \text{ g } H_2O}\right)$$

$$= 0.2619 \text{ mol } H_2O$$

$$X_{C_2H_5OH} = \frac{\text{number of moles of } C_2H_5OH}{\text{total number of moles of } C_2H_5OH + H_2O}$$

$$= \frac{0.003776 \text{ mol}}{0.003776 \text{ mol} + 0.2619 \text{ mol}} = 0.0142$$

6. 0.800m    We calculate the molality of ethanol as follows:

$$? \text{ } m \text{ } C_2H_5OH = \frac{\text{number of moles of } C_2H_5OH}{\text{number of kilograms of } H_2O}$$

$$= \frac{0.003776 \text{ mol } C_2H_5OH}{(4.72 \text{ g } H_2O)\left(\frac{1 \text{ kg}}{10^3 \text{ g}}\right)}$$

$$= 0.800m \text{ } C_2H_5OH$$

7. 2.6 g    Rearranging the equation

$$M \text{ } K_2Cr_2O_7 = \frac{\text{number of moles of } K_2Cr_2O_7}{\text{number of liters of } K_2Cr_2O_7 \text{ soln.}}$$

we find

$$(M \text{ } K_2Cr_2O_7)(\text{number of liters of } K_2Cr_2O_7 \text{ soln.}) = \text{number of moles of } K_2Cr_2O_7$$

and

$(M \ K_2Cr_2O_7)$ (number of liters of $K_2Cr_2O_7$ soln.) = $\dfrac{\text{number of grams of } K_2Cr_2O_7}{\text{molecular weight of } K_2Cr_2O_7}$

Substituting values into the rearranged equation, we find

$(0.15M \ K_2Cr_2O_7)(0.060 \text{ liter } K_2Cr_2O_7 \text{ soln.}) = \dfrac{? \text{ g } K_2Cr_2O_7}{294 \text{ g } K_2Cr_2O_7}$

$? \text{ g } K_2Cr_2O_7 = 2.6 \text{ g } K_2Cr_2O_7$

We could also use a logic chain to solve the problem:

$? \text{ g } K_2Cr_2O_7 =$

$(0.15M \ K_2Cr_2O_7)(60 \text{ ml } K_2Cr_2O_7 \text{ soln.}) \left( \dfrac{1 \text{ liter } K_2Cr_2O_7 \text{ soln.}}{10^3 \text{ ml } K_2Cr_2O_7 \text{ soln.}} \right) \left( \dfrac{294 \text{ g } K_2Cr_2O_7}{1 \text{ mol } K_2Cr_2O_7} \right)$

$= 2.6 \text{ g } K_2Cr_2O_7$

## II. Dilution

See Sections 9.6 and 9.7 of your text if you have any questions on this material.

1. 0.22 liters, or 220 ml

The number of moles of solute in the unknown volume of concentrated solution equals the number of moles of solute in the dilute solution:

$$V_1M_1 = V_2M_2$$

Rearranging the preceding equation and substituting values into it, we find

$$V_1 = \dfrac{V_2M_2}{M_1}$$

$$= \dfrac{(1.0 \text{ liter})(0.10M)}{0.45M}$$

$$= 0.22 \text{ liter, or } 220 \text{ ml}$$

2. 10.7 ml

If we wish to use

$$V_1M_1 = V_2M_2$$

we must first determine the molarity of the 30.5% HCl solution:

$$? \; M = \left(\frac{30.5 \text{ g HCl}}{100 \text{ g HCl soln.}}\right)\left(\frac{1.12 \text{ g HCl soln.}}{1 \text{ ml HCl soln.}}\right)\left(\frac{1 \text{ mol HCl}}{36.46 \text{ g HCl}}\right)\left(\frac{10^3 \text{ ml}}{1 \text{ liter}}\right)$$

$$= 9.369M$$

Rearranging the equation

$$V_1 M_1 = V_2 M_2$$

and substituting values into it, we find

$$V_1 = \frac{V_2 M_2}{M_2}$$

$$= \frac{(500 \text{ ml})(0.200M)}{9.369M}$$

$$= 10.7 \text{ ml}$$

If we use the factor-label method, we can solve the problem in one step:

$$? \text{ ml HCl soln.} =$$

$$0.500 \text{ liter } (0.200M \text{ HCl}) \left(\frac{36.46 \text{ g HCl}}{1 \text{ mol HCl}}\right)\left(\frac{100 \text{ g HCl soln.}}{30.5 \text{ g HCl}}\right)\left(\frac{1 \text{ ml HCl soln.}}{1.12 \text{ g HCl soln.}}\right)$$

$$= 10.7 \text{ ml HCl soln.}$$

3. 10 ml

In general,

$$V_1 C_1 = V_2 C_2$$

in which $V$ is volume and $C$ is concentration. In this problem the concentrations are expressed in mg/ml. If we rearrange the preceding equation and substitute values into it, we find

$$V_1 = \frac{V_2 C_2}{C_1}$$

$$= \frac{(100 \text{ ml})(0.100 \text{ mg/ml})}{(1.00 \text{ mg/ml})} \qquad = 10 \text{ ml}$$

4. 3.0*M*

Rearranging the equation

$$V_1 M_1 = V_2 M_2$$

and substituting values into it, we find

$$M_1 = \frac{V_2 M_2}{V_1}$$

$$? \ M_1 = \frac{(100 \text{ ml})(0.75M)}{(25 \text{ ml})}$$

$$= 3.0M$$

5. 4.1 ml

First we change one of the concentrations so that it has the same units as the other. We determine the molarity of the 0.10 mg/ml solution of vitamin $B_2$:

$$? \ M \ B_2 = \left(\frac{0.10 \text{ mg } B_2}{1 \text{ ml soln.}}\right)\left(\frac{1 \text{ g}}{1000 \text{ mg}}\right)\left(\frac{1000 \text{ ml}}{1 \text{ liter}}\right)\left(\frac{1 \text{ mol } B_2}{376 \text{ g } B_2}\right)$$

$$= 2.66 \times 10^{-4} M \ B_2$$

Then we rearrange the equation

$$V_1 M_1 = V_2 M_2$$

and substitute values into it:

$$V_1 = \frac{V_2 M_2}{M_1}$$

$$V_1 = \frac{(1.0 \text{ liter})(2.66 \times 10^{-4} M \ B_2)}{6.5 \times 10^{-2} M \ B_2}$$

$$= 0.0041 \text{ liter, or } 4.1 \text{ ml}$$

III. Properties of solutions

1. $H_2O$
[9.2]

*Like dissolves like*. Hydrogen fluoride, HF, and water, $H_2O$, are highly polar and have hydrogen bonding ability. Tetrafluoromethane, $CF_4$, and bromine, $Br_2$, are nonpolar.

2. molality
   [9.6]

Molality is defined as the number of moles of solute per kilogram of solvent:

$$m = \frac{\text{number of moles of solute}}{\text{number of kilograms of solvent}}$$

Molality is not defined in terms of volume, as are molarity and normality:

$$M = \frac{\text{number of moles of solute}}{\text{number of liters of solution}}$$

$$N = \frac{\text{number of equivalents of solute}}{\text{number of liters of solution}}$$

Volume changes with a change in temperature. In problems concerning boiling point elevation and freezing point depression, $m$ is used instead of $M$ or $N$.

3. 1.00$m$ AlCl$_3$
   [9.9, 9.13]

If we assume each compound to be a strong electrolyte, each molecule of compound would dissociate into the following number of ions:

| | |
|---|---|
| $AlCl_3$ | 4 |
| $CaCl_2$ | 3 |
| $NaCl$ | 2 |

If we substitute values into the equation

$$\Delta T_f = i K_f m$$

we find that the freezing point depression is largest for $AlCl_3$, for which $i = 4$.

4. endothermic
   [9.5]

As the temperature increases, the amount of $CoSO_4$ that dissolves increases:

$$H_2O + CoSO_4 \rightarrow Co^{2+}(aq) + SO_4^{2-}(aq)$$

The equilibrium shifts therefore to the right. According to Le Chatelier's principle, the application of stress to a system in equilibrium causes the system to react to counteract the stress, thus establishing a new equilibrium state. If a temperature rise (the addition of heat to the system) causes the equilibrium to shift to the right, heat must appear on the left side of the equation:

$$\text{heat} + H_2O + CoSO_4 \rightarrow Co^{2+}(aq) + SO_4^{2-}(aq)$$

If heat is required in the solution process, the process is endothermic.

5. yes
   [9.4]

When a substance dissolves, energy (the lattice energy) is required to break the crystal lattice and energy (the hydration energy) is released by the hydration of the ions.  The solution process for $CoSO_4$ is endothermic, i.e., in the process more energy is required than is released.  Thus, the absolute value of the lattice energy is greater than that of the hydration energy.

6. 0.0374 atm
   [9.8]

Since $Al_2(SO_4)_3$ is a strong electrolyte, in solution a mole of it produces 5 mol of ions:

$$Al_2(SO_4)_3 \rightarrow 2Al^{3+} + 3SO_4^{2-}$$

Assuming an ideal solution is formed, we can use Raoult's law to solve the problem:

$$p_{H_2O} = X_{H_2O}\, p_{H_2O}^o$$

$$= 0.0395 \text{ atm} \left( \frac{9.00 \text{ mol}}{9.00 \text{ mol} + 0.500 \text{ mol}} \right)$$

$$= 0.0374 \text{ atm}$$

Since the solute is nonvolatile,

$$P_{total} = 0.0374 \text{ atm}$$

7. yes, 0.020$N$
   [9.7]

If we substitute values into the equation

$$V_1 N_1 = V_2 N_2$$

or into its rearranged form

$$N_1 = \frac{V_2 N_2}{V_1}$$

we find

$$? N = \frac{(1.0 \text{ ml})(0.10N)}{(5.0 \text{ ml})}$$

$$= 0.020N$$

8.  0.525$M$
    0.535$m$
    [9.6]

When the heating of the solution is complete, 43.7 g of $BaCl_2$ remains as residue. Thus, we can calculate the molarity:

$$? M \; BaCl_2 = \frac{\text{number of moles of } BaCl_2}{\text{number of moles of } BaCl_2 \text{ soln.}}$$

$$= \frac{(43.7 \text{ g } BaCl_2)\left(\dfrac{1 \text{ mol } BaCl_2}{208.2 \text{ g } BaCl_2}\right)}{(400 \text{ ml } BaCl_2 \text{ soln.})\left(\dfrac{1 \text{ liter } BaCl_2 \text{ soln.}}{1000 \text{ ml } BaCl_2 \text{ soln.}}\right)}$$

$$= 0.525M \; BaCl_2$$

Determining the molality is not as straightforward as determining the molarity. Since kg solvent, not kg solution, is needed, we must make use of the density and the fact that the weight of the solution equals the weight of water plus the weight of $BaCl_2$. We first calculate the weight of the solution from the density:

$$? \text{ g soln.} = 400 \text{ ml soln.} \left(\frac{1.09 \text{ g soln.}}{1 \text{ ml soln.}}\right) = 436 \text{ g soln.}$$

Of the 436 g of solution, we know that 43.7 g is $BaCl_2$. Thus, we can calculate the weight of water:

$$? \text{ g } H_2O = 436 \text{ g soln.} - 43.7 \text{ g } BaCl_2 = 392 \text{ g } H_2O$$

We can now calculate the molality of $BaCl_2$:

$$? m \; BaCl_2 = \frac{\text{number of moles of } BaCl_2}{\text{number of kg of water}}$$

$$= \frac{(43.7 \text{ g } BaCl_2)\left(\dfrac{1 \text{ mol } BaCl_2}{208.2 \text{ g } BaCl_2}\right)}{(392 \text{ g } H_2O)\left(\dfrac{1 \text{ kg } H_2O}{1000 \text{ kg } H_2O}\right)}$$

$$= 0.535m \; BaCl_2$$

9. no
   [9.5]

Henry's law, which states that the solubility of a gas in a solution is directly proportional to the partial pressure of that gas above the solution, is valid only for dilute solutions and low pressures. Also, extremely soluble gases usually react with the solvent and thus do not follow Henry's law.

10. NaCl
    [9.9, 9.13 ]

We can calculate the cost of a single mole of ions for each compound:

for $CaCl_2$

$$? \ \$/mol \ ions = \left(\frac{\$88}{2000 \ lb \ CaCl_2}\right)\left(\frac{1 \ lb \ CaCl_2}{454 \ g \ CaCl_2}\right)\left(\frac{111 \ g \ CaCl_2}{1 \ mol \ CaCl_2}\right)\left(\frac{1 \ mol \ CaCl_2}{3 \ mol \ ions}\right)$$

$$= \ 3.6 \times 10^{-3} \ \$/mol \ ions$$

for NaCl

$$? \ \$/mol \ ions = \left(\frac{\$56}{2000 \ lb \ NaCl}\right)\left(\frac{1 \ lb \ NaCl}{454 \ g \ NaCl}\right)\left(\frac{58.4 \ g \ NaCl}{1 \ mol \ NaCl}\right)\left(\frac{1 \ mol \ NaCl}{2 \ mol \ ions}\right)$$

$$= \ 1.8 \times 10^{-3} \ \$/mol \ ions$$

Thus, it is cheaper to use NaCl than $CaCl_2$.

12. increase
    [9.4]

13. $BeCl_2$
    [9.2]

An ion with a larger value of the ratio of ionic charge to ionic radius forms a more stable hydrated ion and thus has a larger enthalpy of hydration, $\Delta H_{hydration}$.

14. $3.64 \times 10^4$
    g/mol
    [9.10]

Rearranging the van't Hoff equation

$$\pi V = nRT$$

we find

$$n = \frac{\pi V}{RT}$$

Since

$$n = \text{number of moles} = \frac{\text{weight in grams}}{\text{molecular weight in grams}}$$

the van't Hoff equation becomes

$$\frac{\text{weight in grams}}{\text{molecular weight in grams}} = \frac{\pi V}{RT}$$

or

$$\text{molecular weight in grams} = \frac{(RT)\,(\text{weight in grams})}{\pi V}$$

Substituting values into the van't Hoff equation, we find

$$\text{molecular weight in grams} = \frac{(0.08206 \text{ liter atm mol}^{-1}\text{K}^{-1})\,(298^\circ\text{K})\,(1.49 \text{ g})}{(1.00 \times 10^{-2} \text{ atm})\,(1.00 \times 10^2 \text{ ml})\,(1 \text{ liter}/10^3 \text{ ml})}$$

$$= 3.64 \times 10^4 \text{ g/mol}$$

15.  $3.08 \times 10^{-2}$ atm    We can use Raoult's law to solve this problem:
 [9.8]

$$p_{H_2O} = X_{H_2O}\, p^o_{H_2O}$$

We must calculate the mole fraction of $H_2O$. First, however, we must determine the number of moles of water and sucrose:

$$? \text{ mol } H_2O = 100 \text{ g } H_2O \left(\frac{1 \text{ mol } H_2O}{18.02 \text{ g } H_2O}\right)$$

$$= 5.55 \text{ mol } H_2O$$

$$? \text{ mol } C_{12}H_{22}O_{11} = 27.4 \text{ g } C_{12}H_{22}O_{11}\left(\frac{1 \text{ mol } C_{12}H_{22}O_{11}}{342.3 \text{ g } C_{12}H_{22}O_{11}}\right)$$

$$= 8.00 \times 10^{-2} \text{ mol } C_{12}H_{22}O_{11}$$

The mole fraction of water is therefore

$$X_{H_2O} = \frac{5.55 \text{ mol}}{5.55 \text{ mol} + 0.800 \text{ mol}}$$

$$= 0.986$$

Substituting values into Raoult's law, we find

$$p_{H_2O} = X_{H_2O}\, p^o_{H_2O}$$

$$= 0.986(3.12 \times 10^{-2} \text{ atm})$$

$$= 3.08 \times 10^{-2} \text{ atm}$$

Since the solute is nonvolatile,

$$P_{total} = P_{H_2O} = 3.08 \times 10^{-2} \text{ atm}$$

16. 121 meq/liter
   [9.7]

Using $V_1N_1 = V_2N_2$, we find

$$N_1 = \frac{V_2N_2}{V_1}$$

$$= \frac{(2.42 \text{ ml})(0.0100 \text{ eq/liter})}{0.200 \text{ ml}}$$

$$= 0.121 \text{ eq/liter}$$

In clinical laboratories the value would be reported in units of milliequivalents per liter:

$$? \text{ meq/liter} = \left(\frac{0.121 \text{ eq}}{1 \text{ liter}}\right)\left(\frac{1000 \text{ meq}}{1 \text{ eq}}\right)$$

$$= 121 \text{ meq/liter}$$

17. [9.5]

According to Henry's law, the amount of gas dissolved in solution increases as the partial pressure of the gas increases. Therefore, since the pressure of oxygen in a hyperbaric chamber is greater than that in the normal atmosphere, more oxygen will dissolve in the body fluids when a patient is in a hyperbaric chamber than when in the normal atmosphere. The greater the amount of oxygen in the body fluids, the greater the inhibition of growth of *Clostridium welchii*.

18. $2.9 \times 10^2$ g NaCl
   [7.4, 9.9]

Atmospheric pressure in Denver is lower than one atmosphere; thus, the boiling point of water in Denver is lower than the normal boiling point, i.e., the temperature at which the vapor pressure of water equals one atmosphere.

We must calculate the amount of NaCl necessary to raise the boiling point of 1000 g $H_2O$ to $100^\circ C$ in Denver. We can calculate the molality of the solution from the equation for boiling point elevation:

$$\Delta T_b = imK_b$$

Rearranging the previous equation and substituting values into it, we find

$$? \ m = \frac{\Delta T_b}{i K_b}$$

$$= \frac{(100 - 95)\,^\circ C}{2(0.512\,^\circ C/m)}$$

$$= 4.9m$$

We then calculate the number of grams of NaCl necessary for the 1000 g $H_2O$:

$$? \ g \ NaCl = \left(\frac{5 \ mol \ NaCl}{1000 \ g \ H_2O}\right) \ (1000 \ g \ H_2O) \ \left(\frac{58.4 \ g \ NaCl}{1 \ mol \ NaCl}\right)$$

$$= 2.9 \times 10^2 \ g \ NaCl$$

The NaCl solution is very concentrated and it is doubtful that NaCl is completely dissociated, i.e., $i$ is less than 2. Although the concentration of NaCl necessary to increase the boiling point by 5°C would be larger than 4.9$m$, we have made a good first approximation.

SELF-TEST

I. Complete the test in 30 minutes:

1. A 0.2070 g sample of an unknown organic acid is dissolved in 100 ml of 0.02560$N$ sodium hydroxide. The resulting solution is titrated to the end point with 36.50 ml of 0.0240$N$ HCl. Calculate the equivalent weight of the unknown acid.

2. A 20.00 ml portion of commercial vinegar, the density of which is 1.01 g/ml, is titrated with 28.75 ml of 0.503$M$ NaOH. What is the weight percentage of acid, $HC_2H_3O_2$, in vinegar?

3. A sample of sodium is added to water to make 500 ml of solution. At STP 33.6 liters of dry hydrogen gas is collected from the reaction of sodium metal with water:

$$2Na(s) + 2H_2O \rightarrow 2NaOH(aq) + H_2(g)$$

What is the normality of the NaOH solution produced by the reaction?

4. What volume of a concentrated hydrochloric acid solution, which is 37% HCl and has a specific gravity of 1.189 g/ml, should be used to prepare 2000 ml of 0.100$M$ HCl solution? Specific gravity is the ratio of the density of a material to the density of water.

5. A solution containing 0.500 g of an unknown non-volatile solute in 10.0 g of camphor has a freezing point of 159.4°C. What is the molecular weight of the solute if the normal freezing point of camphor is 179.0°C and the molal freezing-point depression constant for camphor is 49°C/$m$?

# 10
# Electrochemistry

OBJECTIVES

(a) You should be able to demonstrate your knowledge of the following terms by defining them, describing them, or giving specific examples of them:

ampere [10.1]
anode [10.2]
cathode [10.2]
concentration polarization [10.11]
conductance [10.5]
conduction [10.1, 10.2]
coulomb [10.1]
current [10.1]
electrolysis [10.3]
electrolytic cell [10.2]

electromotive force, emf [10.7]
equivalent conductance [10.5]
faraday [10.4]
Faraday's laws [10.4]
Nernst equation [10.9]
ohm [10.1]
Ohm's law [10.1]
overvoltage [10.11]
resistance [10.1]
standard electrode potential [10.8]
standard emf [10.7]
volt [10.1]
voltaic cell, also called galvanic cell [10.6]

(b) You should be able to use Faraday's laws.

(c) You should be able to determine the conductivity of a solution.

(d) Using a table of standard electrode potentials, you should be able to determine the potential of an electrochemical cell for any set of concentrations of reactants and products.

(e) You should be able to diagram electrochemical cells and determine the electrode signs, the direction of electron movement, and the direction of ion migration.

**EXERCISES**

I. Answer each of the following with *true* or *false*. If a statement is false, correct it.

_____

1. In an electrolytic cell oxidation occurs at the positive electrode.

_____

2. In a galvanic cell oxidation occurs at the positive electrode.

_____

3. The anode of a cell is indicated on the left in the cell notation.

_____

4. In both galvanic and electrolytic cells, reduction occurs at the cathode.

_____

5. A positive cell potential indicates that the reaction should occur spontaneously as written.

_____  6. A battery is an electrolytic cell.

_____  7. One faraday is required to reduce one mole of $Cu^{2+}$ ions to Cu(s).

_____  8. An Avogadro's number of $Ni^{2+}$ ions can be reduced to Ni(s) with two faradays.

_____  9. Four faradays produce 22.4 liters of $O_2$ at STP in the electrolysis of water:

$$2H_2O(l) \rightarrow O_2(g) + 4H^+(aq) + 4e^-$$

_____  10. Electrode potentials are temperature dependent.

_____  11. The best oxidizing agent in Table 10.1 of the study guide is $F_2$.

## Table 10.1   Standard electrode potentials at 25°C

| HALF REACTION | $\mathscr{E}°$ (volts) |
|---|---|
| $Li^+ + e^- \rightleftharpoons Li$ | −3.045 |
| $K^+ + e^- \rightleftharpoons K$ | −2.925 |
| $Ba^{2+} + 2e^- \rightleftharpoons Ba$ | −2.906 |
| $Ca^{2+} + 2e^- \rightleftharpoons Ca$ | −2.866 |
| $Na^+ + e^- \rightleftharpoons Na$ | −2.714 |
| $Mg^{2+} + 2e^- \rightleftharpoons Mg$ | −2.363 |
| $Al^{3+} + 3e^- \rightleftharpoons Al$ | −1.662 |
| $2H_2O + 2e^- \rightleftharpoons H_2 + 2OH^-$ | −0.82806 |
| $Zn^{2+} + 2e^- \rightleftharpoons Zn$ | −0.7628 |
| $Cr^{3+} + 3e^- \rightleftharpoons Cr$ | −0.744 |
| $Fe^{2+} + 2e^- \rightleftharpoons Fe$ | −0.4402 |
| $Cd^{2+} + 2e^- \rightleftharpoons Cd$ | −0.4029 |
| $Ni^{2+} + 2e^- \rightleftharpoons Ni$ | −0.250 |
| $Sn^{2+} + 2e^- \rightleftharpoons Sn$ | −0.136 |
| $Pb^{2+} + 2e^- \rightleftharpoons Pb$ | −0.126 |
| $2H^+ + 2e^- \rightleftharpoons H_2$ | 0 |
| $Cu^{2+} + 2e^- \rightleftharpoons Cu$ | +0.337 |
| $Cu^+ + e^- \rightleftharpoons Cu$ | +0.521 |
| $I_2 + 2e^- \rightleftharpoons 2I^-$ | +0.5355 |
| $Fe^{3+} + e^- \rightleftharpoons Fe^{2+}$ | +0.771 |
| $Ag^+ + e^- \rightleftharpoons Ag$ | +0.7991 |
| $Br_2 + 2e^- \rightleftharpoons 2Br^-$ | +1.0652 |
| $O_2 + 4H^+ + 4e^- \rightleftharpoons 2H_2O$ | +1.229 |
| $Cr_2O_7^{2-} + 14H^+ + 6e^- \rightleftharpoons 2Cr^{3+} + 7H_2O$ | +1.33 |
| $Cl_2 + 2e^- \rightleftharpoons 2Cl^-$ | +1.3595 |
| $MnO_4^- + 8H^+ + 5e^- \rightleftharpoons Mn^{2+} + 4H_2O$ | +1.51 |
| $F_2 + 2e^- \rightleftharpoons 2F^-$ | +2.87 |

[a]Data from A. J. de Bethune and N. A. Swendeman Loud, "Table of Electrode Potentials and Temperature Coefficients," pp. 414-424 in *Encyclopedia of Electrochemistry* (C. A. Hampel, editor), Van Nostrand Reinhold, New York, 1964, and from A. J. de Bethune and N. A. Swendeman Loud, *Standard Aqueous Electrode Potentials and Temperature Coefficients*, 19 pp., C. A. Hampel, publisher, Skokie, Illinois, 1964.

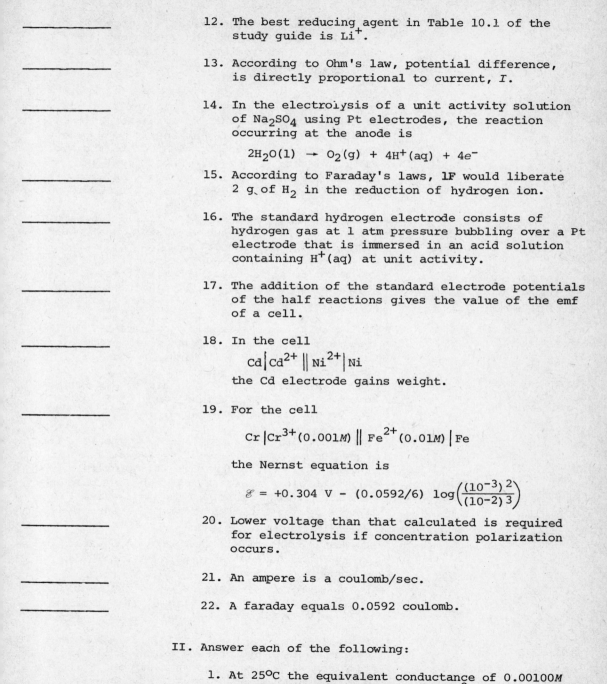

12. The best reducing agent in Table 10.1 of the study guide is $Li^+$.

13. According to Ohm's law, potential difference, is directly proportional to current, $I$.

14. In the electrolysis of a unit activity solution of $Na_2SO_4$ using Pt electrodes, the reaction occurring at the anode is

$$2H_2O(l) \rightarrow O_2(g) + 4H^+(aq) + 4e^-$$

15. According to Faraday's laws, 1F would liberate 2 g of $H_2$ in the reduction of hydrogen ion.

16. The standard hydrogen electrode consists of hydrogen gas at 1 atm pressure bubbling over a Pt electrode that is immersed in an acid solution containing $H^+(aq)$ at unit activity.

17. The addition of the standard electrode potentials of the half reactions gives the value of the emf of a cell.

18. In the cell

$$Cd \mid Cd^{2+} \parallel Ni^{2+} \mid Ni$$

the Cd electrode gains weight.

19. For the cell

$$Cr \mid Cr^{3+}(0.001M) \parallel Fe^{2+}(0.01M) \mid Fe$$

the Nernst equation is

$$\mathscr{E} = +0.304 \text{ V} - (0.0592/6) \log\left(\frac{(10^{-3})^2}{(10^{-2})^3}\right)$$

20. Lower voltage than that calculated is required for electrolysis if concentration polarization occurs.

21. An ampere is a coulomb/sec.

22. A faraday equals 0.0592 coulomb.

II. Answer each of the following:

1. At 25°C the equivalent conductance of 0.00100$M$ acetic acid, $HC_2H_3O_2$, is 49.2 $cm^2$/ohm equivalent.

The equivalent conductance at infinite dilution of $HC_2H_3O_2$ is 390.7 $cm^2$/ohm equivalent. What is the degree of dissociation, $\alpha$, of acetic acid in 0.00100$M$ solution?

2. Corrosion is an oxidation process. Often iron is coated with other metals to protect it from rusting. Determine what would happen to iron if it were coated with tin, as in tin cans, and the coating cracked, exposing the iron to air and water. (Hint: Assume the metal forms an electrochemical cell in solution, the cathode reaction being the reduction of $O_2$ to $OH^-$.)

3. If the iron of problem 2 of this section were galvanized, i.e., coated with zinc, what would happen if the coating cracked and the iron became exposed to air and water. (Hint: Assume the metal forms an electrochemical cell in solution, the cathode reaction being the reduction of $O_2$ to $OH^-$.)

4. During cellular respiration, oxidation-reduction reactions take place. If the free energy of the reaction NAD to FAD (nicotinamide adenine dinucleotide to flavin adenine dinucleotide) is −12.00 kcal/mol and the standard emf is +0.26 V, how many electrons are involved in this oxidation step in the respiratory chain?

5. In the Dow process for obtaining magnesium, the magnesium is precipitated from seawater as the hydroxide, which is dissolved in HCl after purification. The resulting magnesium chloride solution is evaporated, and the magnesium chloride is melted and electrolyzed. Answer the following:

   a. How many hours would it take to obtain 5.0 kg of magnesium metal using $1.0 \times 10^4$ amperes?

   b. What volume of chlorine gas at STP would be evolved in the time it takes to obtain the 5.0 kg of Mg metal using $1.0 \times 10^4$ amperes?

6. Predict whether or not each of the following reactions will occur spontaneously and write a balanced chemical equation for each reaction that is predicted to occur. Assume that each

reactant and product is present at unit activity in aqueous solution at $25^\circ C$.

    a. $Cd^{2+}$ ion reduced to Cd by Ni

    b. Sn oxidized to $Sn^{2+}$ by $Cu^{2+}$ ion

    c. $I_2$ reduced to $I^-$ ion by Ag

    d. $MnO_4^-$ ion reduced to $Mn^{2+}$ ion by $Br^-$ ion

7. Write the shorthand cell notation for the spontaneous reactions in problem 6 of this section.

8. In $0.00500M$ solution, acetic acid is 5.8% dissociated at $25^\circ C$. If $\Lambda_0$ for $HC_2H_3O_2$ is 390.7 $cm^2$/ohm equivalent, what is the specific conductance, $\kappa$, of 0.00500M $HC_2H_3O_2$?

9. Use Table 10.1 of the study guide to predict what will happen when a bar of cadmium is added to a solution of $Fe^{3+}$.

10. The salt $Co_2(SO_4)_3 \cdot 18H_2O$ decomposes when added to water.  Why?

11. If 17.6 liters of $O_2$ at STP are evolved at the anode in an electrochemical cell and the only reaction at the cathode is the reduction of $Cu^{2+}$ to Cu(s), how many grams of Cu are formed?

12. Use the standard electrode potentials of bromine-containing compounds in basic solution

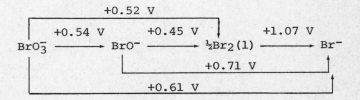

to answer the following:

    a. Does $BrO^-$ disproportionate in basic solution?

    b. Does $Br_2$ disproportionate in basic solution?

13. The potential difference between two hydrogen cells ($H_2 \rightarrow 2H^+ + 2e^-$) is 0.076 $V$. One of the cells contains 1.0$M$ $H^+$ and $H_2(g)$ at 1 atm. The other contains 1 atm $H_2(g)$. What is the concentration of $H^+$ in the second cell?

$$Pt \mid H_2 \mid H^+(?M) \parallel H^+(1.0M) \mid H_2 \mid Pt$$

14. Calculate the concentration of $Fe^{2+}$ and $Fe^{3+}$ in a half cell that contains a total iron concentration of $1.00 \times 10^{-3}M$ if the potential difference of the cell with respect to a standard hydrogen electrode is -0.694 $V$.

$$Pt \mid Fe^{2+}(?M) \mid Fe^{3+}(?M) \parallel H^+(1M) \mid H_2(1 \text{ atm}) \mid Pt$$

15. For 3.00 hours a current of 10.0 amp plated out 13.6 g of a substance. What is the equivalent weight of the substance?

16. In a fuel cell hydrogen and oxygen can react to produce electricity. The process is represented by the cell in Figure 10.1 of the study guide.

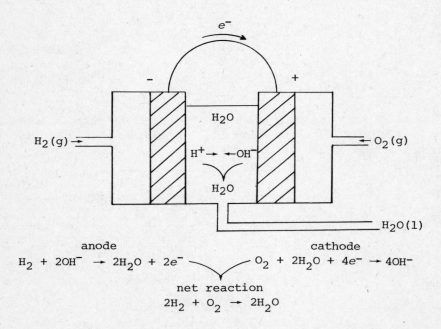

anode
$$H_2 + 2OH^- \rightarrow 2H_2O + 2e^-$$

cathode
$$O_2 + 2H_2O + 4e^- \rightarrow 4OH^-$$

net reaction
$$2H_2 + O_2 \rightarrow 2H_2O$$

Figure 10.1 Fuel cell

In the process hydrogen gas is oxidized to water at the anode and oxygen gas is reduced to $OH^-$ at the cathode.  The excess water is removed. Answer the following:

a. If 47.8 liters of hydrogen at STP react in 10.0 minutes, how many liters of oxygen at STP react in the same time?

b. How many moles of water are produced in 10.0 minutes?

c. What is the average current produced during this time?

17. Wind generators are attractive sources of energy except for the fact that the wind occasionally ceases to blow.  It has been proposed that the generator be used to electrolyze water during off-peak power times and that the hydrogen and oxygen generated during these hours could be used in a fuel cell to generate electricity when the wind stops blowing.  If the efficiency is 100%, i.e., no energy losses occur, what size of tank would be needed to store sufficient hydrogen and oxygen at STP to supply a single residence with 20 amps for 5.0 hours?

III. Do the following for each of the following electro-chemical cells:

a. Label the anode and cathode.

b. Write the reaction occurring at the anode and that occurring at the cathode.

c. Show the direction of the electron movement in the external connection.

d. Indicate the sign of the cathode and that of the anode.

e. Show the direction of ion movement in the cell.

f. Diagram the cell in the shorthand notation de-scribed in Section 10.7 of your text.  Assume all concentrations are $1M$.

g. Determine the half-cell reactions from the standard electrode potentials in Table 10.1 of the study guide.

h. Calculate the cell potential.

1.

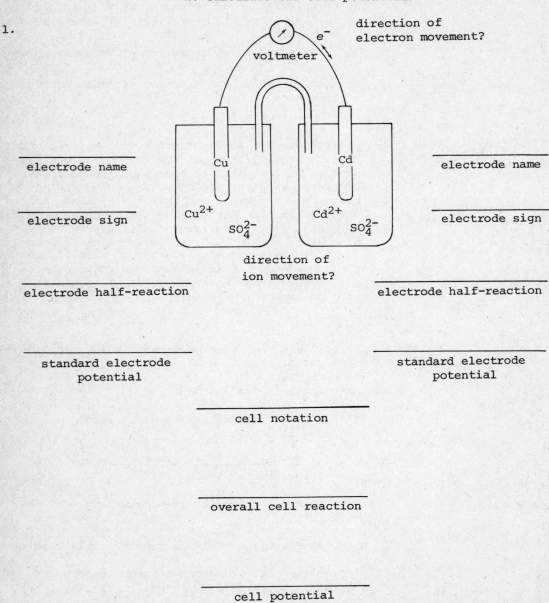

direction of electron movement?

direction of ion movement?

electrode name _____

electrode sign _____

electrode half-reaction _____

standard electrode potential _____

electrode name _____

electrode sign _____

electrode half-reaction _____

standard electrode potential _____

cell notation _____

overall cell reaction _____

cell potential _____

2.

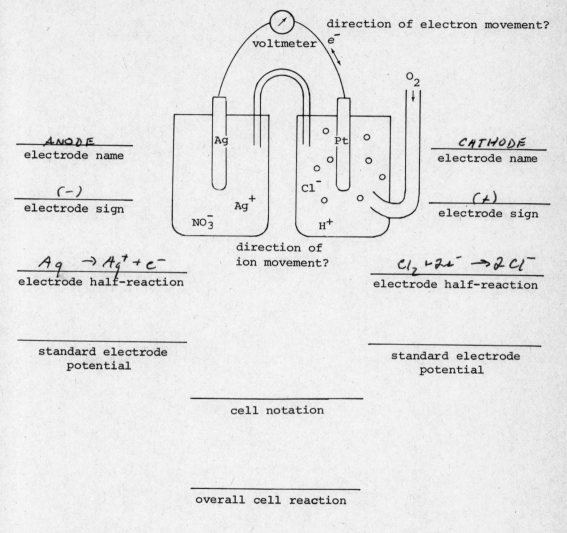

direction of electron movement?

voltmeter   $e^-$

$O_2$

<u>ANODE</u>
electrode name

<u>(-)</u>
electrode sign

Ag

Pt

$Cl^-$

$Ag^+$

$NO_3^-$

$H^+$

<u>CATHODE</u>
electrode name

<u>(+)</u>
electrode sign

direction of
ion movement?

$Ag \rightarrow Ag^+ + e^-$
electrode half-reaction

$Cl_2 + 2e^- \rightarrow 2Cl^-$
electrode half-reaction

standard electrode
potential

standard electrode
potential

cell notation

overall cell reaction

cell potential

3.

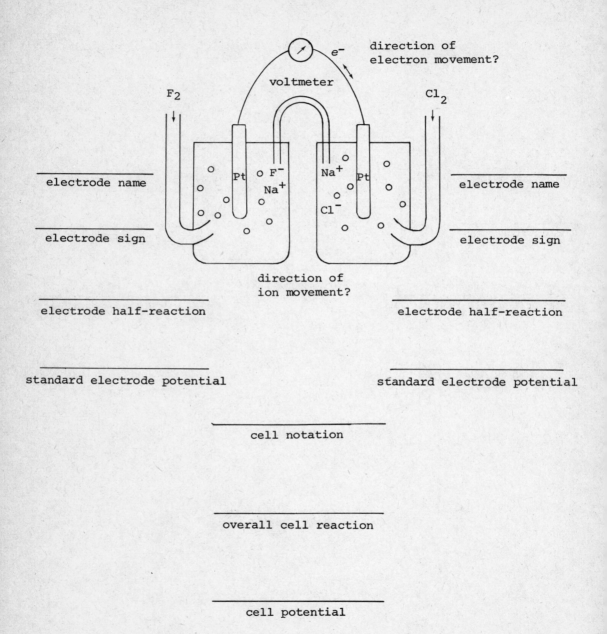

electrode name _____

electrode sign _____

electrode half-reaction _____

standard electrode potential _____

electrode name _____

electrode sign _____

electrode half-reaction _____

standard electrode potential _____

cell notation _____

overall cell reaction _____

cell potential _____

ANSWERS TO
EXERCISES

I. Basic electrochemical principles

1. True [10.2]

2. False [10.2]     In a galvanic cell the positive electrode is the
                    cathode, and reduction occurs at this electrode.  The
                    conventions relating to the terms anode and cathode
                    are outlined in Table 10.2 of the study guide.

Table 10.2  Electrode conventions

|  | CATHODE | ANODE |
|---|---|---|
| ions attracted | cations | anions |
| direction of electron movement | into cell | out of cell |
| half reaction | reduction | oxidation |
| sign |  |  |
|    electrolysis cell | negative | positive |
|    galvanic cell | positive | negative |

You may find it easier to remember the conventions
relating to a galvanic, or voltaic, cell by using
the mnemonic device outlined in Table 10.3 of the
study guide.  Note that when the opposite electrochemical terms are alphabetized, the first term
describes the anode and the second describes the
cathode.

For an electrolysis cell, i.e., a cell in which an
external power supply causes the reaction to proceed in a nonspontaneous direction, the mnemonic
device is valid except that the electrode signs are
reversed.  In an electrolysis cell oxidation occurs
at the positive electrode and electrons move away
from this electrode in the external circuit.  Also,
cell notations are not used for electrolysis cells.

Table 10.3  Electrode conventions for galvanic, or voltaic, cells

|  | ANODE | CATHODE |
|---|---|---|
| ions attracted | anions | cations |
| electrode sign | negative | positive |
| electrode process | oxidation | reduction |
| direction of electron movement | away | toward |
| location in cell notation | left | right |

3. True [10.8]

4. True [10.2]

5. True [10.8]

6. False [10.12]  Batteries are sources of energy and are therefore voltaic cells. Electrolytic cells require an energy source to operate.

7. False [10.4]  The balanced equation

$$Cu^{2+} + 2e^- \rightarrow Cu(s)$$

indicates that 2 mol of electrons, or 2F, are required to produce 1 mol of Cu(s).

8. True [10.4]

9. True [10.8]  The oxygen-producing half reaction

$$2H_2O \rightarrow O_2 + 4H^+ + 4e^-$$

indicates that four faradays are required to produce 1 mol of $O_2$, or 22.4 liters of $O_2$, at STP.

10. True [10.9]  Notice that the Nernst equation

$$\mathscr{E} = \mathscr{E}^\circ - \frac{RT}{nF} \ln Q$$

includes temperature, $T$. The standard electrode potentials in Table 10.1 of the study guide are given for a specific temperature, 25°C.

11. True [10.8]

12. False [10.8]  A positive cell emf indicates that the reaction occurs spontaneously as written. In Table 10.1 of

the study guide oxidizing agents are on the left.
The stronger the oxidizing agent, the more positive
the value of $\mathscr{E}°$.  Reducing agents are on the right.
The stronger the reducing agent the more negative
the value of $\mathscr{E}°$.  Thus, we diagram Table 10.1 of the
study guide as follows:

increasing
strength
of
oxidizing
agent

| | oxidizing<br>agent | reducing<br>agent | |

increasing
strength
of
reducing
agent

The best reducing agent in Table 10.1 of the study
guide is Li.  The $Li^+$ ion is reduced; thus, it is
an oxidizing agent with a $\mathscr{E}°$ of $-3.045$ V.  The lithium
atom is a reducing agent, being oxidized to $Li^+$ with
an $\mathscr{E}°$ of $+3.045$ V.

13. True [10.1]          Ohm's law states

$$\mathscr{E} = IR$$

The proportionality constant $R$ is called the resistance.

14. True [10.3]

15. False [10.4]         According to the reaction

$$2H^+ + 2e^- \rightarrow H_2$$

$1F$, i.e., 1 mol of electrons, would liberate ½ mol
of $H_2$, or ½ mol $H_2(2$ g $H_2/1$ mol $H_2) = 1$ g $H_2$.

16. True [10.8]

17. False [10.8]         The cell emf is the sum of the half-cell potential
for the oxidation reaction and the half-cell
potential for the reduction reaction.  In the table
of standard electrode potentials (Table 10.1 of the
study guide), all reactions are written as reduction
reactions.  Thus, both the sign of one electrode
potential and the corresponding reaction must be
reversed to obtain the necessary half-cell potential
and oxidation reaction.

18. False [10.8]   By convention, the anode is indicated on the left. The cell reaction as diagramed is

$$Cd(s) + Ni^{2+} \rightarrow Cd^{2+} + Ni(s)$$

Thus, Ni is deposited on the electrode, the Ni electrode gains weight, and the Cd electrode loses weight.

19. True [10.9]   The balanced equation

$$2Cr + 3Fe^{2+} \rightarrow 2Cr^{3+} + 3Fe$$

indicates that six electrons are exchanged, $n = 6$, and that

$$Q = \frac{[Cr^{3+}]^2}{[Fe^{2+}]^3} = \frac{[10^{-3}]^2}{[10^{-2}]^3}$$

The metals Cr and Fe have unit activity and are not included in $Q$. The value of $\mathscr{E}^\circ$ is +0.304 V:

$$3\left[Fe^{2+} + 2e^- \rightleftharpoons Fe\right] \qquad \mathscr{E}^\circ = -0.4402 \text{ V}$$

$$2\left[Cr \rightleftharpoons Cr^{3+} + 3e^-\right] \qquad \mathscr{E}^\circ = +0.744 \text{ V}$$

$$\overline{2Cr + 3Fe^{2+} \rightarrow 2Cr^{3+} + 3Fe \qquad \mathscr{E}^\circ = +0.304 \text{ V}}$$

20. False [10.11]   Higher voltage must be applied if concentration polarization occurs. As electrolysis proceeds, there is a buildup or deficit of ions around the electrode, thus setting up a concentration cell about the electrode. The concentration cell produces a back emf, opposite to the applied voltage.

21. True [10.1]

22. False [10.4]   One faraday equals 96,487 coulombs.

## II. Electrochemical calculations

1. 0.126
   [10.5]   At infinite dilution, weak electrolytes are theoretically 100% ionized. Thus, $\Lambda_o$ for acetic acid is 390.7 cm²/ohm equivalent. Since we know the equivalent conductance of the 0.00100$M$ acetic acid solution, we can obtain the degree of dissociation, $\alpha$, for the solution by using the relationship

$$\alpha = \frac{\Lambda}{\Lambda_o}$$

$$= \frac{49.2 \ cm^2/ohm \ equivalent}{390.7 \ cm^2/ohm \ equivalent}$$

$$= 0.126$$

Thus, 0.00100$M$ acetic acid is 12.6% ionized at 25$^{o}$C.

2. [10.8]    For the tin-coated iron, possible oxidation reactions (see Table 10.1 of the study guide) are

$$Fe \ \rightarrow \ Fe^{2+} + 2e^{-} \qquad \mathscr{E}^{o} = +0.4402 \ V$$
$$Sn \ \rightarrow \ Sn^{2+} + 2e^{-} \qquad \mathscr{E}^{o} = +0.136 \ V$$
$$2H_2O \ \rightarrow \ O_2 + 4H^{+} + 4e^{-} \qquad \mathscr{E}^{o} = -1.229 \ V$$

The reduction reaction is

$$O_2 + 2H_2O + 4e^{-} \ \rightarrow \ 4OH^{-} \qquad \mathscr{E}^{o} = +0.401 \ V$$

Since iron is the best of the three reducing agents, it will be oxidized, i.e., it will rust, when not protected.

3. [10.8]    For galvanized iron, possible oxidation reactions (see Table 10.1 of study guide) are

$$Fe \ \rightarrow \ Fe^{2+} + 2e^{-} \qquad \mathscr{E}^{o} = +0.4402 \ V$$
$$Zn \ \rightarrow \ Zn^{2+} + 2e^{-} \qquad \mathscr{E}^{o} = +0.7628 \ V$$
$$2H_2O \ \rightarrow \ O_2 + 4H^{+} + 4e^{-} \qquad \mathscr{E}^{o} = -1.229 \ V$$

The reduction reaction is

$$O_2 + 2H_2O + 4e^{-} \ \rightarrow \ 4OH^{-} \qquad \mathscr{E}^{o} = +0.401 \ V$$

The iron is protected since zinc is a better reducing agent than iron.  The zinc will be oxidized by the $O_2$ and $H_2O$.

4. 2    We use the relationship between standard Gibbs free
   [10.7]   energy and standard emf

$$\Delta G^{o} = -nF \mathscr{E}^{o}$$

$$n = \frac{-\Delta G^{o}}{F \mathscr{E}^{o}}$$

$$n = \frac{-(-12.00 \ kcal/mol)}{(23.061 \ kcal/V)(+0.26 \ V)}$$

$$n = 2.0$$

5. a. 1.1 hr
   [10.4]

The reduction reaction is

$$Mg^{2+} + 2e^- \rightarrow Mg$$

Thus, 2 mol of electrons, or $2F$, is required per mole of Mg. We can use a logic chain to solve the problem:

$$? \text{ hr} = 5.0 \text{kg Mg} \left(\frac{10^3 g}{1 \text{ kg}}\right)\left(\frac{1 \text{ mol Mg}}{24.3 \text{ g Mg}}\right)\left(\frac{2F}{1 \text{ mol Mg}}\right)\left(\frac{96,500 \text{ coulombs}}{1 F}\right)$$

$$\left(\frac{1}{1.0 \times 10^4 \text{ coulomb/sec}}\right)\left(\frac{1 \text{ min}}{60 \text{ sec}}\right)\left(\frac{1 \text{ hr}}{60 \text{ min}}\right)$$

$$= 1.1 \text{ hr}$$

   b. $4.6 \times 10^3$
      liters $Cl_2$

The oxidation reaction is

$$2Cl^- \rightarrow Cl_2 + 2e^-$$

Thus, for every mole of Mg produced, one mole of $Cl_2$, or 22.4 liters of $Cl_2$, is obtained. Therefore,

$$? \text{ liters } Cl_2 = 5.0 \text{ kg Mg} \left(\frac{10^3 g}{1 \text{ kg}}\right)\left(\frac{1 \text{ mol Mg}}{24.3 \text{ g Mg}}\right)\left(\frac{1 \text{ mol } Cl_2}{1 \text{ mol Mg}}\right)\left(\frac{22.4 \text{ liters } Cl_2}{1 \text{ mol } Cl_2}\right)$$

$$= 4.6 \times 10^3 \text{ liters } Cl_2$$

We can also solve the problem as follows:

$$? \text{ liters } Cl_2 = 1.1 \text{ hr} \left(\frac{60 \text{ min}}{1 \text{ hr}}\right)\left(\frac{60 \text{ sec}}{1 \text{ min}}\right)\left(\frac{1.0 \times 10^4 \text{ coulombs}}{1 \text{ sec}}\right)$$

$$\left(\frac{1 F}{96,500 \text{ coulombs}}\right)\left(\frac{1 \text{ mol } Cl_2}{2 F}\right)\left(\frac{22.4 \text{ liters } Cl_2}{1 \text{ mol } Cl_2}\right)$$

$$= 4.6 \times 10^3 \text{ liters } Cl_2$$

6. [10.8]

Spontaneous reactions are reactions with a positive cell potential.

Reaction (a) is not spontaneous since $\mathscr{E}^°_{cell} = -0.153$ V:

$$Cd^{2+} + 2e^- \rightarrow Cd \qquad\qquad \mathscr{E}^° = -0.4029 \text{ V}$$

$$Ni \rightarrow Ni^{2+} + 2e^- \qquad\qquad \mathscr{E}^° = +0.250 \text{ V}$$

$$\overline{Cd^{2+} + Ni \rightarrow Cd + Ni^{2+} \qquad\qquad \mathscr{E}^°_{cell} = -0.153 \text{ V}}$$

Reaction (b) is spontaneous since $\mathscr{E}^°_{cell} = +0.473$ V.

$$Sn \rightarrow Sn^{2+} + 2e^- \qquad\qquad \mathscr{E}^\circ = +0.136 \text{ V}$$
$$\underline{Cu^{2+} + 2e^- \rightarrow Cu \qquad\qquad \mathscr{E}^\circ = +0.337 \text{ V}}$$
$$Sn + Cu^{2+} \rightarrow Sn^{2+} + Cu \qquad \mathscr{E}^\circ_{cell} = +0.473 \text{ V}$$

Reaction (c) is not spontaneous since $\mathscr{E}^\circ_{cell} = -0.2636$ V:

$$I_2 + 2e^- \rightarrow 2I^- \qquad\qquad \mathscr{E}^\circ = +0.5355 \text{ V}$$
$$\underline{2Ag \rightarrow 2Ag^+ + 2e^- \qquad\qquad \mathscr{E}^\circ = -0.7991 \text{ V}}$$
$$I_2 + 2Ag \rightarrow 2I^- + 2Ag^+ \qquad \mathscr{E}^\circ_{cell} = -0.2636 \text{ V}$$

Note that the half reaction for the oxidation of Ag must be multiplied by 2 before addition so that the electrons lost and gained in the half reactions will cancel. The $\mathscr{E}^\circ$ for the Ag/Ag$^+$ electrode, however, is *not* multiplied by 2. The $\mathscr{E}^\circ$ valve of a half reaction is independent of the number of electrons lost or gained.

Reaction (d) is spontaneous since $\mathscr{E}^\circ_{cell} = +0.45$ V:

$$2\left[MnO_4^- + 8H^+ + 5e^- \rightarrow Mn^{2+} + 4H_2O\right] \qquad \mathscr{E}^\circ = +1.51 \text{ V}$$
$$\underline{5\left[2Br^- \rightarrow Br_2 + 2e^-\right] \qquad\qquad\qquad\qquad \mathscr{E}^\circ = -1.0652 \text{ V}}$$
$$2MnO_4^- + 16H^+ + 10Br^- \rightarrow 2Mn^{2+} + 8H_2O + 5Br_2 \quad \mathscr{E}^\circ_{cell} = +0.45 \text{V}$$

7. [10.8]

In standard cell notation the anode is listed first. Oxidation always occurs at the anode. Thus, for reaction (b) the cell is noted

$$Sn\,\big|\,Sn^{2+}\,\big\|\,Cu^{2+}\,\big|\,Cu$$

and for reaction (d) the cell is noted

$$Pt\,\big|\,Br^-\,\big|\,Br_2\,\big\|\,MnO_4^-\,\big|\,Mn^{2+}\,\big|\,Pt$$

The inert platinum electrodes are noted in the cell notation for reaction (b) since they are necessary to make an external electrical connection.

8. $1.1 \times 10^{-4}$ ohm$^{-1}$ cm$^{-1}$

[10.5]

To calculate the specific conductane, $\kappa$, we use the equation

$$\Lambda = \frac{1000\kappa}{N}$$

First, however, we must calculate the equivalent

conductance, $\Lambda$, of the 0.00500$M$ acetic acid solution:

$$\Lambda = \alpha \Lambda_o$$

$$\Lambda = (0.058)(390.7 \text{ cm}^2/\text{ohm equivalent})$$

$$\Lambda = 22.7 \text{ cm}^2/\text{ohm equivalent}$$

Thus,

$$\Lambda = \frac{1000\kappa}{N}$$

$$\kappa = \frac{\Lambda N}{1000}$$

$$= \frac{(22.7 \text{ cm}^2/\text{ohm equivalent})(5.00 \times 10^{-3} \text{ equivalents/cm}^3)}{1000}$$

$$= 1.1 \times 10^{-4} \text{ ohm}^{-1} \text{ cm}^{-1}$$

9.  [10.8]      From Table 10.1 of the study guide we choose reactions that contain any of the possible reactants, Cd, $Fe^{3+}$, and $H_2O$:

oxidation reactions

$$Cd \rightarrow Cd^{2+} + 2e^- \qquad\qquad \mathscr{E}° = +0.4029 \text{ V}$$

$$2H_2O \rightarrow O_2 + 4H^+ + 4e^- \qquad\qquad \mathscr{E}° = -1.229 \text{ V}$$

reduction reactions

$$Fe^{3+} + e^- \rightarrow Fe^{2+} \qquad\qquad \mathscr{E}° = +0.771 \text{ V}$$

$$Fe^{2+} + 2e^- \rightarrow Fe \qquad\qquad \mathscr{E}° = -0.4402 \text{ V}$$

$$2H_2O + 2e^- \rightarrow H_2 + 2OH^- \qquad\qquad \mathscr{E}° = -0.82806 \text{ V}$$

The only pair of reactions that has a positive cell potential is

$$Cd \rightarrow Cd^{2+} + 2e^- \qquad\qquad \mathscr{E}° = +0.4029 \text{ V}$$

$$\underline{2\left[Fe^{3+} + e^- \rightarrow Fe^{2+}\right] \qquad\qquad \mathscr{E}° = +0.771 \text{ V}}$$

$$Cd + 2Fe^{3+} \rightarrow Cd^{2+} + 2Fe^{2+} \qquad \mathscr{E}°_{cell} = +1.174 \text{ V}$$

10.  [10.8]     In Appendix D of your text the reduction potential of $Co^{3+}$ is given:

$$Co^{3+} + e^- \rightarrow Co^{2+} \qquad\qquad \mathscr{E}° = +1.808 \text{V}$$

The oxidation potential of water can be obtained from Table 10.1 of the study guide:

$$2H_2O \rightarrow O_2 + 4H^+ + 4e^- \qquad \mathscr{E}^\circ = -1.229 \text{ V}$$

Thus, when we couple the reduction of $Co^{3+}$ and the oxidation of water, we find that $\mathscr{E}^\circ$ for the oxidation-reduction reaction is positive:

$$4\left[Co^{3+} + e^- \rightarrow Co^{2+}\right] \qquad \mathscr{E}^\circ = +1.808 \text{ V}$$
$$2H_2O \rightarrow O_2 + 4H^+ + 4e^- \qquad \mathscr{E}^\circ = -1.229 \text{ V}$$
$$4Co^{3+} + 2H_2O \rightarrow 4Co^{2+} + O_2 + 4H^+ \quad \mathscr{E}^\circ = +0.579 \text{ V}$$

Thus, when the $Co^{3+}$ salt is added to water, the reduction of $Co^{3+}$ and the oxidation of water occur spontaneously.

11. 99.9 g Cu
[10.4]

The balanced equation is

$$2Cu^{2+} + 2H_2O \rightarrow 2Cu + O_2 + 4H^+$$

Using a logic chain, we find

$$? \text{ g Cu} = 17.6 \text{ liters } O_2 \left(\frac{1 \text{ mol } O_2}{22.41 \text{ liters } O_2}\right)\left(\frac{2 \text{ mol Cu}}{1 \text{ mol } O_2}\right)\left(\frac{63.55 \text{ g Cu}}{1 \text{ mol Cu}}\right)$$
$$= 99.9 \text{ g Cu}$$

12. [10.8]
a. yes

$$2\left[BrO^- + 2e^- + 2H^+ \rightarrow Br^- + H_2O\right] \qquad \mathscr{E}^\circ = +0.71 \text{ V}$$
$$BrO^- + 2H_2O \rightarrow BrO_3^- + 4H^+ + 4e^- \qquad \mathscr{E}^\circ = -0.54 \text{ V}$$
$$3 BrO^- \rightarrow 2Br^- + BrO_3^- \qquad \mathscr{E}^\circ = +0.17 \text{ V}$$

b. yes

$$5\left[\tfrac{1}{2}Br_2 + e^- \rightarrow Br^-\right] \qquad \mathscr{E}^\circ = +1.07 \text{ V}$$
$$\tfrac{1}{2}Br_2 + 6OH^- \rightarrow BrO_3^- + 3H_2O + 5e^- \qquad \mathscr{E}^\circ = -0.52 \text{ V}$$
$$3Br_2 + 6OH^- \rightarrow 5Br^- + BrO_3^- + 3H_2O \qquad \mathscr{E}^\circ = +0.55 \text{ V}$$

13. $5.2 \times 10^{-2}M$
[10.9]

The reaction taking place is

$$H_2(1 \text{ atm}) + 2H^+(1.0M) \rightarrow H_2(1 \text{ atm}) + 2H^+(?M) \quad \mathscr{E} = +0.076V$$

Two faradays are involved in the cell reaction and therefore $n = 2$. Since the same electrode is in each half cell, $\mathscr{E}^\circ$ for the cell is zero.

We can obtain the unknown $[H^+]$ from the Nernst equation:

$$\mathscr{E} = \mathscr{E}^\circ - \frac{0.0592}{n} \log Q$$

$$0.076 = 0 - \frac{0.0592}{2} \log\left(\frac{(1)\ [H^+]^2}{(1)\ (1)^2}\right)$$

$$\log [H^+]^2 = 0.076 \left(\frac{-2}{0.0592}\right)$$

$$= -2.57$$

$$[H^+]^2 = \text{antilog } -2.57$$

$$= 2.7 \times 10^{-3}$$

$$[H^+] = 5.2 \times 10^{-2} M$$

14. [10.9]
$[Fe^{2+}] = 9.52 \times 10^{-4} M$
$[Fe^{3+}] = 4.8 \times 10^{-5} M$

Since the total iron concentration is $1.00 \times 10^{-3} M$,

$$[Fe^{2+}] + [Fe^{3+}] = 1.00 \times 10^{-3} M$$

If we let $x = [Fe^{2+}]$, then $\left((1.00 \times 10^{-3}) - x\right) M = [Fe^{3+}]$

We can now use the Nernst equation to solve the problem:

$$\mathscr{E} = \mathscr{E}^\circ - \frac{0.0592}{n} \log Q$$

$$\log Q = \mathscr{E} - \mathscr{E}^\circ \left(\frac{-n}{0.0592}\right)$$

$$\log \frac{\left((1.00 \times 10^{-3}) - x\right) M}{x} = -0.694 - (-0.771) \left(\frac{-1}{0.0592}\right)$$

$$x = 9.52 \times 10^{-4} M = [Fe^{2+}]$$

$$\left((1.00 \times 10^{-3}) - x\right) M = 4.8 \times 10^{-5} M = [Fe^{3+}]$$

15. 12.1 g/equivalent
[10.4]

We can use Faraday's laws and a logic chain to solve this problem:

$$? \text{ equivalents} = 3.00 \text{ hr} \left(\frac{60 \text{ min}}{1 \text{ hr}}\right)\left(\frac{60 \text{ sec}}{1 \text{ min}}\right)(10.0 \text{ coulomb/sec})$$

$$\left(\frac{1F}{96,487 \text{ coulombs}}\right)\left(\frac{1 \text{ equivalent}}{1F}\right)$$

$$= 1.12 \text{ equivalents}$$

$$\text{equivalent weight} = \frac{\text{weight}}{\text{number of equivalents}}$$

$$= \frac{13.6 \text{ g}}{1.12 \text{ equivalents}}$$

$$= 12.1 \text{ g/equivalent}$$

16. a. 23.9 liters $O_2$
[10.4]

The net reaction shows that 1 mol of $O_2$ reacts with 2 mol of $H_2$; therefore, half as much $O_2$ as $H_2$, or 23.9 liters of $O_2$, is required at STP:

$$? \text{ liters } O_2 = 47.8 \text{ liters } H_2 \left(\frac{1 \text{ mol } O_2}{2 \text{ mol } H_2}\right)\left(\frac{22.4 \text{ liters } O_2/1 \text{ mol } O_2}{22.4 \text{ liters } H_2/1 \text{ mol } H_2}\right)$$

$$= 23.9 \text{ liters } O_2$$

b. 2.13 mol $H_2O$

From the net reaction we know that 2 mol of $H_2$ reacts with 1 mol of $O_2$ to produce 2 mol of $H_2O$. Thus,

$$? \text{ mol } H_2O = 47.8 \text{ liters } H_2 \left(\frac{2 \text{ mol } H_2O}{2 \text{ mol } H_2}\right)\left(\frac{1 \text{ mol } H_2}{22.4 \text{ liters } H_2}\right)$$
$$= 2.13 \text{ mol } H_2O$$

c. 686 coulombs/sec

To determine the current, expressed in amperes or coulombs/sec, we use Faraday's laws and a logic chain:

$$? \text{ coulomb/sec} = \left(\frac{47.8 \text{ liters } H_2}{10 \text{ min}}\right)\left(\frac{1 \text{ mol } H_2}{22.4 \text{ liters } H_2}\right)\left(\frac{2F}{1 \text{ mol } H_2}\right)$$

mol $H_2$ reacted per min

$$\left(\frac{96{,}500 \text{ coulombs}}{1 F}\right)\left(\frac{1 \text{ min}}{60 \text{ sec}}\right)$$

current in coulombs/sec

$$= 686 \text{ coulombs/sec}$$

17. 42 liters $H_2$
21 liters $H_2$
[10.4]

To determine the size of the tanks, we use Faraday's laws and set up a logic chain:

$$? \text{ liters } H_2 = 20 \text{ coulombs/sec} \left(\frac{60 \text{ sec}}{1 \text{ min}}\right)\left(\frac{60 \text{ min}}{1 \text{ hr}}\right)\left(\frac{22.4 \text{ liters } H_2}{1 \text{ mol } H_2}\right)\left(\frac{1 \text{ mol } H_2}{2F}\right)$$

$$\left(\frac{1F}{96,500 \text{ coulombs}}\right)(5 \text{ hr})$$

$$= 42 \text{ liters } H_2$$

Since the cell reaction

$$2H_2(g) + O_2(g) \rightarrow 2H_2O(l)$$

indicates that 1 mol of $O_2$ reacts with 2 mol of $H_2$, only 21 liters of $O_2$ are required.

## III. Electrochemical cells

1. In solving problems such as this, check the table of standard electrode potentials (Table 10.1 of the study guide) first to determine which species will be oxidized and which will be reduced. The couple with the more positive electrode potential will occur as written, a reduction; the other will occur in reverse, an oxidation.

   In this cell $Cu^{2+}$ is reduced and Cd is oxidized. The electrons travel toward the positive electrode. The ions that are reduced, the cations, migrate toward the cathode, and the anions migrate toward the anode. In the cell notation the anode is represented to the left of the salt bridge, which is represented by the double bar, and the cathode is represented to the right of the salt bridge. The cell potential is

   $$\mathscr{E}^\circ_{\text{right}} - \mathscr{E}^\circ_{\text{left}} = +0.740 \text{ V}$$

1.

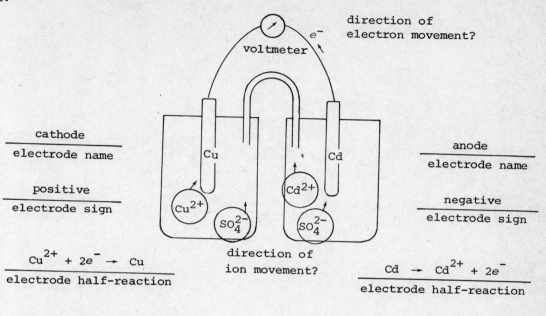

direction of
electron movement?

$e^-$

voltmeter

cathode
_____
electrode name

anode
_____
electrode name

positive
_____
electrode sign

negative
_____
electrode sign

Cu

Cd

$Cd^{2+}$

$Cu^{2+}$

$SO_4^{2-}$

$SO_4^{2-}$

direction of
ion movement?

$Cu^{2+} + 2e^- \rightarrow Cu$
_____
electrode half-reaction

$Cd \rightarrow Cd^{2+} + 2e^-$
_____
electrode half-reaction

$+0.337V$
_____
standard electrode
potential

$+0.4029V$
_____
standard electrode
potential

$Cd \mid Cd^{2+} \parallel Cu^{2+} \mid Cu$
_____
cell notation

$Cd + Cu^{2+} \rightarrow Cd^{2+} + Cu$
_____
overall cell reaction

$+0.740V$
_____
cell potential

2.

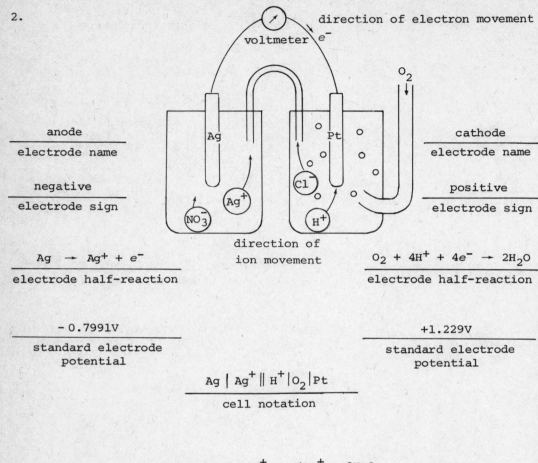

direction of electron movement

voltmeter $e^-$

$O_2$

Ag

Pt

anode
___
electrode name

negative
___
electrode sign

cathode
___
electrode name

positive
___
electrode sign

direction of
ion movement

$Ag \rightarrow Ag^+ + e^-$
___
electrode half-reaction

$O_2 + 4H^+ + 4e^- \rightarrow 2H_2O$
___
electrode half-reaction

$-0.7991V$
___
standard electrode
potential

$+1.229V$
___
standard electrode
potential

$Ag \mid Ag^+ \parallel H^+ \mid O_2 \mid Pt$
___
cell notation

$4Ag + O_2 + 4H^+ \rightarrow 4Ag^+ + 2H_2O$
___
overall cell reaction

$+0.430V$
___
cell potential

3.

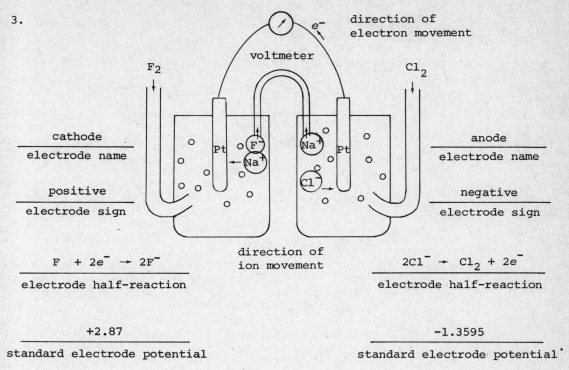

direction of
electron movement

voltmeter

$F_2$    $Cl_2$

| cathode | anode |
|---------|-------|
| electrode name | electrode name |

| positive | negative |
|----------|----------|
| electrode sign | electrode sign |

direction of
ion movement

$F + 2e^- \rightarrow 2F^-$

electrode half-reaction

$2Cl^- \rightarrow Cl_2 + 2e^-$

electrode half-reaction

+2.87

standard electrode potential

-1.3595

standard electrode potential

$Pt \,|\, Cl_2 \,|\, Cl^- \,\|\, F^- \,|\, F_2 \,|\, Pt$

cell notation

$F_2 + 2Cl^- \rightarrow Cl_2 + 2F^-$

overall cell reaction

+1.51V

cell potential

SELF-TEST          Complete the test in 20 minutes:

I. Do the following:

1. Calculate the emf of the following cell at 25°C:

$$Cd(s) \,\vert\, Cd^{2+}(0.050M) \,\Vert\, Ag^{+}(0.50M) \,\vert\, Ag(s)$$

The pertinent electrode potentials are

$$2e^{-} + Cd^{2+} \rightleftharpoons Cd(s) \qquad \mathscr{E}^{\circ} = -0.40 \text{ V}$$

$$e^{-} + Ag^{+} \rightleftharpoons Ag(s) \qquad \mathscr{E}^{\circ} = +0.80 \text{ V}$$

The Nernst equation is

$$\mathscr{E} = \mathscr{E}^{\circ} - \frac{0.0592}{n} \log Q$$

2. What weight of Mg is obtained in 10.0 minutes from the electrolysis of dry, molten $MgCl_2$ using a current of 10.0 amp?  The atomic weight of Mg is 24.3 and that of Cl is 35.5.

3. Given the electrode potentials

$$Mg^{2+} + 2e^{-} \rightleftharpoons Mg \qquad \mathscr{E}^{\circ} = -2.36 \text{ V}$$

$$Cu^{2+} + 2e^{-} \rightleftharpoons Cu \qquad \mathscr{E}^{\circ} = +0.34 \text{ V}$$

answer each of the following:

a. What is the best oxidizing agent of the chemical species in the half reactions?

b. In a voltaic cell consisting of standard $Mg^{2+}/Mg$ and $Cu^{2+}/Cu$ half cells, which half cell is the cathode?

c. Is the Cu electrode of the voltaic cell in part (b) of this problem positive or negative?

d. Will Mg metal spontaneously reduce $Cu^{2+}$ ions in $1.0M$ concentration at 25°C?

e. In an electrolytic cell is the cathode positive or negative?

4. At $25^{\circ}C$ the specific conductance, $\kappa$, of a $0.10M$ solution of a weak acid HX is $1.8 \times 10^{-4}$/ohm cm. The equivalent conductance at infinite dilution of HX, $\Lambda_o$, is 400 $cm^2$/ohm equivalent.  What is the degree of dissociation, $\alpha$, of 0.10M HX at $25^{\circ}C$?

II. Answer each of the following.  The questions refer to a galvanic cell that involves the reaction

$$2VO^{2+} + 4H^+ + Ni(s) \rightarrow 2V^{3+} + Ni^{2+} + 2H_2O$$

Pertinent electrode potentials are

$$2e^- + Ni^{2+} \rightleftharpoons Ni \qquad\qquad \mathscr{E}° = -0.25\ V$$

$$e^- + 2H^+ + VO^{2+} \rightleftharpoons V^{3+} + H_2O \qquad \mathscr{E}° = +0.36\ V$$

1. The standard emf of the cell is
   a.  0.11V          c.  0.97V
   b.  0.61V          d.  0.47V

2. The emf of the cell could be increased by
   a.  increasing $Ni^{2+}$ concentration
   b.  lowering the pH
   c.  reducing $VO^{2+}$ concentration
   d.  increasing $V^{3+}$ concentration

3. The anode could be made of
   a.  Ni             c.  Pt
   b.  V              d.  $H_2$

4. The cathode could be made of
   a.  Ni             c.  Pt
   b.  V              d.  $H_2$

5. The cell must be constructed so that the mixing of the following is avoided:
   a. $V^{3+}$ and $Ni^{2+}$
   b. $V^{3+}$ and Ni(s)
   c. $VO^{2+}$ and $V^{3+}$
   d. $VO^{2+}$ and Ni(s)

6. The best oxidizing agent is
   a. $Ni^{2+}$          c. $VO^{2+}$ and $H^+$
   b. Ni(s)              d. $V^{3+}$

# 11
# The Nonmetals, Part I

OBJECTIVES

(a) You should be familiar with the physical properties of the group 0 elements (the noble gases), the group VII A elements (the halogens), and the group VI A elements (except oxygen, see Chapter 8 of your text). The elements discussed in Chapter 11 of your text are shown in the shaded area of Figure 11.1 of the study guide.

(b) You should be familiar with the abundance of these elements in nature, the ways in which they can be obtained from natural sources, and their uses.

(c) You should know the fundamental chemistry of these elements, i.e., the types of compounds they form, how such compounds are prepared, and how such compounds react.

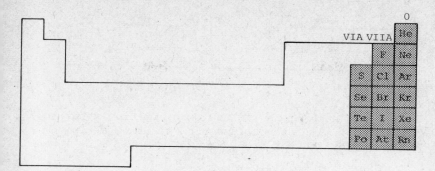

Figure 11.1   Periodic table

(d) You should be able to apply the basic information of all previous chapters of your text toward an understanding of the chemistry of these elements.

EXERCISES

I. Answer the following:

1. Chlorine can be prepared by the reaction of concentrated HCl with solid $MnO_2$. Write the balanced equation for the reaction:

   $MnO_2(s)$ + HCl(aq) →

2. Use the ion-electron method to balance the equation for the reaction of iodate ion and sulfite ion to form iodine and sulfate ion in acid solution.

3. What is the physical state at STP of each of the following: fluorine, chlorine, bromine, and iodine?

4. Write the electronic configuration of each of the following:
   (a) Br                       c. Cl

   (b) $Br^-$                     d. $Cl^-$

5. From the following standard electrode potential diagram of chlorine, chloride, and the chlorine oxyanions, predict which compounds are stable in basic solution:

$$\text{ClO}_4^- \xrightarrow{+0.36 \text{ V}} \text{ClO}_3^- \xrightarrow{+0.33 \text{ V}} \text{ClO}_2^- \xrightarrow{+0.66 \text{ V}} \text{ClO}^- \xrightarrow{+0.40 \text{ V}} \text{Cl}_2 \xrightarrow{+1.36 \text{ V}} \text{Cl}^-$$

+0.63 V

+0.50 V          +0.89 V

6. Complete each of the following by writing a balanced equation. If not reaction occurs, write *NR* in the space provided for products.

a. $SiO_2(s)$ + 6HF(aq) $\longrightarrow$

b. $Br_2(l)$ + $2I^-$(aq) $\longrightarrow$

c. $S(s)$ + $H_2(g)$ $\longrightarrow$

d. $Ba^{2+}$(aq) + $SO_4^{2-}$ (aq) $\longrightarrow$

e. $Cl_2(g)$ + Bi(s) $\longrightarrow$

f. $Cl_2(g)$ + Hg(l) $\longrightarrow$

g. Se(s) + Hg(l) $\longrightarrow$

h. $CaF_2(s)$ + $H_2SO_4(l)$ $\xrightarrow{\text{heat}}$

i. $Br_2(g)$ + $H_2(g)$ $\longrightarrow$

j. $Br_2(l)$ + $2Cl^-$(aq) $\longrightarrow$

7. What is the orbital hybridization of sulfur in each of the following? Draw structures of the molecules.
   a. $SF_4$          b. $SF_6$

8. Predict the shape of each of the following interhalogens:

   a. $ClF_2^+$  b. $ICl_2^-$  c. $ClF_3$  d. $ICl_4^-$

9. How are pure neon, argon, krypton, and xenon obtained commercially

10. What is the commercial source of pure helium?

11. Why have $XeF_2$, $XeF_4$, and $XeF_6$ been prepared, but not $XeF_3$ and $XeF_5$?

12. In addition to the xenon fluoride compounds mentioned in problem 11 of this section, some krypton fluoride compounds have been prepared. What are the chances of preparing compounds containing fluorine and either He, Ne, or Ar?

13. Draw the molecular orbital energy-level diagram for diatomic fluorine and explain why $F_2$ is more stable than two free fluorine atoms.

14. Hydrofluoric acid, HF, is a weak electrolyte, but the others of the series HF, HCl, HBr, and HI are strong. Why?

15. What is the color and the physical state of each of the following at STP: oxygen, sulfur, selenium, and tellurium?

16. Write the electronic configuration of tellurium.

17. Draw a Lewis structure of $S_8$.

18. Write the formula of each of the following compounds in the space provided:

_____ a. selenous acid
_____ b. perbromic acid
_____ c. tellurium dioxide
_____ d. xenon(VIII) oxide
_____ e. calcium fluoride
_____ f. sodium bisulfate

_____ g. pyrosulfuric acid
_____ h. hypoiodous acid
_____ i. phosphorus trichloride
_____ j. peroxydisulfuric acid
_____ k. sodium perxenate
_____ l. sodium thiosulfate

19. Determine the oxidation number of iodine in each of the following compounds. Name each compound in the space provided.

_____ a. HOI
_____ b. $IF_7$
_____ c. $I_2$
_____ d. NaI
_____ e. $I_2O_5$
_____ f. $H_5IO_6$

_____ g. AgI
_____ h. $ICl_3$
_____ i. $I_2O_4$
_____ j. $As_2I_4$
_____ k. $I_2Cl_6$

20. Draw a Lewis structure of the thiosulfate ion, $S_2O_3^{2-}$ .

21. How many grams of $MnO_2$ are needed to prepare 1.00 liter of dry $Cl_2$ gas measured at $25^{\circ}C$ and 1.00 atm by the following reaction?

$$MnO_2(s) + 4HCl(aq) \rightarrow Cl_2(g) + MnCl_2(aq) + 2H_2O(l)$$

22. What would you expect the shape of the periodate ion $IO_6^{5-}$ to be? Draw the Lewis structure.

23. Chlorine is contained in DDT, $C_{14}H_9Cl_5$, the common insecticide. What is the percentage of chlorine in DDT?

## ANSWERS TO EXERCISES

I. To determine the amount of descriptive chemistry you should know, rely on your class notes and the comments of your instructor.

1. [8.5, 11.3]    $MnO_2(s) + 4HCl(aq) \rightarrow Cl_2(g) + MnCl_2(aq) + 2H_2O(l)$

2. [8.5, 11.3]    $2IO_3^- + 2H^+ + 5SO_3^{2-} \rightarrow 5SO_4^{2-} + I_2 + H_2O$

3. [11.2]

| ELEMENT | COLOR | PHYSICAL STATE AT STP |
|---|---|---|
| fluorine | pale yellow | gas |
| chlorine | green-yellow | gas |
| bromine | red | liquid |
| iodine | black | solid |

4. 2.16, 2.17, 11.2

a. Cl : $1s^2\ 2s^2\ 2p^6\ 3s^2\ 3p^5$

b. Cl$^-$: $1s^2\ 2s^2\ 2p^6\ 3s^2\ 3p^6$

c. Br : $1s^2\ 2s^2\ 2p^6\ 3s^2\ 3p^6\ 3d^{10}\ 4s^2\ 4p^5$

d. Br$^-$: $1s^2\ 2s^2\ 2p^6\ 3s^2\ 3p^6\ 3d^{10}\ 4s^2\ 4p^6$

5. 11.7

The chloride ion, $Cl^-$, and the perchlorate ion, $ClO_4^-$, are stable to disproportionation in basic solution. The perchlorate ion is stable because it can not be further oxidized, and the chloride ion is stable because it can not be further reduced. Neither of these ions can disproportionate because neither can undergo oxidation and reduction spontaneously and simultaneously.  The chlorate ion, $ClO_3^-$, however, can disproportionate:

$$6e^- + 3H_2O + ClO_3^- \rightarrow Cl^- + 6OH^- \qquad \mathscr{E}° = +0.63 \text{ V}$$

$$3\left[2OH^- + ClO_3^- \rightarrow ClO_4^- + H_2O + 2e\right] \qquad \mathscr{E}° = -0.36 \text{ V}$$

$$4\,ClO_3^- \rightarrow Cl^- + 3ClO_4^- \qquad \mathscr{E}° = +0.27 \text{ V}$$

Also, the chlorite ion, $ClO_2^-$, can disproportionate:

$$2e^- + ClO_2^- + H_2O \rightarrow ClO^- + 2OH^- \qquad \mathscr{E}° = +0.60 \text{ V}$$

$$ClO_2^- + 2OH^- \rightarrow ClO_3^- + H_2O + 2e^- \qquad \mathscr{E}° = +0.33 \text{ V}$$

$$2ClO_2^- \rightarrow ClO^- + ClO_3^- \qquad \mathscr{E}° = +0.27 \text{ V}$$

Both of the products of the disproportionation of $ClO_2^-$ can disproportionate.  The $ClO^-$ can disproportionate as follows:

$$2\left[H_2O + ClO^- + 2e^- \rightarrow Cl^- + 2OH^-\right] \qquad \mathscr{E}° = +0.89 \text{ V}$$

$$ClO^- + 4OH^- \rightarrow ClO_3^- + 2H_2O + 4e^- \qquad \mathscr{E}° = -0.50 \text{ V}$$

$$3ClO^- \rightarrow 2Cl^- + ClO_3^- \qquad \mathscr{E}° = +0.39 \text{ V}$$

Chlorine gas, $Cl_2$, disproportionates as follows:

$$2OH^- + Cl_2 \rightarrow ClO^- + Cl^- + H_2O \qquad \mathscr{E}° = +0.96 \text{ V}$$

6. a. [11.5]

$$SiO_2(s) + 6HF(aq) \rightarrow 2H^+(aq) + SiF_6^{2-}(aq) + 2H_2O(l)$$

   b. [11.3]

$$Br_2(l) + 2I^-(aq) \rightarrow 2Br^-(aq) + I_2(s)$$

You should learn some of the important halogen reactions summarized in Table 11.1 of the study guide.

Table 11.1  Some reactions of the halogens ($X_2$ = $F_2$, $Cl_2$, $Br_2$, or $I_2$)

| GENERAL REACTION | REMARKS |
|---|---|
| $nX_2 + 2M \rightarrow 2MXn$ | $F_2$, $Cl_2$ with practically all metals; $Br_2$, $I_2$ with all except noble metals |
| $X_2 + H_2 \rightarrow 2HX$ | |
| $3X_2 + 2P \rightarrow 2PX_3$ | with excess P; similar reactions with As, Sb, and Bi |
| $5X_2 + 2P \rightarrow 2PX_5$ | with excess $X_2$, but not with $I_2$; $SbF_5$, $SbCl_5$, $AsF_5$, $AsCl_5$, and $BiF_5$, may be similarly prepared |
| $X_2 + 2S \rightarrow S_2X_2$ | with $Cl_2$, $Br_2$ |
| $X_2 + H_2O \rightarrow H^+ + X^- + HOX$ | not with $F_2$ |
| $2X_2 + 2H_2O \rightarrow 4H^+ + 4X^- + O_2$ | $F_2$ rapidly; $Cl_2$, $Br_2$ slowly in sunlight |
| $X_2 + H_2S \rightarrow 2HX + S$ | |
| $X_2 + CO \rightarrow COX_2$ | $Cl_2$, $Br_2$ |
| $X_2 + SO_2 \rightarrow SO_2X_2$ | $F_2$, $Cl_2$ |
| $X_2 + 2X'^- \rightarrow X_2' + 2X^-$ | $F_2 > Cl_2 > Br_2 > I_2$ |
| $X_2 + X_2' \rightarrow 2XX'$ | formation of the interhalogen compounds (all except lF) |

c. [ 11.9 ]  $S(s) + H_2(g) \rightarrow H_2S(g)$

You should learn some of the important sulfur, selenium, and tellurium reactions summarized in Table 11.2 of the study guide.

d. [ 11.12 ]  $Ba^{2+}(aq) + SO_4^{2-}(aq) \rightarrow BaSO_4(s)$

The compounds $BaSO_4$, $SrSO_4$, $PbSO_4$, and $Hg_2SO_4$ are fairly insoluble in water; $CaSO_4$ and $AgSO_4$ are slightly soluble.  All other known metal sulfates are water soluble.

e. [11.3]  $3Cl_2(g) + 2Bi(s) \rightarrow 2BiCl_3(s)$ ( See Table 11.1.)

f. [ 11.3 ]  $Cl_2(g) + Hg(l) \rightarrow HgCl_2(s)$ ( See Table 11.1.)

g. [ 11.9 ]  $Se(s) + Hg(l) \rightarrow HgSe(s)$ ( See Table 11.2.)

Table 11.2 Some reactions of sulfur, selenium, and tellurium

| REACTION OF SULFUR | REMARKS |
| --- | --- |
| $nS + mM \rightarrow M_mS_n$ | Se, Te react similarly with many metals (not noble metals) |
| $nS + S^{2-} \rightarrow S_{n+1}^{2-}$ | for S and Te, $n = 1$ to 5; for Se $n = 1$ to 4 |
| $S + H_2 \rightarrow H_2S$ | S > Se > Te; elevated temperatures; compounds are better prepared by actions of dilute HCl on sulfides, selenides, or tellurides |
| $S + O_2 \rightarrow SO_2$ | S > Se > Te; dioxides of Se and Te are easier to prepare with a mixture of $O_2 + NO_2$ |
| $S + 3F_2 \rightarrow SF_6$ | S, Se, Te with excess $F_2$ |
| $S + 2F_2 \rightarrow SF_4$ | S, Se; $TeF_4$ is made indirectly $(TeF_6 + Te)$ |
| $2S + X_2 \rightarrow S_2X_2$ | S, Se; $X_2 = Cl_2$ or $Br_2$ |
| $S + 2X_2 \rightarrow SX_4$ | S, Se, Te with excess $Cl_2$; Se, Te with excess $Br_2$; Te with excess $I_2$ |
| $S_2Cl_2 + Cl_2 \rightarrow 2SCl_2$ | $SCl_2$ only, $SBr_2$ unknown; $SeCl_2$, $SeBr_2$ (only in vapor state); $TeCl_2$, $TeBr_2$ are made by thermal decomposition of higher halides |
| $S + 4HNO_3 \rightarrow SO_2 + 4NO_2 + 2H_2O$ | hot, concentrated nitric acid; S yields mixtures of $SO_2$ and $SO_4^{2-}$; Se yields $H_2SeO_3(SeO_2 \cdot H_2O)$; Te yields $2TeO_2 \cdot HNO_3$ |

h. [11.5]    $CaF_2(s) + H_2SO_4(l) \xrightarrow{\text{heat}} CaSO_4(s) + 2HF(g)$

Hydrogen fluoride is prepared commercially in this way.

i. [11.5]    $Br_2(g) + H_2(g) \rightarrow 2HBr(l)$

j. [11.3]    $Br_2(l) + 2Cl^-(aq) \rightarrow NR$ (See Table 11.1.)

7.  a. [4.2., 4.3]

$dsp^3$

b.

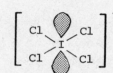

$d^2sp^3$

8. [4.2., 4.3]

a.

angular

b.

linear

c.

T-shaped

d.

square planar

9. [11.1]    These gases are obtained by fractional distillation of liquefied air.

10. [11.1]    Helium is obtained from natural gas deposits.

11. [11.1]    The compounds $XeF_3$ and $XeF_5$ would each have an odd number of valence electrons. Most molecules that are easily prepared have an even number of valence electrons.

12. [11.1]    The atoms of the elements of group 0 increase in size from He to Xe.  The valence electrons are also less tightly held from He to Xe, i.e., the ionization potentials decrease down the group.

13. [4.4, 4.5]    See Figure 11.2 of the study guide for the molecular orbital diagram of $F_2$.

14. [8.10, 11.5]    The bond of HF is stronger than that of any other hydrogen halide; therefore, HF is less easily dissociated.  When dissociation does occur, $HF_2^-$(aq) species form.  Hydrogen bonding stabilizes HF in solution.

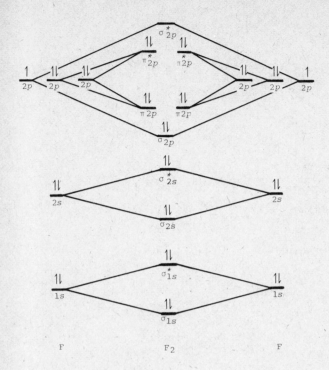

Figure 11.2   Molecular orbital diagram of $F_2$

15. [ 11.8 ]

| ELEMENT | COLOR | PHYSICAL STATE AT STP |
|---|---|---|
| oxygen | colorless | gas consisting of $O_2$ molecules |
| sulfur | yellow | solid consisting of $S_8$ rings |
| selenium | red to black | solid consisting of $Se_8$ rings and/or $Se_n$ chains |
| tellurium | silver to white | solid consisting of $Te_n$ chains |

16. [ 2.16, 2.17 ]   Te: $1s^2\ 2s^2\ 2p^6\ 3s^2\ 3p^6\ 3d^{10}\ 4s^2\ 4p^6\ 4d^{10}\ 5s^2\ 5p^4$

17. [ 11.9 ]

or

| 18. | | | |
|---|---|---|---|
| a. | $H_2SeO_3$ | selenous acid | [ 11.11 ] |
| b. | $HBrO_4$ | perbromic acid | [ 11.7 ] |
| c. | $TeO_2$ | tellurium dioxide | [ 11.11 ] |
| d. | $XeO_4$ | xenon(VIII) oxide | [ 11.1 ] |
| e. | $CaF_2$ | calcium fluoride | [ 11.6 ] |
| f. | $NaHSO_4$ | sodium bisulfate | [ 11.12 ] |
| g. | $H_2S_2O_7$ | pyrosulfuric acid | [ 11.12 ] |
| h. | $HOI$ | hypiodous acid | [ 11.7 ] |
| i. | $PCl_3$ | phosphorus trichloride | [ 11.6 ] |
| j. | $H_2S_2O_8$ | peroxydisulfuric acid | [ 11.12 ] |
| k. | $Na_4XeO_6$ | sodium perxenate | [ 11.1 ] |
| l. | $Na_2S_2O_3$ | sodium thiosulfate | [ 11.12 ] |

19. [ 3.10 ]     Write the formula of the compound in the space
                 provided:

| | OXIDATION NUMBER | NAME | FORMULA |
|---|---|---|---|
| a. | +1 | hypoiodous acid | _____ |
| b. | +7 | iodine heptafluoride | _____ |
| c. | 0 | iodine | _____ |
| d. | -1 | sodium iodide | _____ |
| e. | +5 | diiodine pentoxide | _____ |
| f. | +7 | periodic acid | _____ |
| g. | -1 | silver(I) iodide or silver iodide | _____ |
| h. | +3 | iodine trichloride | _____ |
| i. | +4 | diiodine tetraoxide | _____ |
| j. | -1 | diarsenic tetraiodide | _____ |
| k. | +6 | diiodine hexachloride | _____ |

20. [ 11.12 ]    The thiosulfate anion has the same structure as the
                 sulfate ion except that one oxygen atom is replaced
                 by a sulfur atom:

$$\left[\begin{array}{c} \ddot{:O}: \\ | \\ :\ddot{O} - S - \ddot{O}: \\ | \\ :\ddot{S}: \end{array}\right]^{2-}$$

21. [ 6.9, 11.3 ]     We calculate the number of grams of $MnO_2$:

$$?g\ MnO_2 = 1.00\ \text{liter}\ Cl_2 \left(\frac{1\ \text{mol}\ Cl_2}{22.4\ \text{liters}\ Cl_2}\right)\left(\frac{1\ \text{mol}\ MnO_2}{1\ \text{mol}\ Cl_2}\right)\left(\frac{86.94\ g\ MnO_2}{1\ \text{mol}\ MnO_2}\right)$$

$$= 3.88\ g\ MnO_2$$

22. [ 3.6, 11.7 ]     The Lewis structure is

$$\left[\begin{array}{ccc} \ddot{:O} & & \ddot{O}: \\ :\ddot{O} - & I & - \ddot{O}: \\ \ddot{:O} & & \ddot{O}: \end{array}\right]^{5-}$$

From a consideration of electron pair repulsions, we predict the ion to be octahedral:

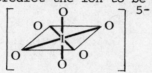

23. [ 5.4 ]     First we determine the molecular weight of DDT:

    14 carbon atoms   : 14(12.011) = 168.15

     9 hydrogen atoms:  9(1.0079) =   9.0711

     5 chlorine atoms:  5(35.453) = 177.26
    _____

        molecular weight of DDT = 354.48

Then we determine the percentage of chlorine in DDT:

$$\%\ Cl = \left(\frac{177.26\ g/mol\ DDT}{354.48\ g/mol\ DDT}\right)100$$

$$= 50.006$$

SELF-TEST          I. We suggest that you go through your lecture notes and
                   compile a list of chemical reactions , formulas, and
                   names of compounds.  Check your material to make sure
                   that you did not copy it incorrectly and then prepare
                   your own self-test.  Write a series of incomplete re-
                   actions with either the reactants or products given

and a list of formulas without names or vice versa.
The next day complete the reactions and give either
the name or formula of the compound without referring
to any material.

If your instructor has emphasized the necessity of
learning all chemical reactions in Chapter 11 of
your text, complete the following test in 15 minutes:

1. Complete and balance each of the following equations. If no reaction occurs, write *NR*.

   a. $OH^-(aq) + Cl_2(aq) \rightarrow$

   b. $FeS(s) + H^+(aq) + Cl^-(aq) \rightarrow$

   c. $Br_2(g) + Ag_2O(s) + H_2O(l) \rightarrow$

   d. $PCl_3(l) + H_2O(l) \rightarrow$

   e. $Cl^-(aq) + H_2O(l) \xrightarrow[\text{heat}]{\text{electrolysis}}$

   f. $Cl^-(aq) + H^+(aq) + H_2O(l) \rightarrow$

   g. $Cl_2(g) + I^-(aq) \rightarrow$

   h. $Ba^{2+}(aq) + SO_4^{2-}(aq) \rightarrow$

   i. $C_{12}H_{22}O_{11}(s) + H_2SO_4(conc.) \rightarrow$
      sucrose

   j. $S_2O_3^{2-}(aq) + H^+(aq) \rightarrow$

# 12
# The Nonmetals, Part II

OBJECTIVES

(a) You should be familiar with the physical properties of the elements of groups III A, IV A, and V A. These elements are shown in the shaded area of the periodic table of Figure 12.1 of the study guide.

(b) You should be familiar with the abundance of these elements in nature, the ways in which they can be obtained in pure form, and their uses.

(c) You should know the fundamental chemistry of these elements, i.e., the types of compounds they form, how such compounds are prepared, and how such compounds react.

241

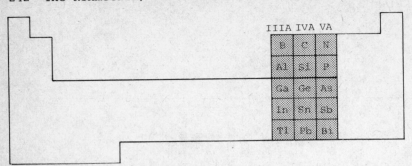

Figure 12.1  Periodic table

(d) You should be able to apply the basic information of all previous chapters of your text toward an understanding of the chemistry of these elements.

EXERCISES

I.  Answer the following:

1. What is the color and physical state of each element of group V A at STP?

2. Write the electronic configuration of each of the following:

   a. P    b. $P^{3-}$    c. As    d. $As^{3+}$   e. $As^{5+}$

3. Which element of group V A is most abundant on Earth?

4. What is the oxidation number of nitrogen in each of the following compounds?  Name each compound.

_____ a. $N_2O_4$            _____ c. $N_2O$

_____ b. NO                  _____ d. $N_2O_5$

_____ e. $N_2O_3$

5. Write the formula of each of the following in the space provided:

_____ a. ammonia              _____ g. triphosphoric acid
_____ b. aluminum carbide     _____ h. calcium cyanamide
_____ c. hydroxylamine        _____ i. phosphorus(III) oxide
_____ d. antimony(III) sulfide _____ j. hydrazine
_____ e. calcium nitride      _____ k. tetranitrogen tetrasulfide
_____ f. phosphorus(V) oxide  _____ l. phosphine
                            _____ m. nitric oxide

6. Complete and balance the following equations. If no reaction occurs, write *NR* in the space provided for the products.

    a. $PCl_3(l) + Cl_2(g) \longrightarrow$

    b. $NaN_3(s) \xrightarrow{heat}$

    c. $CaCO_3(s) \xrightarrow{heat}$

    d. $NH_3(g) + O_2(g) \xrightarrow[Pt]{heat}$

    e. $P_4O_{10}(s) + H_2O(l) \longrightarrow$

    f. $Hg(CN)_2(s) \xrightarrow{heat}$

    g. $CaC_2(s) + H_2O(l) \longrightarrow$

    h. $Sb_2S_3(s) + O_2(g) \longrightarrow$

    i. $NH_4NO_3(s) \xrightarrow{heat}$

    j. $AsCl_3(l) + H_2O(l) \longrightarrow$

    k. $Sb(s) + O_2(g) \xrightarrow{heat}$

    l. $PCl_5(s) + Cl_2(g) \longrightarrow$

    m. $B(s) + N_2(g) \xrightarrow{heat}$

7. Predict the structure of $NO_3^-$. Draw the resonance forms of the ion and include the formal charges.

8. Given the following electrode potential diagram for common nitrogen compounds in acid solution, predict which compounds disproportionate:

$$NO_3^- \xrightarrow{+0.94\ V} HNO_2 \xrightarrow{+1.00\ V} NO \xrightarrow{+1.59\ V} N_2O \xrightarrow{+1.77\ V} N_2 \xrightarrow{+0.27\ V} NH_4^+$$

(+0.96 V above HNO₂→NO; +1.12 V below NO₃⁻→NO)

9. Draw the Lewis structure of each of the following compounds. Predict the shape of each.

    a. $BF_3$    b. $BF_4^-$

10. List two homopolar molecules that are isoelectronic with the cyanide ion.

11. Draw the Lewis structure of cyanogen.

12. Which element of group IV A has the highest melting point?

13. How many grams of water must be added to 1.00 g $P_4O_{10}$ to produce tetrametaphosphoric acid?

$$P_4O_{10}(s) + 2H_2O(l) \xrightarrow{0^{\circ}C} H_4P_4O_{10}(s)$$

14. Which of the following compounds is most basic?

    a. $PH_3$   b. $AsH_3$   c. $SbH_3$   d. $BiH_3$

15. Which of the following solids is the best electrical conductor?

    a. $W_2N$   b. $S_4N_4$   c. BN   d. AlN

16. Which of the following elements forms a three-center bond in some of its compounds?

    a. Te   b. As   c. Si   d. B

17. Which of the following would not oxidize when heated in air?

    a. CO   b. $P_4$   c. $N_2$   d. C

18. Which of the following compounds has the lowest boiling point?

    a. $NH_3$   b. $PH_3$   c. $AsH_3$   d. $SbH_3$

# ANSWERS TO EXERCISES

## I. Properties of nonmetals

1. [12.1]

| ELEMENT | COLOR | PHYSICAL STATE (STP) |
|---|---|---|
| nitrogen | colorless | gas consisting of $N_2$ molecules |
| phosphorus | white, red, black | solid consisting of $P_4$ (white) or $P_n$ (black) |
| arsenic | gray metallic, yellow | solid consisting of $As_4$ (yellow) or $As_n$ (gray metallic) |
| antimony | gray metallic, yellow | solid consisting of $Sb_4$ (yellow) or $Sb_n$ (gray metallic) |
| bismuth | gray metallic | solid consisting of $Bi_n$ |

2. [12.1]

a. P  : $1s^2\ 2s^2\ 2p^6\ 3s^2\ 3p^3$

b. $P^{3-}$ : $1s^2\ 2s^2\ 2p^6\ 3s^2\ 3p^6$

c. As  : $1s^2\ 2s^2\ 2p^6\ 3s^2\ 3p^6\ 3d^{10}\ 4s^2\ 4p^3$

d. $As^{3+}$: $1s^2\ 2s^2\ 2p^6\ 3s^2\ 3p^6\ 3d^{10}\ 4s^2$

e. $As^{5+}$: $1s^2\ 2s^2\ 2p^6\ 3s^2\ 3p^6\ 3d^{10}$

3. nitrogen [12.2]

4. [12.2]   After checking the answers, write the formulas of the compounds in the space provided:

| OXIDATION NUMBER OF NITROGEN | NAME | FORMULA |
|---|---|---|
| a. | +4 | dinitrogen tetraoxide | _____ |
| b. | +2 | mononitrogen monoxide, or nitric oxide | _____ |
| c. | +1 | dinitrogen monoxide, or nitrous oxide | _____ |
| d. | +5 | dinitrogen pentoxide | _____ |
| e. | +3 | dinitrogen trioxide | _____ |

5.    After checking the answers, write the name of the compound in the space provided:

a. $NH_3$ [12.4] _____     g. $H_5P_3O_{10}$ [12.8] _____

b. $Al_4C_3$ [12.11] _____     h. CaNCN [12.4] _____

c. $NH_2OH$ [12.4] _____     i. $P_4O_6$ [12.8] _____

d. $Sb_2S_3$ [12.2] _____     j. $N_2H_4$ [12.4] _____

e. $Ca_3N_2$ [12.3] _____     k. $N_4S_4$ [12.6] _____

f. $P_4O_{10}$ [12.8] _____     l. $PH_3$ [12.4] _____

m. NO [12.4] _____

6.

a. $PCl_3(l) + Cl_2(g) \longrightarrow PCl_5(s)$ [12.5]

b. $2NaN_3(s) \xrightarrow{heat} 2Na(l) + 3N_2(g)$ [12.2]

c. $CaCO_3(s) \xrightarrow{heat} CaCO(s) + CO_2(g)$ [12.12]

d. $4NH_3(g) + 5O_2(g) \xrightarrow[Pt]{heat} 4NO(g) + 6H_2O(g)$ [12.4]

e. $P_4O_{10}(s) + 6H_2O(l) \longrightarrow 4H_3PO_4(aq)$ [12.8]

f. $Hg(CN)_2(s) \xrightarrow{heat} Hg(l) + C_2N_2(g)$ [12.14]

g. $CaC_2(s) + 2H_2O(l) \longrightarrow Ca(OH)_2(aq) + C_2H_2(g)$ [12.11]

h. $2Sb_2S_3(s) + 9O_2(g) \xrightarrow{heat} Sb_4O_6(g) + 6SO_2(g)$ [12.2]

i. $NH_4NO_3(s) \xrightarrow{heat} N_2O(g) + 2H_2O(g)$ [12.7]

j. $AsCl_3(l) + 3H_2O(l) \longrightarrow H_3AsO_3(aq) + 3H^+(aq) + 3Cl^-(aq)$ [12.5]

k. $4Sb(s) + 3O_2(g) \xrightarrow{heat} Sb_4O_6(s)$ [12.8]

l. $PCl_5(s) + Cl_2(g) \longrightarrow NR$ [12.5]

m. $2B(s) + N_2(g) \xrightarrow{heat} 2BN(s)$ [12.17]

7. [12.7]    We predict the molecule to be triangular planar. The resonance structures are

8. [12.7]

NO$_3^-$   : stable

HNO$_2$ : disproportionates:

$$3HNO_2 \rightarrow 2NO + H_2O + NO_3^- + H^+ \qquad \mathscr{E}° = +0.06 \text{ V}$$

NO   : disproportionates

$$4NO + H_2O \rightarrow N_2O + 2HNO_2 \qquad \mathscr{E}° = +0.59 \text{ V}$$

Write another disproportionation reaction giving stable products.

N$_2$O   : disproportionates

$$5N_2O + H_2O \rightarrow 4N_2 + 2NO_3^- + 2H^+ \qquad \mathscr{E}° = +0.65 \text{ V}$$

N$_2$   : stable

NH$_4^+$ : stable

9. [12.17]

a.

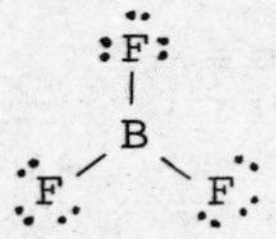

b.

triangular
planar

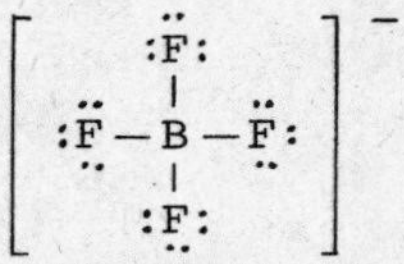

tetrahedral

10. N$_2$, CO

[12.14]

11. :N≡C—C≡N:

[12.14]

12. carbon

[12.9]

13. 0.127 g        We calculate the number of grams of water:

$$? \text{ g } H_2O = 1.00 \text{ g } P_4O_{10} \left( \frac{1 \text{ mol } P_4O_{10}}{283.9 \text{ g } P_4O_{10}} \right) \left( \frac{2 \text{ mol } H_2O}{1 \text{ mol } P_4O_{10}} \right) \left( \frac{18.02 \text{ g } H_2O}{1 \text{ mol } H_2O} \right)$$

$$= 0.127 \text{ g } H_2O$$

14. $PH_3$
    [12.4]

15. $W_2N$                  Ionic nitrides are better conductors than covalent
    [12.3]                  nitrides.

16. B
    [12.16]

17. $N_2$                   Air is composed of 21% oxygen and 78% nitrogen by
    [8.1]                   volume.  Oxygen and nitrogen do not react with each
                            other.

18. $PH_3$
    [8.9]

SELF-TEST          I. We suggest that you go through your lecture notes and
                      compile a list of chemical reactions, formulas, and
                      names of compounds.  Check your material to make sure
                      that you did not copy any material incorrectly and
                      then prepare your own self-test.  Write a series of
                      incomplete reactions with either the reactants or prod-
                      ucts given and a list of formulas without names or
                      vice versa.  The next day complete the reactions and
                      give either the name or formula of the compound with-
                      out referring to any material.  You may wish to ex-
                      change tests with your friends.

                      If your instructor has emphasized the necessity of
                      learning all chemical reactions in Chapter 12 of your
                      text, complete the following test in 10 minutes:

                      1. Draw the resonance structures of nitric acid.

                      2. Complete the following reactions:

                          a. $Sb_2S_3(s) + O_2(g) \rightarrow$
                          b. $CaO(s) + C(s) \rightarrow$
                          c. $Ca_3N_2(s) + H_2O(l) \rightarrow$
                          d. $As_4O_6(s) + C(s) \rightarrow$
                          e. $PBr_3(l) + H_2O(l) \rightarrow$
                          f. $PBr_3(l) + O_2(g) \rightarrow$
                          g. $PI_3(s) + I_2(s) \rightarrow$

3. Hypophosphorous acid is a

   a. monoprotic acid
   b. diprotic acid
   c. tripotic acid
   d. base

# 13

# Chemical Kinetics and Chemical Equilibrium

OBJECTIVES

(a) You should be able to demonstrate your knowledge of the following terms by defining them, describing them, or giving specific examples of them:

    activated complex [13.2]
    Arrhenius equation [13.6]
    catalyst [13.3]
    energy of activation [13.2]
    equilibrium [13.8]
    law of mass action [13.5]
    Le Chatelier's principle [13.10]
    order [13.5]
    rate constant [13.5]
    rate-determining step [13.7]

reaction intermediates [13.7]
reaction mechanism [13.1]
steady-state approximation [13.7]

(b) Given experimental data, you should be able to formulate the rate equation for a reaction.

(c) You should be able to evaluate the specific rate constant, $k$, and the energy of activation, $E_a$, for a reaction.

(d) You should understand the effect that a change in conditions such as temperature and concentration will have on the reaction rate, as well as the effect that the presence of a catalyst will have.

(e) You should be able to propose a reaction mechanism based on experimental data.

(f) Given a chemical equation, you should be able to write the expression of the equilibrium constant of the reaction.

(g) Given the concentrations of species at equilibrium, you should be able to evaluate the equilibrium constant, $K$, for a reaction.

(h) You should be able to use an equilibrium constant to determine the concentration of a species at equilibrium.

(i) You should be able to use Le Chatelier's principle to determine the effect that a change in conditions will have on an equilibrium system.

(j) Given the equilibrium constants for a reaction at two temperatures, you should be able to evaluate $\Delta H$ for that reaction.

(k) Given the values of $\Delta H$ and $K$ for a reaction at a temperature, $T_1$, you should be able to calculate the value of $K$ for that reaction at another temperature, $T_2$.

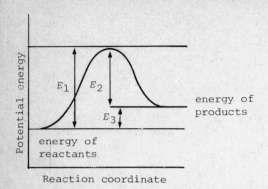

Figure 13.1 Energy diagram

**EXERCISES**

I. Answer each of the following with *true* or *false*. If a statement is false, correct it.

_____. 1. The reaction represented in the energy diagram of Figure 13.1 of the study guide is exothermic.

_____ 2. In the energy diagram of Figure 13.1 of the study guide, $E_1$ is the activation energy for the forward reaction.

_____ 3. In the energy diagram of Figure 13.1 of the study guide, $E_2$ is $\Delta E$ for the reaction.

_____ 4. A catalyst changes the path of a reaction.

_____ 5. A catalyst changes the position of equilibrium.

_____ 6. If the equilibrium constant for a reaction is large, the reaction will proceed rapidly.

_____ 7. If the specific rate constant for a reaction is large, the reaction will proceed rapidly.

_____ 8. Rate constants are independent of temperature.

_____ 9. Reaction intermediates appear in the net equation of a reaction.

10. For the reaction

$$3Fe(s) + 4H_2O(g) \rightarrow Fe_3O_4(s) + 4H_2(g)$$

the expression for the equilibrium constant is

$$K = \frac{[H_2]^4}{[H_2O]^4}$$

11. Equilibrium constants are not affected by pressure.

12. When a system reaches a state of dynamic equilibrium, the reaction stops.

II. Do the following:

1. For the reaction

$$A(g) + B(g) \rightarrow C(g)$$

the following data were obtained from three experiments:

| EXPERIMENT | [A] | [B] | RATE (FORMATION OF C) |
|---|---|---|---|
| 1 | 0.20M | 0.20M | $3.0 \times 10^{-4} M/min$ |
| 2 | 0.60M | 0.60M | $81.0 \times 10^{-4} M/min$ |
| 3 | 0.60M | 0.20M | $9.0 \times 10^{-4} M/min$ |

Use the data to answer the following:

a. What is the rate equation for the reaction?

b. What is the order of the reaction?

c. What is the numerical value of the specific rate constant?

d. If the volume of the reaction vessel were halved, what would be the effect on the reaction rate?

e. Is the reaction as written a one-step process?

2. Assume that the rate-determining step of a reaction is

$$2A(g) + B(g) \rightarrow C(g)$$

and that 2 mol of A and 1 mol of B are mixed in a liter container.  Compare the following to the initial reaction rate of the mixture:

   a. rate when half of A and half of B have been consumed.

   b. rate when two-thirds of A and two-thirds of B have been consumed

   c. initial reaction rate of a mixture of 2 mol of A and 2 mol of B in a 1 liter container.

   d. initial reaction rate of a mixture of 4 mol of A and 2 mol of B in a 1 liter container.

3. For the reaction

$$H_2(g) + I_2(g) \rightarrow 2 HI(g)$$

The slope of the plot of log $k$ vs. $\frac{1}{T}$ is $-8880^\circ K$. What is the energy of activation for the reaction?

4. For the reaction

$$A(g) + B(g) \rightarrow C(g)$$

The specific rate constants of the forward and backward reactions are 0.16 $M^{-1}$ sec$^{-1}$ and 4.0 x $10^4$ sec$^{-1}$ respectively.  If each reaction occurs in one step, what is the value of the equilibrium constant?

5. Write the expression for the equilibrium constant, $K$, for each of the following reactions:

   a. $H_2(g) + I_2(g) \rightleftharpoons 2HI(g)$

   b. $PCl_5(g) \rightleftharpoons PCl_3(g) + Cl_2(g)$

c. $N_2(g) + 3H_2(g) \rightleftharpoons 2NH_3(g)$

d. $C(s) + CO_2(g) \rightleftharpoons 2CO(g)$

e. $2NOBr(g) \rightleftharpoons 2NO(g) + Br_2(g)$

f. $2Cl_2(g) + 2H_2O(g) \rightleftharpoons 4HCl(g) + O_2(g)$

g. $2Hg(g) + O_2(g) \rightleftharpoons 2HgO(s)$

6. A 1.00 liter vessel initially contains 2.01 mol
   $N_2$ and 2.08 mol $H_2$ at 500°C. After the system
   reaches equilibrium, 0.50 mol $NH_3$ has formed.
   Answer the following:
   a. What are the equilibrium concentrations of
      $N_2$ and $H_2$?
   b. What is the equilibrium constant, $K$, at
      500°C?
   c. What is $K_p$ at 500°C?

7. For the equilibrium

   $A(g) + B(g) \rightleftharpoons C(g)$

   $K$ is 1.00 x $10^{-3}$ liter/mol at 100°C and 1.00 x
   $10^{-7}$ liter/mol at 200°C. Answer the following:
   a. Is the reaction endothermic or exothermic?
   b. What is the value of $\Delta H$ for the reaction?

8. At temperature $T_1$ and a total pressure of 1.00atm,
   $N_2O_4$ is 20.0% dissociated:

   $N_2O_4(g) \rightleftharpoons 2NO_2(g)$

   Assume that 1.00 mol of $N_2O_4$ is present initially
   and answer the following:
   a. How many moles of $N_2O_4(g)$ and $NO_2(g)$ are
      present at equilibrium?
   b. What is the total number of moles of gas
      present at equilibrium?

  c. What are the equilibrium pressures of
     $N_2O_4(g)$ and $NO_2(g)$?
  d. What is the value of $K_p$ at temperature $T_1$?

9. At $100^{\circ}C$ the equilibrium constant, $K$, for the
   reaction

   $$CO(g) + Cl_2(g) \rightleftharpoons COCl_2(g)$$

   is $4.57 \times 10^9$ liter/mol.  If 1.00 mol of $COCl_2$
   is placed in a 1.00 liter container, what is the
   concentration of CO after equilibrium has been
   established?(Note that the subtraction of a very
   small number from a large number may be neglected.)

10. The equilibrium constant, $K_p$, for the reaction

    $$C_4H_{10}(g) \rightleftharpoons 2H_2(g) + C_4H_6(g)$$

    is $1.0 \times 10^{-6}atm^2$ at $600^{\circ}C$.  If 2.0 mol $H_2$ and
    1.0 mol $C_4H_{10}$ are put into a 1.00 liter flask at
    $600^{\circ}C$, how many moles of $C_4H_6$ will be formed?

11. Consider the equilibrium

    $$2SO_2(g) + O_2(g) \rightleftharpoons 2SO_3(g) + heat$$

    What effect would each of the following stresses
    have on the reaction?
    a. addition of $SO_2$ at constant $V$ and $T$
    b. reduction of volume at constant temperature
    c. increase in temperature

12. At $400^{\circ}C$ equilibrium for the reaction

    $$27 \text{ kcal} + 2Cl_2(g) + 2H_2O(g) \rightleftharpoons 4HCl(g) + O_2(g)$$

    is established.  What would be the effect on the
    number of moles of $Cl_2(g)$ present if each of the
    following changes were made?
    a. increase of temperature to $600^{\circ}C$
    b. addition of $O_2(g)$

c. removal of $H_2O(g)$
d. reduction of the volume of the container
e. introduction of a catalyst

13   What is the concentration of HI at equilibrium at
357$^\circ$C if 6.22$M$ $H_2$ and 5.71$M$ $I_2$ are in a container
and $K$ for the reaction

$$H_2(g) + I_2(g) \rightleftharpoons 2HI(g)$$

is 71.3?

ANSWERS TO
EXERCISES
1. False
   [13.2]

I.   Principles of chemical kinetics and equilibrium

The reaction is endothermic.  The energy of the
products is higher than that of the reactants; thus,
the addition of energy is necessary for the reaction
to occur.

2. True
   [13.2]

3. False
   [13.2]

Since $E_2$ is the difference between the energy of the
products and that of the activated complex, it is
the activation energy for the backward reaction.
The $\Delta E$ for the forward reaction is $E_3$, the difference
between the energy of the products and that of the
reactants.

4. True
   [13.3]

5. False
   [13.3, 13.10]

A catalyst has no effect on the position of equilibrium since it affects the rate of the forward reaction and that of the backward reaction to an equal
extent.

6. False
   [13.8]

The equilibrium constant indicates nothing about how
a reaction will occur.  It only gives a ratio of
products to reactants.

7. True
   [13.5]

8. False
   [13.6]

The rate constant, $k$, varies with temperature, $T$, in a manner described by the Arrhenius equation

$$k = Ae^{-E_a/RT}$$

9. False
   [13.7]

Reaction intermediates are not final products; thus, they would not appear in the net equation.

10. True
    [13.9]

11. True
    [13.10]

12. False
    [13.8]

In a state of dynamic equlibrium the reaction continues, but the forward and reverse rates are equal.

II.   Chemical kinetics and equilibrium calculations

1. a. rate = $k[A][B]^2$
      [13.5]

Comparing experiments 1 and 3, we find

[B] = constant

[A] changes

$$\text{rate} = k[A]^x[B]^y$$

Since $k$ and, in this case, $[B]^y$ are constants, the rate is proportional to $[A]^x$. In experiment 3, [A] is 3 times greater than that in experiment 1. The rate in experiment 3 is also 3 times greater than that in experiment 1. The rate is directly proportional to the change in [A]; thus,

$$x = 1$$

and

$$\text{rate} = k[A][B]^y$$

Comparing experiments 2 and 3 we find

[A] = constant

[B] changes

$$\text{rate} = k[A]^x[B]^y$$

Since $k$ and, in this case, [A] are constants, the rate is proportional to $[B]^y$.  In experiment 2, [B] is 3 times greater than that in experiment 3.  The rate in experiment 2 is 9 times greater than that in experiment 3.  The rate is directly proportional to the square of the change in [B]; thus,

$$y = 2$$

and

$$rate = k[A][B]^2$$

b. 3
[13.5]

The reaction order is the sum of the exponents of the concentrations appearing in the equation.  Thus,

$$reaction\ order = sum\ of\ exponents$$

$$= 1 + 2$$

$$= 3$$

c. 0.038 $M^{-2}$ $min^{-1}$

We can use any set of data in the equation

$$rate = k[A][B]^2$$

Using the data from experiment 1, we find

$$k = \frac{rate}{[A][B]^2}$$

$$= \frac{3.0 \times 10^{-4} M/min}{(0.20M)(0.20M)^2}$$

$$= 3.8 \times 10^{-2}\ liter^2\ mol^{-2}\ min^{-1},$$

$$or\ 3.8 \times 10^{-2}\ M^{-2}\ min^{-1}$$

d. The rate would be 8 times greater than that of the original.
[13.5]

The rate equation is

$$rate = k[A][B]^2$$

If the volume of the reaction vessel were halved, all concentrations would be doubled.  For example, if the concentrations were originally 1 mol/1 liter, they would be 1 mol/0.5 liter, or $2M$, when the volume were halved.  Thus, the rate of the reaction when the volume were halved would be 8 times greater than that of the original:

$$\text{rate} = k[A][B]^2$$

$$= k(2M)(2M)^2$$

$$= k(8M^3)$$

e. no
[13.7]

If this were a one-step process, the reaction as written would be the rate-determining step and the exponents of the concentrations of the reactants in the rate equation would be the same as the coefficients in the chemical reaction:

$$\text{rate} = k[A]^1[B]^1$$

2. a. one-eighth of the original rate
[13.5, 13.7]

The rate-determining step is

$$2A(g) + B(g) \rightarrow C(g)$$

Thus,

$$\text{rate} = k[A]^2[B]$$

and initially

$$\text{rate} = k[A]^2[B]$$

$$= k(2M)^2(1M)$$

$$= k(4M^3)$$

When half of each reactant has been consumed,

$$\text{rate} = k[A]^2[B]$$

$$= k(1M)^2(0.5M)$$

$$= k(0.5M^3)$$

Thus, the rate when half of each reactant has been consumed is one-eighth of the initial rate.

b. eight twenty-sevenths of the initial rate [13.5]

When two-thirds of each reactant has been consumed,

$$\text{rate} = k\,[\text{A}]^2[\text{B}]$$

$$= k\,(4/3M)^2(2/3M)$$

$$= k\,(16/9M^2)(2/3M)$$

$$= k\,(32/27M^3)$$

$$= k\,(4M^3)(8/27)$$

Thus, the rate when two-thirds of reactant has been consumed is eight twenty-sevenths of the initial rate.

c. The initial rate of the mixture of 2 mol of A and 2 mol of B is twice as great as that of the mixture of 2 mol of A and 1 mol of B. [13.5]

The initial rate of a mixture of 2 mol of A and 2 mol of B in a 1 liter container is

$$\text{rate} = k\,[\text{A}]^2[\text{B}]$$

$$= k\,(2M)^2(2M)$$

$$= k\,(8M^3)$$

$$= k\,(4M^3)(2)$$

Thus, the initial rate of a mixture of 2 mol of A and 2 mol of B in a 1 liter container is 2 times as great as that of a mixture of 2 mol of A and 1 mol of B in a 1 liter container.

d. The initial rate of the mixture of 4 mol of A and 2 mol of B is 8 times as great as that of the mixture of 2 mol of A and 1 mol of B. [13.5]

The initial rate of a mixture of 4 mol of A and 2 mol of B in a 1 liter container is

$$\text{rate} = k\,[\text{A}]^2[\text{B}]$$

$$= k\,(4M)^2(2M)$$

$$= k\,(32M^3)$$

$$= k\,(4M^3)(8)$$

Thus, the initial rate of a mixture of 4 mol of A and 2 mol of B in a 1 liter container is 8 times as great as the initial rate of a mixture of 2 mol of A and 1 mol of B in a 1 liter container.

3. 40.6 kcal/mol
   [13.6]

The Arrhenius equation is

$$k = Ae^{-E_a/RT}$$

If we take the natural log of the Arrhenius equation, we find

$$\ln k = \ln A - \left(\frac{E_a}{RT}\right)$$

If we change the preceding equation to common logs, we find

$$2.303 \log k = 2.303 \log A - \left(\frac{E_a}{RT}\right)$$

$$\log k = \log A - \frac{E_a}{2.303RT}$$

$$\log k = \frac{-E_a}{2.303RT} + \log A$$

The preceding equation is the equation of a straight line, $y = mx + b$. The slope of a plot of $\log k$ vs. $\frac{1}{T}$ is $-E_a/2.303R$. Thus,

$$m = - \frac{E_a}{2.303R}$$

$$E_a = -2.303Rm$$

$$= (-2.303)(1.987 \text{ cal}/^\circ\text{K mol})(-8880^\circ\text{K})$$

$$= 40.6 \text{ kcal/mol}$$

4. 4.0 x $10^{-6}M^{-1}$
   [13.8]

The relationship between the equilibrium constant and the rates of the forward and backward reactions is

$$K = \frac{k_f}{k_b}$$

$$= \frac{0.16\text{M}^{-1}\text{sec}^{-1}}{4.0 \text{ x } 10^4\text{sec}^{-1}}$$

$$= 4.0 \text{ x } 10^{-6}M^{-1}$$

5. [13.9]

a. $K = \dfrac{[HI]^2}{[H_2][I_2]}$

b. $K = \dfrac{[PCl_3][Cl_2]}{[PCl_5]}$

c. $K = \dfrac{[NH_3]^2}{[N_2][H_2]^3}$

d. $K = \dfrac{[CO]^2}{[CO_2]}$

e. $K = \dfrac{[NO]^2[Br_2]}{[NOBr]^2}$

f. $K = \dfrac{[HCl]^4[O_2]}{[Cl_2]^2[H_2O]^2}$

g. $K = \dfrac{1}{[Hg]^2[O_2]}$

The exponent of each species is equal to the coefficient of that species in the equation of the chemical reaction.

The concentrations of solids at constant temperature are constant and are included in the value of $K$; therefore, they are not written in the expression for the equilibrium constant.

6. [13.9]

a. $[N_2] = 1.76M$

$[H_2] = 1.33M$

At equilibrium

$[NH_3] = 0.50$ mol/1.00 liter $= 0.50M$

The balanced equation

$N_2(g) + 3H_2(g) \rightleftharpoons 2NH_3(g)$

indicates that for every 2 mol of $NH_3$ formed, 1 mol of $N_2$ is lost. Thus, at equilibrium

$$[N_2] = \dfrac{2.01 \text{ mol } N_2 - (0.50 \text{ mol } NH_3 \text{ formed})\left(\dfrac{1 \text{ mol } N_2 \text{ lost}}{2 \text{ mol } NH_3 \text{ formed}}\right)}{1.00 \text{ liter soln.}}$$

$= 1.76M$

Similarly, for every 2 mol of $NH_3$ formed, 3 mol of $H_2$ is lost:

$$[H_2] = \frac{2.08 \text{ mol } H_2 - (0.50 \text{ mol } NH_3 \text{ formed}) \left( \dfrac{3 \text{ mol } H_2 \text{ lost}}{2 \text{ mol } NH_3 \text{ formed}} \right)}{1.00 \text{ liter soln}}$$

$$= 1.33M$$

b. $K =$
$6.0 \times 10^{-2} M^{-2}$

[13.9]

We substitute concentrations into the expression for the equilibrium constant:

$$K = \frac{[NH_3]^2}{[N_2][H_2]^3}$$

$$= \frac{(0.50M)^2}{(1.76M)(1.33M)^3}$$

$$= 6.0 \times 10^{-2} M^{-2}, \text{ or } 6.0 \times 10^{-2} \text{ liter}^2 \text{ mol}^{-2}$$

c. $K_p =$
$1.5 \times 10^{-5} \text{ atm}^{-2}$

The relationship between $K_p$ and $K$ is

$$K_p = K(RT)^{\Delta n}$$

For the reaction

$$\Delta n = -2$$

and

$$K_p = K(RT)^{\Delta n}$$

$$= (6.0 \times 10^{-2} \text{ liter}^2 \text{ mol}^{-2}) \left[ (0.0821 \text{ liter atm } {}^\circ K^{-1} \text{ mol}^{-1})(773 {}^\circ K) \right]^{-2}$$

$$= 1.5 \times 10^{-5} \text{ atm}^{-2}$$

7. [13.10]

a. exothermic

The expression for the equilibrium constant is

$$K = \frac{[C]}{[A][B]}$$

The equilibrium constant, $K$, decreases with increasing temperature, a fact that implies that the numerator of the equilibrium expression is becoming smaller. Thus, the equilibrium is shifting to the left. Since an increase in temperature shifts the euilibrium to the left, heat must appear on the right side; therefore, heat is released and the reaction is exothermic.

b. $-3.22 \times 10^{-4}$ cal mol$^{-1}$

Rearranging the equation

$$\log \frac{K_2}{K_1} = \frac{\Delta H}{2.303R} \left( \frac{T_2 - T_1}{T_1 T_2} \right)$$

and substituting values into it, we find

$$\Delta H. = 2.303R \left( \frac{T_1 T_2}{T_2 - T_1} \right) \log \left( \frac{K_2}{K_1} \right)$$

$$= 2.303(1.987 \text{ cal}^\circ\text{K}^{-1}\text{mol}^{-1}) \left( \frac{(373^\circ\text{K})(473^\circ\text{K})}{(473^\circ\text{K} - 373^\circ\text{K})} \right) \log \left( \frac{1.00 \times 10^{-7}}{1.00 \times 10^{-3}} \right)$$

$$= -3.22 \times 10^4 \text{ cal mol}^{-1}$$

8. a. 0.80 mol $N_2O_4$

0.40 mol $NO_2$

[13.10]

The $N_2O_4$ is 20.0% dissociated; therefore, 0.80 mol of $N_2O_4$ remains:

? mol $N_2O_4$ remaining = 1.00 mol $N_2O_4$ - 0.200(1.00) mol $N_2O_4$

$$= 0.80 \text{ mol } N_2O_4$$

Since 0.20 mol of $N_2O_4$ is dissociated, we know from the balanced equation that 0.40 mol of $NO_2$ is formed:

? mol $NO_2$ formed = 0.20 mol $N_2O_4$ lost $\left( \dfrac{2 \text{ mol } NO_2 \text{ formed}}{1 \text{ mol } N_2O_4 \text{ lost}} \right)$

$$= 0.40 \text{ mol } NO_2$$

b. 1.20 mol gas     We calculate the total number of moles of gas present at equilibrium:

$$\text{moles of gas} = 0.80 \text{ mol gas} + 0.40 \text{ mol gas}$$

$$= 1.20 \text{ mol gas}$$

c. $p_{N_2O_4} =$

    0.667 atm

$p_{NO_2} =$

    0.333 atm

We determine the partial pressure of each gas:

$$p_{N_2O_4} = \left( \frac{n_{N_2O_4}}{n_{N_2O_4} + n_{NO_2}} \right) P$$

$$= \left( \frac{0.80 \text{ mol}}{0.80 \text{ mol} + 0.40 \text{ mol}} \right) 1.00 \text{ atm}$$

$$= 0.667 \text{ atm}$$

$$p_{NO_2} = \left( \frac{n_{NO_2}}{n_{N_2O_4} + n_{NO_2}} \right) P$$

$$= \left( \frac{0.40 \text{ atm}}{0.80 \text{ atm} + 0.40 \text{ atm}} \right) 1.00 \text{ atm}$$

$$= 0.333 \text{ atm}$$

d. $1.66 \times 10^{-1}$ atm   We substitute partial pressures into the expression for $K_P$:

$$K_P = \frac{(p_{NO_2})^2}{p_{N_2O_4}}$$

$$= \frac{(0.333 \text{ atm})^2}{(0.667 \text{ atm})}$$

$$= 0.166 \text{ atm}$$

9. [CO]= $1.48 \times 10^{-5} M$   At 100 °C the equilibrium constant is
   [13.10]

$$K = 4.57 \times 10^9 \text{ liter/mol} = \frac{[COCl_2]}{[CO] [Cl_2]}$$

We make the following assumptions:

| SPECIES | ORIGINAL CONCENTRATION | EQUILIBRIUM CONCENTRATION |
|---|---|---|
| CO | O | $(x)M$ |
| $Cl_2$ | O | $(x)M$ |
| $COCl_2$ | $1.00M$ | $(1.00 - x = 1.00)M$ |

We substitute concentrations into the expression for the equilibrium constant:

$$K = 4.57 \times 10^9 \text{ liter/mol} = \frac{[COCl_2]}{[CO][Cl_2]}$$

$$4.57 \times 10^9 \text{ liter/mol} = \frac{1.00M}{x^2}$$

$$x^2 = 2.19 \times 10^{-10}M^2$$

$$x = 1.48 \times 10^{-5}M$$

Note that $1.48 \times 10^{-5}M$ is negligible compared to $1.00M$.

10. $4.8 \times 10^{-11}$ mol $C_4H_6$ [13.9]

The change in the number of moles, $\Delta n$, is +2. Therefore,

$$K = K_P(RT)^{-\Delta n}$$

$$= 1.0 \times 10^{-6} \text{ atm}^2 \left[(0.0821 \text{ liter atm } ^{\circ}K^{-1} \text{ mol}^{-1})(873^{\circ}K)\right]^{-2}$$

$$= 1.9 \times 10^{-10}M^2$$

The equilibrium expression is

$$K = \frac{[H_2]^2[C_4H_6]}{[C_4H_{10}]}$$

From the reaction equation we know that for every 1 mol of $C_4H_6$ formed, 2 mol of $H_2$ is formed. Also, for every 1 mol of $C_4H_6$ formed, 1 mol of $C_4H_{10}$ is lost. Therefore, we make the following assumptions:

| SPECIES | ORIGINAL CONCENTRATION | EQUILIBRIUM CONCENTRATION |
|---|---|---|
| $C_4H_6$ | 0 | $(x)M$ |
| $H_2$ | $2.0M$ | $(2.0 + 2x = 2.0)M$ |
| $C_4H_{10}$ | $1.0M$ | $(1.0 - x = 1.0)M$ |

We substitute concentrations into the expression for the equilibrium constant:

$$K = 1.9 \times 10^{-10}M^2 = \frac{(2.0M)^2(x)}{(1.0M)}$$

$$x = \frac{1.9 \times 10^{-10}M^2}{4.0M}$$

$$x = 4.8 \times 10^{-11}M$$

Note that $4.8 \times 10^{-11}M$ is negligible compared to $1.0M$ and $9.6 \times 10^{-11}M$ is negligible compared to $2.0M$.

11. [13.10]

a. shift to right   A shift to the right occurs since the stress is on the left.

b. shift to right   A shift to the right occurs since the stress is on the left. The two moles of gas on the right would occupy less volume than the three moles of gas on the left.

c. shift to left   A shift to the left occurs since the stress is on the right. The reaction is exothermic; thus, the shift is toward the left when the temperature is increased.

12.  [13.10]

    a. decrease          The number of moles of $Cl_2$ would decrease.  The reaction is endothermic, and the increase in temperature would shift the reaction to the side without heat.

    b. increase          The number of moles of $Cl_2$ would increase.  An increase in the number of moles of $O_2$ would shift the reaction away from the side with $O_2$.

    c. increase          The number of moles of $Cl_2$ would increase.  The removal of $H_2O$ would cause a shift to the left.

    d. increase          The number of moles of $Cl_2$ would increase.  There are fewer moles of gas on the left side than on the right.  Thus, to alleviate the stress of volume reduction, the equilibrium would shift to the side that contains fewer molecules since such molecules collectively would occupy less volume.

    e. no change        There would be no change in the number of moles of $Cl_2$ since a catalyst would have no effect on the position of equilibrium.

13.  [HI] = 9.55$M$     The expression for the equilibrium constant is
     [13.10]

$$K = 71.3 = \frac{[HI]^2}{[H_2][I_2]}$$

From the balanced equation we know that for every 2 mol of HI formed, 1 mol of $H_2$ and 1 mol of $I_2$ are lost, or for every 1 mol of HI formed, $\frac{1}{2}$ mol of $H_2$ and $\frac{1}{2}$ mol of $I_2$ are lost.  Therefore, we make the following assumptions:

| SPECIES | ORIGINAL CONCENTRATION | EQUILIBRIUM CONCENTRATION |
|---------|------------------------|---------------------------|
| HI | 0 | $(x)M$ |
| $H_2$ | 6.22$M$ | $(6.22 - \frac{x}{2})M$ |
| $I_2$ | 5.71$M$ | $(5.71 - \frac{x}{2})M$ |

We substitute concentrations into the expression for the equilibrium constant and use the quadratic equation

$$x = \frac{-b \pm \sqrt{b^2 - 4ac}}{2a}$$

to solve for $x$:

$$K = 71.3 = \frac{[HI]^2}{[H_2][I_2]}$$

$$= \frac{x^2}{(6.22 - \frac{x}{2})(5.71 - \frac{x}{2})}$$

$$= \frac{x^2}{35.5 - 2.86x - 3.11x + \frac{x^2}{4}}$$

$$x^2 = 2531 - 426x + 17.8x^2$$

$$0 = 16.8x^2 - 426x + 2531$$

$$x = \frac{-(-426) \pm \sqrt{(-426)^2 - 4(16.8)(2531)}}{2(16.8)}$$

$$= 9.55 \text{ or } 15.8$$

Since the concentration of $H_2$ and that of $I_2$ originally present are smaller than $\frac{x}{2}$ when $x = 15.8$, the solution is $x = 9.55$. Therefore,

$$HI = 9.55M$$

SELF-TEST

Complete the test in 45 minutes.

Answer each of the following:

_____

1. The chemical equation for the reaction between oxalate ion and mercuric chloride in aqueous solution is

$$C_2O_4^{2-}(aq) + 2HgCl_2(aq) \rightarrow 2CO_2(g) + Hg_2Cl_2(s) + 2Cl^-(aq)$$

At temperature $T$ the following kinetic data were obtained:

| EXPERIMENT | [$HgCl_2$] | [$C_2O_4^{2-}$] | RATE (FORMATION OF $CO_2$) |
|---|---|---|---|
| 1 | 0.080$M$ | 0.200$M$ | $1.5 \times 10^{-5} M\ min^{-1}$ |
| 2 | 0.080$M$ | 0.400$M$ | $6.0 \times 10^{-5} M\ min^{-1}$ |
| 3 | 0.040$M$ | 0.400$M$ | $3.0 \times 10^{-5} M\ min^{-1}$ |

Calculate the kinetic order for $HgCl_2$, the kinetic order for $C_2O_4^{2-}$, and the experimental rate constant with the appropriate units.

2. The bromination of acetone in aqueous solution is given by the equation

$$H_3C - \overset{O}{\underset{\|}{C}} - CH_3(aq) + Br_2(aq) \rightarrow H_3C - \overset{O}{\underset{\|}{C}} - CH_2Br(aq) + HBr(aq)$$

acetone

and has an experimental rate law

rate = $k[\text{acetone}]^1$

Answer each of the following with *increases*, *decreases*, or *remains the same*:

a. As the concentration of acetone is increased, the experimental rate constant _____.
b. As the concentration of acetone is increased, the rate _____.
c. As the concentration of $Br_2$ is increased, the rate _____.
d. If a suitable positive catalyst is introduced, the overall activation energy for the reaction _____.
e. If the temperature is increased, the rate constant for the rate-determining step _____.

3. The reaction of gaseous chlorine and water

$$2Cl_2(g) + 2H_2O(g) \rightleftharpoons 4HCl(g) + O_2(g)$$

is endothermic. The four reactant gases are
mixed in a reaction vessel and allowed to attain
an equilibrium state. Indicate the effect (i.e.,
*increase, decrease,* or *no change*) of the following operations described in the left column on
the equilibrium value of the quantity in the
right column. Each operation is to be considered
separately. Temperature and volume are constant
except when the contrary is indicated.

|               | OPERATION                          | QUANTITY                     |
|---------------|------------------------------------|------------------------------|
| _____   | a. increase in volume of container | number of moles of HCl       |
| _____   | b. increase in volume of container | $K_P$                        |
| _____   | c. addition of $O_2(g)$            | $K_P$                        |
| _____   | d. addition of $O_2(g)$            | number of moles of $O_2$     |
| _____   | e. increase in temperature         | $K_P$                        |

4. The decomposition of $CaSO_3(s)$

$$CaSO_3(s) \rightleftharpoons CaO(s) + SO_2(g) \qquad .$$

is endothermic. At equilibrium there are significant quantities of all three compounds in the
reaction vessel. Suggest two ways in which the
yield of $SO_2(g)$ might be appreciably increased.

5. At $269.9^\circ K$ nitric oxide and bromine are mixed in
a vessel. Initially the partial pressures are

$$p_{NO} = 0.129 \text{ atm}$$

$p_{Br_2} = 0.0543$ atm

The following equilibrium is established:

$$2NO(g) + Br_2(g) \rightleftharpoons 2NOBr(g)$$

The total pressure at equilibrium is 0.145 atm.
Calculate $K_p$.

6. At temperature $T$ 1.00 mol of NOCl is sealed in a
   1.00 liter container.  At equilibrium 20.0% of
   the NOCl is dissociated:

$$2NOCl(g) \rightleftharpoons 2NO(g) + Cl_2(g)$$

What is the value of the equilibrium constant?
Use molar concentrations.

# 14

# Elements
# of Chemical
# Thermodynamics

OBJECTIVES

(a) You should be able to demonstrate your knowledge of
the following terms by defining them, describing
them, or giving specific examples of them:

   bond energy [14.3]
   enthalpy [14.2]
   entropy [14.4]
   first law of thermodynamics [14.1]
   Gibbs free energy [14.5]
   heat of atomization, also called bond
      dissociation energy [14.3]
   internal energy [14.1]
   second law of thermodynamics [14.4]

state function [14.1]
third law of thermodynamics [14.8]

(b) You should be able to calculate $\Delta H$ for a reaction
from heats of combustion, enthalpies of formation,
bomb calorimeter data, and heats of atomization.

(c) You should be able to calculate $K$, $\Delta G$, and $\Delta S$ of an
electrochemical cell reaction from the values of $\Delta H$,
$\mathscr{E}$, and $T$.

(d) Using tabulated values of the Gibbs free energy of
formation, $\Delta G_f^0$, you should be able to calculate $\Delta G^0$
of a reaction.

**EXERCISES**

I. Answer each of the following with *true* or *false*. If
the statement is false, correct it.

_____

1. Consider a 1 ml sample of pure water at 25°C and
   1 atm (state A). The sample is cooled to 1°C
   and then the pressure is reduced to 0.1 atm
   (state B). It takes 5 hours to carry out the
   change from state A to state B. The sample is
   then heated and the pressure is raised to 1 atm.
   In 20 sec the water is at 25°C (a return to
   state A). The internal energy change in going
   from A to B is equal to, but opposite in sign
   from, the change from B to A.

_____

2. The work done in changing from A to B and that
   done in changing from B to A in statement 1 of
   this section are numerically the same but
   opposite in sign.

_____

3. At constant pressure the amount of heat absorbed
   or evolved by a system is called the enthalpy
   change, $\Delta H$.

_____

4. If volume does not change, the amount of heat
   released during a change of state of a system is
   equal to the decrease in internal energy of that
   system.

_____

5. If $\Delta S$ for a change in a system is positive, the
   change is said to be thermodynamically spontaneous.

_____ 6. The more positive the value of $\Delta S_{total}$ for a change, the more rapid the change.

_____ 7. For a spontaneous change the Gibbs free energy change is less than zero.

_____ 8. As the reversible emf of a cell reaction becomes more negative, $\Delta G^O$ becomes more negative.

_____ 9. The standard free energy of formation of any element in its standard state is zero.

_____ 10. The value of $\Delta G$ of a reaction is independent of pressure.

_____ 11. The reaction of acetylene, $H-C \equiv C-H$, with oxygen is spontaneous.

_____ 12. The value of Gibbs free energy of formation is positive for every compound.

_____ 13. The entropy of a mole of solid sodium is less than that of a mole of gaseous sodium.

_____ 14. More heat is released to the surroundings when a mole of $H_2$ is burned at constant pressure (an open flame) than when it is burned at constant volume (a bomb calorimeter).

II. Complete each of the following statements with an entry from the list on the right. An entry may be used more than once.

1. The energy of the universe is constant, but the _____ tends toward a maximum.

2. The _____ of a reaction can be measured indirectly with a bomb calorimeter.

3. The SI unit of pressure is the _____ which equals 1 N/m$^2$.

4. The amount of work done by a process carried out at constant volume is _____.

5. By convention, in the equation describing the change in internal

a. enthalpy
b. entropy
c. zero
d. negative
e. positive
f. gram
g. atmosphere
h. newton
i. pascal
j. state
k. Gibbs free energy

energy of a system, $\Delta E = q - w$,
work done by the system is always
_____.

6. Entropy, enthalpy, and internal
   energy are _____ functions.

7. For any thermodynamically spontaneous change the total change in
   entropy must be a _____
   value.

8. For any thermodynamically spontaneous change the Gibbs free energy
   must be a _____ value.

9. For a system at equilibrium $\Delta G^\circ$
   equals _____.

10. The thermodynamic function that
    does not necessitate a considera-
    tion of the changes in surroundings
    is _____.

11. The $\Delta S$ of dissolution of NaCl in
    $H_2O$ is a _____ value.

III. Answer each of the following:

1. Use the information in Table 14.1 of your text
   and the heat of combustion of $CH_4$ at 25°C and
   1 atm, $\Delta H = -192.0$ kcal/mol [or 803.3 kJ/mol],
   to calculate $\Delta H$ for the following reaction at
   25°C and 1 atm:

   $$CO_2(g) + 2H_2(g) \rightarrow 2H_2O(g) + C(s)$$

2. What is $\Delta E$ for the reaction described in problem
   1 of this section?

3. An unknown reaction is carried out in a bomb
   calorimeter containing only carbon and water.
   The heat of reaction at 25°C is found to be 31.4
   kcal/1 mol C [or 131.4 kJ/1 mol C]. Which of the
   following reactions occurred?

   $$C(s) + 2H_2O(g) \rightarrow CO_2(g) + 2H_2(g)$$

   $$C(s) + H_2O(g) \rightarrow CO(g) + H_2(g)$$

4. Use the data in Table 14.3 of your text to determine the value of $\Delta H$ for each reaction in problem 3 of this section. The heat of atomization of C(s) is 172 kcal/mol [or 720 kJ/mol] and the C≡O bond energy is 257 kcal/mol [or 1080 kJ/mol]. Compare values calculated from average bond energies with those calculated from values of $\Delta H_f^0$ (see problem 3 of this section) and offer an explanation of any differences observed.

5. Determine the C=O bond energy from values given in your text and the study guide.

6. Use values of $\Delta G_f^0$ to determine $\Delta G$ for the oxidation of aluminum at 25°C and 1 atm:

$$4Al(s) + 3O_2(g) \rightarrow 2Al_2O_3(s)$$

7. Calculate $\mathscr{E}°$ for the reaction

$$2Al(s) + 3/2\ O_2(g) \rightarrow Al_2O_3(s)$$

8. Calculate $K_p$ for the reaction

$$C(s) + H_2O(g) \rightarrow CO(g) + H_2(g)$$

9. Use the data in Table 14.3 of your text to determine the standard heat of formation of $CF_4(g)$. The heat of atomization of C(s) is 172 kcal/mol [or 720 kJ/mol].

10. Determine the average C—H bond energy from standard heat of formation data in Table 14.1 of your text. The heat of atomization of C(s) is 172 kcal/mol [or 720 kJ/mol].

11. Use the results of the calculation in problem 10 of this section and standard heat of formation data in Table 14.1 of your text to predict the C—C bond energy in ethane,

```
     H  H
     |  |
 H—C—C—H
     |  |
     H  H
```

Compare the calculated value with the C—C bond energy given in Table 14.3 of your text.

12. Use the values of $\Delta H_f^0$ and $\Delta G_f^0$ given in Tables 14.1 and 14.3 of your text to determine the

change in entropy for the following reaction at 25°C and 1 atm:

$$CH_4(g) + 2O_2(g) \rightarrow CO_2(g) + 2H_2O(g)$$

13. Compare the value of $\Delta S$ calculated in problem 12 of this section with the value computed from the data in Table 14.7 of your text.

14. Calculate the equilibrium constant for the reaction

$$H_2O \rightleftharpoons H^+(aq) + OH^-(aq)$$

from the following electrode potentials:

$$2H_2O + 2e^- \rightleftharpoons H_2(g) + 2OH^-(aq) \qquad \mathscr{E}° = -0.828 \text{ V}$$

$$2H^+(aq) + 2e^- \rightleftharpoons H_2(g) \qquad \mathscr{E}° = 0.000 \text{ V}$$

15. When 1.46 g of adipic acid, $C_6H_{10}O_4$, is burned at 25°C and constant pressure, the products are $CO_2(g)$ and $H_2O(l)$. If 6.690 kcal [or 27.991 kJ] is released in the reaction, what is $\Delta H$ for the reaction of one mole of adipic acid?

| | |
|---|---|
| **ANSWERS TO EXERCISES** | I. Concepts of chemical thermodynamics |
| 1. True [14.1] | Internal energy is a state function. The difference in the internal energies of the two states is independent of the path taken between states. |
| 2. False [14.1] | Work depends on the path, i.e., the method of going from one state to another. |
| 3. True [14.2] | The relationship between $\Delta H$ and $\Delta E$ is $$\Delta H = \Delta E + P\Delta V$$ At constant volume the amount of heat absorbed or evolved by a system is the internal energy change, $\Delta E$. |
| 4. True [14.2] | At constant volume $$\Delta H = \Delta E$$ |

and

$$q_V = q_p$$

5. False
   [14.4]

For a spontaneous change $\Delta S_{total}$ is positive. Since $\Delta S_{total} = \Delta S_{system} + \Delta S_{surroundings}$, it is possible to have a negative value of $\Delta S_{system}$ and still have a thermodynamically spontaneous reaction.

6. False
   [14.4]

A positive value of $\Delta S_{total}$ indicates that the change is thermodynamically favorable. The change, however, may not occur at an observable rate.

7. True
   [14.5]

8. False
   [14.9]

The relationship between $\Delta G^O$ and $\mathscr{E}^\circ$ is

$$\Delta G^O = -nF\,\mathscr{E}^\circ$$

The value of $\Delta G^O$ becomes more positive as $\mathscr{E}^\circ$ becomes more negative.

9. True
   [14.7]

10. False
    [14.6]

The values of $\Delta G$ and $\Delta S$ and the amount of work done by the system are not independent of pressure.

11. True
    [14.7]

The reaction is exothermic, as evidenced by the oxyacetylene flames used in welding.

12. False
    [14.7]

Values of $\Delta G_f$ can be positive or negative.

13. True
    [14.4]

The solid is more ordered than the gas.

14. True
    [14.2]

At constant volume the heat released is

$$\Delta E = q_V$$

No pressure-volume work is done. At constant pressure

$$q_p = \Delta E + P\Delta V$$

Since the $\Delta V$ for the reaction is negative, the heat

released to the surroundings by the burning of $H_2$ is more at constant pressure than at constant volume:

$$H_2(g) + \tfrac{1}{2}O_2(g) \rightarrow H_2O(g)$$

## II. Conventions and terms of thermodynamics

1. entropy [14.4]
2. enthalpy [14.2]
3. pascal [14.2]
4. zero [14.2]
5. positive [14.1]
6. state [14.4]
7. positive [14.4]
8. negative [14.5]
9. zero [14.5]
10. Gibbs free energy [14.5]
11. positive [14.4]

## III. Chemical thermodynamic calculations

1. $\Delta H = -21.3$ kcal [or -89.1 kJ] [14.2]

We add the appropriate equations and the corresponding enthalpy values:

$$CH_4(g) \rightarrow C(s) + 2H_2(g) \qquad \Delta H = 17.89 \text{ kcal}$$

$$CO_2(g) + 2H_2O(g) \rightarrow 2O_2(g) + CH_4(g) \qquad \Delta H = 192.0 \text{ kcal}$$

$$4H_2(g) + 2O_2(g) \rightarrow 4H_2O(g) \qquad \Delta H = 4(-57.80) \text{ kcal}$$

$$\overline{CO_2(g) + 2H_2(g) \rightarrow 2H_2O(g) + C(s)} \qquad \overline{\Delta H = -21.3 \text{ kcal}}$$

2. $\Delta E = -20.7$ kcal [or -86.6 kJ] [14.2]

The relationship between $\Delta E$ and $\Delta H$ is

$$\Delta E = \Delta H - \Delta n \; RT$$

Substituting values into the preceding equation, we find

$$\Delta E = \Delta H - \Delta n RT$$

$$= -21.3 \text{ kcal} - \frac{(-1 \text{ mol})(1.987 \text{ cal}/^\circ\text{K mol})(298^\circ\text{K})}{1000 \text{ cal}/1 \text{ kcal}}$$

$$= -20.7 \text{ kcal}$$

3. $C(s) + 2H_2O(g) \rightarrow$
   $CO_2(g) + 2H_2(g)$
   [14.2]

We use information in Table 14.1 of your text to calculate the theoretical value of $\Delta H$ for each reaction and the information in the problem to calculate the experimental value of $\Delta H$ for each reaction:

For the reaction

$$C(s) + 2H_2O(g) \rightarrow CO_2(g) + 2H_2(g)$$

We first determine the value of $\Delta H$ from values of $\Delta H_f^o$:

$$\Delta H = \Delta H_f^o(CO_2) - 2\Delta H_f^o(H_2O)$$

$$= -94.05 \text{ kcal} - 2(-57.80 \text{ kcal})$$

$$= 21.55 \text{ kcal}$$

Then we determine the value of $\Delta H$ from the experimental data:

$$\Delta H = \Delta E + \Delta nRT$$

$$= 31.4 \text{ kcal} + (1 \text{ mol}) \left( \frac{(1.987 \text{ kcal/}^oK \text{ mol})(298^oK)}{(1000 \text{ cal/1 kcal})} \right)$$

$$= 32.0 \text{ kcal}$$

For the reaction

$$C(s) + H_2O(g) \rightarrow CO(g) + H_2(g)$$

we determine the value of $\Delta H$ from values of $\Delta H_f^o$:

$$\Delta H = \Delta H_f^o(CO) - \Delta H_f^o(H_2O)$$

$$= (-26.42 \text{ kcal}) - (-57.80 \text{ kcal})$$

$$= 31.38 \text{ kcal}$$

We then determine the value of $\Delta H$ from experimental data:

$$\Delta H = \Delta E + \Delta nRT$$

$$= 31.4 \text{ kcal} + (1 \text{ mol}) \left( \frac{(1.987 \text{ kcal/}^oK \text{ mol})(298^oK)}{(1000 \text{ cal/1 kcal})} \right)$$

$$= 32.0 \text{ kcal}$$

The values of $\Delta H$ determined from values of $\Delta H_f^o$ and from experimental data for the reaction

$$C(s) + 2H_2O(g) \rightarrow CO_2(g) + 2H_2(g)$$

are in closer agreement than those values of $\Delta H$ determined for the reaction

$$C(s) + H_2O(g) \rightarrow CO(g) + H_2(g)$$

Therefore, the water and carbon react in the calorimeter to form carbon dioxide and hydrogen.

4. 33 kcal
   [or 138 kJ]
   [14.3]

We add the appropriate equations and the corresponding enthalpy values.  For the reaction

$$C(s) + 2H_2O(g) \rightarrow CO_2(g) + 2H_2(g)$$

we find

| | |
|---|---|
| $2H_2O(g) \rightarrow 4H(g) + 2O(g)$ | $\Delta H = 4(111)$ kcal |
| $C(s) \rightarrow C(g)$ | $\Delta H = 172$ kcal |
| $C(g) + 2O(g) \rightarrow \ddot{O}=C=\ddot{O}(g)$ | $\Delta H = 2(-169)$ kcal |
| $4H(g) \rightarrow 2H_2(g)$ | $\Delta H = 2(-104)$ kcal |
| $C(s) + 2H_2O(g) \rightarrow CO_2(g) + 2H_2(g)$ | $\Delta H = +70$ kcal |

Note that 4 O—H bonds are broken, and two C=O and two H—H bonds are formed.

The value of $\Delta H$ determined from average bond energies is higher by 45 kcal than that determined from values of $\Delta H_f^o$.  The discrepancy probably occurs because the breaking of the second C=O bond in $CO_2$ requires more energy than the breaking of the first.  Thus, the average bond energy value of C=O in Table 14.3 of your text is not the average bond energy of the C=O bonds in $CO_2$.

For the reaction

$$C(s) + H_2O(g) \rightarrow CO(g) + H_2(g)$$

we find

$$H_2O(g) \rightarrow 2H(g) + O(g) \qquad \Delta H = 2(111) \text{ kcal}$$

$$C(s) \rightarrow C(g) \qquad \Delta H = 172 \text{ kcal}$$

$$C(g) + O(g) \rightarrow CO(g) \qquad \Delta H = -257 \text{ kcal}$$

$$2H(g) \rightarrow H_2(g) \qquad \Delta H = -104 \text{ kcal}$$

$$\overline{C(s) + H_2O(g) \rightarrow CO(g) + H_2(g)} \qquad \overline{\Delta H = 33 \text{ kcal}}$$

The value of $\Delta H$ calculated from average bond energies is higher by 2 than that calculated from values of $\Delta H_f^O$.

5. +192 kcal/mol
[or 803 kJ/mol]
[14.3]

The enthalpy of formation of a C$=$O bond in $CO_2$ is one-half the enthalpy of formation of one mole of $CO_2$(g), i.e., one-half the enthalpy of the reaction:

$$C(g) + 2O(g) \rightarrow CO_2(g)$$

We can determine $\Delta H_f$ of $CO_2$(g) as follows:

$$C(s) + O_2(g) \rightarrow CO_2(g) \qquad \Delta H_f^O = -94.05 \text{ kcal}$$

$$C(g) \rightarrow C(s) \qquad \Delta H_{fusion} = -172 \text{ kcal}$$

$$2O(g) \rightarrow O_2(g) \qquad \Delta H = -118 \text{ kcal}$$

$$\overline{C(g) + 2O(g) \rightarrow CO_2(g)} \qquad \overline{\Delta H = -384 \text{ kcal}}$$

Since there are 2 C$=$O bonds in $CO_2$, the average bond energy of C$=$O in $CO_2$ is

$$\frac{+384 \text{ kcal/mol}}{2} = +192 \text{ kcal/mol}$$

Note the difference in this value and the one given in Table 14.3 of your text.

6. $\Delta G = -753.54$ kcal
[or -3152.8 kJ]
[14.7]

The equation

$$4Al(s) + 3O_2(g) \rightarrow 2Al_2O_3(s)$$

represents the formation of 2 mol of $Al_2O_3$ at 25°C and 1 atm. From Table 14.6 of your text we note

$$2Al(s) + 3/2\ O_2(g) \rightarrow Al_2O_3(s) \quad \Delta G_f^O = -376.77 \text{ kcal}$$

Thus,

$$4Al(s) + 3O_2(g) \rightarrow 2Al_2O_3(s) \qquad \Delta G_f^O = -753.54 \text{ kcal}$$

7. 2.7230 **V**
   [14.7]

The relationship between $\Delta G^O$ and $\mathscr{E}°$ is $\Delta G^O = -nF\mathscr{E}°$ Rearranging the preceding equation and substituting values into it, we find

$$\mathscr{E}° = -\frac{\Delta G^O}{nF}$$

$$= -\frac{(-753.54 \text{ kcal})}{(12)(23.061 \text{ kcal/V})}$$

$$= 2.7230 \text{ v}$$

8. $K_p = 10^{-16}$
   [14.9]

Since

$$\Delta G^O = -2.303\ RT \log K_p$$

we must determine $\Delta G^O$ for the reaction:

$$\Delta G^O = \Delta G_f^O(CO) - \Delta G_f^O(H_2O)$$

$$= -32.81 \text{ kcal} - (-54.64 \text{ kcal})$$

$$= 21.83 \text{ kcal}$$

Rearranging

$$\Delta G^O = -2.303\ RT \log K_p$$

and substituting values into it, we find

$$\log K_p = -\frac{\Delta G^O}{2.303\ RT}$$

$$= -\frac{21.83 \text{ kcal/mol}}{(2.303)(1.987 \times 10^{-3} \text{ kcal/°K mol})(298°K)}$$

$$= -16.0$$

and

$$K_p = 10^{-16}$$

9. −218 kcal
   [or −912 kJ]
   [14.3]

We add the appropriate equations and corresponding values of $\Delta H$:

$$C(s) \rightarrow C(g) \qquad \Delta H = 172 \text{ kcal}$$

$$2F_2(g) \rightarrow 4F(g) \qquad \Delta H = 2(37) \text{ kcal}$$

$$C(g) + 4F(g) \rightarrow CF_4(g) \qquad \Delta H = 4(-116) \text{ kcal}$$

$$\overline{C(s) + 2F_2(g) \rightarrow CF_4(g)} \qquad \overline{\Delta H = -218 \text{ kcal}}$$

The value of $\Delta H_f^O$ of one mole of $CF_4(g)$ from Table 14.1 of your text is -218.3 kcal.

10. +99.5 kcal/mol
    [or 416 kJ/mol]
    [14.3]

We add the appropriate equations and corresponding values of $\Delta H$:

$$C(s) + 2H_2(g) \rightarrow CH_4(g) \qquad \Delta H = -17.89 \text{ kcal}$$

$$C(g) \rightarrow C(s) \qquad \Delta H = -172 \text{ kcal}$$

$$4H(g) \rightarrow 2H_2(g) \qquad \Delta H = 2(-104) \text{ kcal}$$

$$\overline{C(g) + 4H(g) \rightarrow CH_4(g)} \qquad \overline{\Delta H = -398 \text{ kcal}}$$

The enthalpy of formation of $CH_4(g)$ is -398 kcal/mol. In each molecule of $CH_4(g)$ there are 4 C—H bonds. Thus, the enthalpy of formation of one C—H bond would be one fourth of the enthalpy of formation of $CH_4$: ¼ x -398 kcal/mol = -99.5 kcal/mol.

Therefore, +99.5 kcal/mol is required to break a C—H bond.

11. +79 kcal
    [or -330 kJ]
    [14.3]

We add the appropriate equation and corresponding values of $\Delta H$ to find the enthalpy of formation of ethane:

$$2C(s) + 3H_2(g) \rightarrow C_2H_6(g) \qquad \Delta H = -20.24 \text{ kcal}$$

$$2C(g) \rightarrow 2C(s) \qquad \Delta H = 2(-172) \text{ kcal}$$

$$6H(g) \rightarrow 3H_2(g) \qquad \Delta H = 3(-104) \text{ kcal}$$

$$\overline{2C(g) + 6H(g) \rightarrow C_2H_6(g)} \qquad \overline{\Delta H = -676 \text{ kcal}}$$

Since -676 kcal/mol is $\Delta H$ for the formation of 6 C—H bonds and 1 C—C bond, we subtract the enthalpy for the formation of 6 C—H bonds from the enthalpy of formation of $C_2H_6$ to find the enthalpy for the formation of a C—C bond:

$$\Delta H (C\!-\!C \text{ bond}) = \Delta H_f(C_2H_6) - 6\Delta H(C\!-\!H \text{ bond})$$

$$= -676 \text{ kcal/mol} - 6(-99.5 \text{ kcal/mol})$$

$$= -79 \text{ kcal/mol}$$

**12.** -1.21 cal/$^{O}$K
[or -5.06 J/$^{O}$K]
[14.5]

First we determine $\Delta H^O$ for the reaction:

$$\Delta H = \Delta H_f^O(CO_2) + 2\Delta H_f^O(H_2O) - \Delta H_f^O(CH_4)$$

$$= -94.05 \text{ kcal} + 2(-57.80 \text{ kcal}) - (-17.89 \text{ kcal})$$

$$= -191.76 \text{ kcal}$$

Then we determine $\Delta G^O$ for the reaction:

$$\Delta G = \Delta G_f^O(CO_2) + 2\Delta G_f^O(H_2O) - \Delta G_f(CH_4)$$

$$= -94.26 \text{ kcal} + 2(-54.64 \text{ kcal}) - (-12.14 \text{ kcal})$$

$$= -191.40 \text{ kcal}$$

Finally we calculate $\Delta S$ for the reaction:

$$\Delta S = \frac{\Delta H - \Delta G}{T}$$

$$= \frac{(-191.76 \text{ kcal}) - (-191.40 \text{ kcal})}{298^O K}$$

$$= -1.21 \times 10^{-3} \text{ kcal/}^O K$$

$$= -1.21 \text{ cal/}^O K$$

**13.** -1.226 cal/$^{O}$K
[or -5.130 J/$^{O}$K]
[14.8]

We calculate $\Delta S^O$:

$$\Delta S = \left[ S^O(CO_2) + 2S^O(H_2O) \right] - \left[ S^O(CH_4) + 2S^O(O_2) \right]$$

$$= \left[ 51.06 \text{ cal/}^O K + 2 \text{ mol}(45.11 \text{ cal/}^O K \text{ mol}) \right] -$$

$$\left[ 44.50 \text{ cal/}^O K \text{ mol} + 2 \text{ mol}(49.003 \text{ cal/}^O K \text{ mol}) \right]$$

$$= -1.226 \text{ cal/}^O K$$

The values of $\Delta S$ computed in problems 12 and 13 of this section are in good agreement.

**14.** $K_p = 10^{-14}$
[14.7, 14.9]

We add the following half reactions and corresponding values of $\mathscr{E}^O$ to determine the value of $\mathscr{E}^O$ for the ionization of water:

$$2H_2O + 2e^- \rightarrow H_2 + 2OH^- \qquad \mathscr{E}^\circ = -0.828 \text{ V}$$

$$H_2 \rightarrow 2H^+ + 2e^- \qquad \mathscr{E}^\circ = 0.000 \text{ V}$$

$$\overline{2H_2O \rightarrow 2H^+ + 2OH^- \qquad \mathscr{E}^\circ = -0.828 \text{ V}}$$

Since

$$\Delta G^\circ = -nF\mathscr{E}^\circ$$

and

$$\Delta G^\circ = -2.303RT \log K_p$$

then

$$-nF\mathscr{E}^\circ = -2.303RT \log K_p$$

and

$$\log K_p = \frac{nF\mathscr{E}^\circ}{2.303RT}$$

Substituting values into the preceding equation, we find

$$\log K_p = \frac{nF\mathscr{E}^\circ}{2.303RT}$$

$$= \frac{1(23.061 \text{ kcal/V})(-0.828 \text{ V})}{2.303(1.987 \times 10^{-3} \text{ kcal/}^\circ\text{K mol})(298^\circ\text{K})}$$

$$= -14.0$$

15. -669 kcal
    [or -2800 kJ]
    [14.2]

First we write a balanced equation for the reaction

$$2C_6H_{10}O_4(s) + 13O_2(g) \rightarrow 10H_2O(l) + 12CO_2(g)$$

Then we compute the number of moles of $C_6H_{10}O_4$ in 1.46 g $C_6H_{10}O_4$:

$$? \text{ mol } C_6H_{10}O_4 = 1.46 \text{ g } C_6H_{10}O_4\left(\frac{1 \text{ mol } C_6H_{10}O_4}{146.1 \text{ g } C_6H_{10}O_4}\right)$$

$$= 1.00 \times 10^{-2} \text{ mol } C_6H_{10}O_4$$

We know that 6.690 kcal is released when $1.00 \times 10^{-2}$ mol of $C_6H_{10}O_4$ is burned. Therefore, we can calculate the amount of heat released by 1 mol of $C_6H_{10}O_4$:

$$\Delta H = \left( \frac{-6690 \text{ cal}}{0.01 \text{ mol } C_6H_{10}O_4} \right) (1 \text{ mol } C_6H_{10}O_4) \left( \frac{1 \text{ kcal}}{1000 \text{ cal}} \right)$$

$$= -669 \text{ kcal}$$

**SELF-TEST**

Complete the test in 35 minutes:

1. The heat of combustion of oxalic acid, $H_2C_2O_4$, is −60.4 kcal/mol [or −253 kJ/mol] at 25°C and constant volume. What is $\Delta H$?

2. Using standard electrode potentials at 25°C (Appendix D of your text), calculate $K_p$ for

   $$Cu^{2+}(aq) + Sn(s) \rightarrow Cu(s) + Sn^{2+}(aq)$$

3. Using standard entropies and enthalpies of formation, determine if the reaction

   $$CO(g) + H_2O(g) \rightarrow CO_2(g) + H_2(g)$$

   is spontaneous as written and calculate $K_p$.

# 15
# Acids and Bases

OBJECTIVES

(a) You should be able to demonstrate your knowledge of the following terms by defining them, describing them, or giving specific examples of them:

acid [15.1, 15.2, 15.3]
amphiprotic [15.3]
anhydride [15.1]
Arrhenius concept [15.1]
base [15.1, 15.2, 15.3]
Brønsted-Lowry concept [15.3]
conjugate pair [15.3]
electrophilic [15.7]
hydrolysis [15.5]

leveling effect [15.4]
Lewis concept [15.7]
neutralization [15.1, 15.2]
nucleophilic [15.7]
solvent system concept [15.2]

(b) You should be able to identify acids and bases in given reactions and deduce relative strengths of each from displacement of equilibrium and/or from a qualitative knowledge of factors that influence the strength of each.


EXERCISES

I.  Fill in each of the statements with the appropriate entry or entries from the following list.  An entry may be used more than once.

a. ammonium ion
b. chloride ion
c. hydroflouric acid
d. hydroiodic acid
e. sodium acetate
f. water
g. zinc(II) ion

1. In glacial acetic acid, i.e., pure acetic acid, _____ would be a strong base.

2. The strongest acid that can exist in liquid ammonia is _____.

3. _____ is the weakest hydro acid of the elements of group VII A.

4. _____ are species that can act as Brønsted acids in water.

5. _____ are species that can act as Brønsted bases in water.

6. _____ is a species that can act as a Lewis acid.

7. _____ are species that can act as Lewis bases.

8. The strongest electrophilic species is _____.

9. The conjugate acid of ammonia is _____.

10. _____ is an amphiprotic material.

11. If 0.1 mol of _____ were added to 1 liter of water, a basic solution would be formed. Assume that the species are added as chloride salts of cations and sodium salts of anions.

12. If 0.1 mol of _____ were added to 1 liter of water, a neutral solution would be formed. Assume that the species are added as chloride salts of cations and sodium salts of anions.

II. Answer each of the following. Assume that each equilibrium is displaced to the right.

1. The strongest Lewis acid in the reaction

$$HCl + H_2O \rightleftharpoons Cl^- + H_3O^+$$

is _____.

2. The strongest Brønsted base in the reaction

$$H_2SO_4 + F^- \rightleftharpoons HSO_4^- + HF$$

is _____.

3. The strongest Brønsted acid in the reaction

$$HF + \left[ H-\overset{\overset{\displaystyle O}{\|}}{C}-\overset{\ominus}{O} \right]^- \rightleftharpoons F^- + H-\overset{\overset{\displaystyle O}{\|}}{C}-OH$$

is _____.

III. Use qualitative knowledge of factors that influence the strength of an acid to answer the following:

For each pair of compounds, determine which compound is the stronger acid:

1. HI and HF

2. 

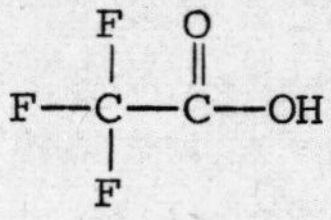

$$\underset{\underset{F}{|}}{\overset{\overset{F}{|}}{F-C}}-\overset{O}{\overset{\|}{C}}-OH \quad \text{and} \quad \underset{\underset{H}{|}}{\overset{\overset{H}{|}}{H-C}}-\overset{O}{\overset{\|}{C}}-OH$$

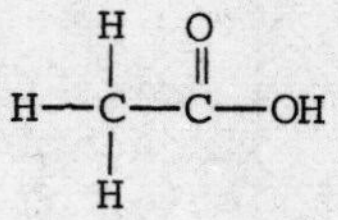

3. 

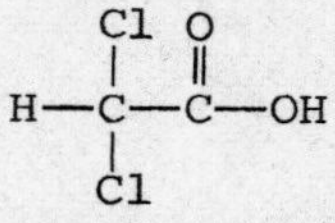

$$\underset{\underset{Cl}{|}}{\overset{\overset{Cl}{|}}{H-C}}-\overset{O}{\overset{\|}{C}}-OH \quad \text{and} \quad \underset{\underset{Cl}{|}}{\overset{\overset{Cl}{|}}{Cl-C}}-\overset{O}{\overset{\|}{C}}-OH$$

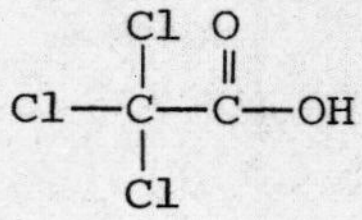

4. $HClO_4$ and $HClO_3$

5. HOI and HOCl


IV. Use qualitative knowledge of factors that influence the strength of an acid or base to answer the following. Assume that each equilibrium is displaced to the right.

1. List all Bronsted acids in the following reactions in order of decreasing strength:

   a. $HOBr + CN^- \rightleftharpoons OBr^- + HCN$

   b. $HCl + H_2O \rightleftharpoons H_3O^+ + Cl^-$

   c. $HI + BrO_3^- \rightleftharpoons HBrO_3 + I^-$

   d. $OH^- + HOBr \rightleftharpoons H_2O + OBr^-$

   e. $H_3O^+ + IO_3^- \rightleftharpoons HIO_3 + H_2O$

2. Which acids in problem 1 of this section are strong acids?

3. List the Brønsted bases in problem 1 of this section in order of decreasing strength.

4. Identify the Lewis acid in the following reaction:

   $PCl_3 + Cl_2 \rightarrow PCl_5$

ANSWERS TO
EXERCISES                    I. Acid-base systems

1. sodium          The acetate ion is the strongest base that can exist
   acetate         in acetic acid.  The basicity of strongly basic
   [15.4]          materials in acetic acid is leveled to that of the
                   acetate ion.

2. ammonium ion
   [15.4]

3. hydrofluoric    There is an increase in acid strength of the hydro
   acid            acids, HX, of the elements in any group of the periodic
   [15.4]          table with increasing atomic size of the electro-
                   negative element, X.

4. ammonium ion    A Brønsted acid is a species that can donate a proton:
   hydrofluoric acid
   hydroiodic acid      $NH_4^+$    +    $OH^-$   ⇌    $NH_3$    +    $H_2O$
   water           **Brønsted**   Brønsted      Brønsted      Brønsted
   [15.3]            acid           base          base          acid


                       HF      +    $H_2O$   ⇌    $F^-$    +    $H_3O^+$
                   Brønsted       Bronsted      Brønsted      Brønsted
                     acid           base          base          acid


                       HI      +    $H_2O$   ⇌    $I^-$    +    $H_3O^+$
                   Brønsted       Brønsted      Brønsted      Brønsted
                     acid           base          base          acid


                      $H_2O$   +    $H^-$    ⇌    $H_2$    +    $OH^-$
                   Brønsted       Brønsted      Brønsted      Brønsted
                     acid           base          base          acid

5. chloride ion    A Bronsted base is a species that can accept a proton:
   sodium acetate
   water
   [15.3]

$$Cl^- \quad + \quad HI \quad \rightleftharpoons \quad HCl \quad + \quad I^-$$
Brønsted base $\quad$ Brønsted acid $\quad$ Brønsted acid $\quad$ Brønsted base

$$C_2H_5O_2^- \quad + \quad HCl \quad \rightleftharpoons \quad C_2H_5O_2H \quad + \quad Cl^-$$
Brønsted base $\quad$ Brønsted acid $\quad$ Brønsted acid $\quad$ Brønsted base

$$H_2O \quad + \quad HBr \quad \rightleftharpoons \quad H_3O^+ \quad + \quad Br^-$$
Brønsted base $\quad$ Brønsted acid $\quad$ Brønsted acid $\quad$ Brønsted base

6. Zn(II) ion
   [15.7]

   A Lewis acid *accepts* an electron pair to form a covalent bond.

7. chloride ion
   water
   [15.7]

   A Lewis base *donates* an electron pair to form a covalent bond.

8. Zn(II) ion
   [15.7]

   The word *electrophilic* is used to describe the ability of a species to attract an electron pair.

9. ammonium ion
   [15.3]

   $$H_2O \quad + \quad NH_3 \quad \rightleftharpoons \quad NH_4^+ \quad + \quad OH^-$$
   acid$_1$ $\quad$ base$_1$ $\quad$ acid$_2$ $\quad$ base$_1$

10. water
    [15.3]

    Water can function as an acid and a base:

    $$H_2O + NH_2^- \rightleftharpoons NH_3 + OH^-$$
    $$H_2O + HBr \rightleftharpoons Br^- + H_3O^+$$

11. sodium acetate
    [15.5]

    Acetate ion hydrolyzes:

    $$H_2O + C_2H_3O_2^- \rightleftharpoons HC_2H_3O_2 + OH^-$$

12. chloride ion
    water
    [15.1]

II. Strengths of acids and bases determined from equilibrium displacement

1. H$^+$
   [15.7]

   The hydrogen ion, H$^+$, from HCl is the electron acceptor.

2. F⁻
[15.3]
                The fluoride ion, $F^-$, is the proton acceptor.

3. HF
[15.3]
                Hydrogen fluoride, HF, is the proton donor.


III. Strengths of acids and bases determined from qualitative
knowledge of factors that influence strength


1. HI
[15.6]

2.
$$F-\overset{\overset{\displaystyle F}{|}}{\underset{\underset{\displaystyle F}{|}}{C}}-\overset{\overset{\displaystyle O}{\|}}{C}-OH$$
[15.6]

3.
$$Cl-\overset{\overset{\displaystyle Cl}{|}}{\underset{\underset{\displaystyle Cl}{|}}{C}}-\overset{\overset{\displaystyle O}{\|}}{C}-OH$$
[15.6]

4. $HClO_4$
[15.6]

5. HOCl
[15.6]


IV. Relative strengths of acids and bases determined from
information on equilibrium displacement and qualitative
knowledge of factors that influence strength


1. [15.6]
HI  _ _ _ _ _ _
HCl _ _ _ _ _ _ _
                The order of acid strength of the hydrogen halides is
                    HI > HBr > HCl > HF

$H_3O^+$
                Hydrochloric acid, HCl, is stronger than the hydronium
                ion, $H_3O^+$ (see reaction c).

                _ _ _ _ _ _ Iodic acid, $HIO_3$, is weaker than the hydronium ion,
$HBrO_3$ _ _ _ _       $H_3O^+$ (see reaction f).  Bromic acid, $HBrO_3$, is expected
$HIO_3$ _ _            to be more acidic than iodic acid, $HIO_3$, and less acidic
     _ _ _ _ _ _ than hydroiodic acid, HI (see reaction d).

HOBr

Hypobromous acid, HOBr, is stronger than HCN (see reaction a) and $H_2O$ (see reaction e).

HCN

$H_2O$

2. HCl and HI
[15.4]

Hydrochloric acid and hydrobromic acid are leveled by water:

$$HCl + H_2O \rightleftharpoons Cl^- + H_3O^+$$

$$HI + H_2O \rightleftharpoons I^- + H_3O^+$$

3. [15.6]
OH⁻

CN⁻

OBr⁻

$IO_2^-$

$IO_3^-$

$BrO_3^-$

$H_2O$

Cl⁻

I⁻

Increasing acid strength parallels decreasing base strength of the conjugate base. Thus, the conjugate base of the strongest acid is the weakest base. Bases weaker than water are not normally considered to be basic species in aqueous solution.

4. $PCl_3$
[15.7]

Trichlorophosphine, $PCl_3$, is a Lewis acid and accepts an electron pair from each chlorine atom of $Cl_2$ to form $PCl_5$.

# SELF-TEST

Complete the test in 10 minutes:

_____

1. The reaction of ammonia and the hydride ion goes to completion:

$$NH_3 + H^- \rightarrow H_2 + NH_2^-$$

What is the strongest Brønsted acid in the reaction? What is the strongest Brønsted base?

_____

2. What is the conjugate acid of HS⁻?

_____

3. The following reaction may be interpreted as an acid-base reaction in a $SO_2$ solvent system:

$$Cs_2SO_3 + SOCl_2 \rightarrow 2CsCl + 2SO_2$$

What is the acid?

_____

4. What is the Lewis base in the following reaction?

$$BF_3 + F^- \rightarrow BF_4^-$$

_____

5. What is the Lewis acid in the following reaction?

$$\left[:\ddot{S}:\right]^{2-} + \ddot{S}: \rightarrow S_2^{2-}$$

_____

6. The following equilibrium is displaced to the right:

$$HC_2H_3O_2 + HS^- \rightleftharpoons H_2S + C_2H_3O_2^-$$

What is the strongest Brønsted base in the reaction??
What is the strongest Brønsted acid?

_____

7. What is the conjugate base of $H_2PO_4^-$ ?

# 16
## Ionic Equilibria, Part I

OBJECTIVES

(a) You should be able to demonstrate your knowledge
of the following terms by defining them, describing
them, or giving specific examples of them:

buffer [16.7]
common-ion effect [16.6]
hydronium ion [16.1]
indicator [16.5]
pH [16.3]
p$K$ [16.7]
polyprotic acids [16.8]
pOH [16.3]
strong electrolyte [16.1]
weak electrolyte [16.1]

(b) You should be able to calculate the pH and the concentration of all species in a solution containing either strong acids or bases, weak acids or bases, buffers, or polyprotic acids.

(c) You should be able to determine the ionization constant of a weak acid from potentiometric titration data or from concentrations of all species in solution.

(d) You should be able to determine the percent ionization of a weak acid from the ionization constant and the pH of a solution containing a known amount of that acid.

EXERCISES    I.    Do the following:

1.  The parietal cells of the human stomach secrete $0.155M$ HCl.  What is the pH of the secretion?

2.  What is the pH of a solution with a $[OH^-]$ of $4.8 \times 10^{-4} M$?

3.  A certain enzyme loses its ability to catalyze a reaction below a pH of 5.7.  To what $[H^+]$ does the pH correspond?

4.  What are the pH and pOH of a $0.020M$ solution of KOH?

5.  Formic acid, HCOOH, has an ionization constant of $1.77 \times 10^{-4}$.  What is the pH of $0.10M$ formic acid?

6.  What are the concentrations of $H^+$, $ClO^-$, and HClO in $0.10M$ hypochlorous acid?  The ionization constant of hypochlorous acid is $3.2 \times 10^{-8}$

7.  How many moles of nitrous acid, $HNO_2$, must be added to water to prepare 1.0 liter of solution with a pH of 3.0?  The ionization constant of nitrous acid is $4.5 \times 10^{-4}$

8.  What is the concentration of a solution in which HX is 1.0% dissociated?  The ionization constant of HX is $5.0 \times 10^{-4}$

9. In metabloic acidosis the $HCO_3^-/CO_2$ ratio in the blood may be 16/1, a ratio that corresponds to a pH of 7.3. What is the p$K$ of carbonic acid?

10. In metabolic alkalosis the $HCO_3^-/CO_2$ ratio may be 40/1. What is the pH of blood under such conditions? Use the value of the ionization constant of carbonic acid calculated in problem 9 of this section.

11. A cell consisting of a hydrogen electrode and a normal calomel electrode contains a solution of unknown pH. The potential is 0.833 V at $25^\circ$C. Calculate the following:
    a. the pH of the solution
    b. the $[H^+]$ of the solution

12. The indicator bromothymol blue has an ionization constant of $5.0 \times 10^{-8}$. The acid color is yellow, and the alkaline color is blue. The acid color is visible when the ratio of yellow to blue is 20 to 1, and the blue is visible when the ratio of blue to yellow is 2 to 1. What is the pH range of color change of bromothymol blue?

13. A solution is prepared by adding 0.040 mol of solid NaCNO, sodium cyanate, to 250 ml of 0.025$M$ cyanic acid, HCNO. Assume that no volume change occurs. Calculate the pH of the solution. The ionization constant of cyanic acid is $1.2 \times 10^{-4}$.

14. Strychnine is a weak base, and its aqueous ionization is represented by

    $$S(aq) + H_2O \rightleftharpoons SH^+(aq) + OH^-(aq)$$

    A 1.0$M$ solution of strychnine has a pH of 11. What is the ionization constant of strychnine?

15. What is the pH of a buffer that is 0.15$M$ in NaZ and 3.0$M$ in HZ if the ionization constant of the weak acid HZ is $5.0 \times 10^{-6}$?

16. How many moles of solid $NH_4Cl$ must be dissolved in 500.0 ml of a solution that is originally

0.250$M$ in $NH_3$ in order to have a final pH of
10.0? Assume that the volume remains 500.0
ml. The ionization constant of $NH_3$ is 1.8 x
$10^{-5}$.

17. Calculate the pH of a solution composed of 50
ml of 1.0$M$ $NaC_2H_3O_2$ and 50 ml of 1.0$M$ $HC_2H_3O_2$.
Assume the volume after mixing to be 100 ml.
The ionization constant of acetic acid, $HC_2H_3O_2$,
is 1.8 x $10^{-5}$.

18. What is the $[H^+]$ in a 0.036$M$ $H_2S$ solution?
For $H_2S$ the ionization constant of the primary
ionization is 1.1 x $10^{-7}$. Assume that effects
due to secondary ionization are negligible.

19. Vitamin C, ascorbic acid, is a diprotic acid
with the formula $C_6H_8O_6$. Calculate the $[H+]$,
$[C_6H_7O_6^-]$, and $[C_6H_6O_6^{2-}]$ of a 0.10$M$ solution of
ascorbic acid. The ionization constants of
ascorbic acid are $K_1$ = 7.9 x $10^{-5}$ and $K_2$ =
1.6 x $10^{-12}$.

20. A buffer solution is prepared by mixing 500 ml
of 1.00$M$ $HC_2H_3O_2$ and 500 ml of 1.00$M$ $NaC_2H_3O_2$.
The ionization constant of acetic acid, $HC_2H_3O_2$,
is 1.8 x $10^{-5}$. Calculate the following:
   a. the pH of the buffer
   b. the pH of the buffer solution after the
   addition of 0.005 mol of HCl

## ANSWERS TO EXERCISES

I. Ionic equilibrium calculations

1. pH = 0.810
   [16.3]

The definition of pH is

$$pH = -\log [H+]$$

Substituting the value of $[H+]$ into the preceding
equation, we find

$$pH = -\log [H+]$$
$$= -\log (1.55 \times 10^{-1})$$
$$= (-\log 1.55) + (-\log 10^{-1})$$
$$= -0.190 + 1$$
$$= 0.810$$

Read Appendix B of your text if you need help in understanding and working with logarithms.

2. pH = 10.7
   [16.3]

The water constant is

$$[H^+] [OH^-] = 1.0 \times 10^{-14}$$

If we solve the preceding equation for $[H^+]$ and substitute values into the equation, we find

$$[H^+] = \frac{1.0 \times 10^{-14}}{4.8 \times 10^{-4}}$$

$$= 2.1 \times 10^{-11} M$$

We determine pH:

$$pH = \log [H^-]$$
$$= -\log (2.1 \times 10^{-11})$$
$$= (-\log 2.1) + (-\log 10^{-11})$$
$$= -0.30 + 11$$
$$= 10.7$$

3. $[H^+] = 2.0 \times 10^{-6} M$
   [16.3]

We can determine $[H^+]$ as follows:

$$- \log [H^+] = pH$$
$$\log [H^+] = -pH$$
$$= -5.7$$
$$= 0.30 - 6.0$$
$$[H^+] = (\text{antilog } 0.30)(\text{antilog } -6.0)$$
$$= 2.0 \times 10^{-6} M$$

We can also use another method of determining $[H^+]$:

$$pH = - \log [H^+]$$
$$[H^+] = 10^{-pH}$$
$$= 10^{-5.7}$$
$$= (10^{+0.3})(10^{-6})$$
$$= 2.0 \times 10^{-6} M$$

4. pH = 12.3
   pOH = 1.7
   [16.3]

Potassium hydroxide is a strong electrolyte.

Thus,

$$[OH^-] = 2.0 \times 10^{-2} M$$

and

$$pOH = -\log [OH^-]$$

$$= -\log (2.0 \times 10^{-2})$$

$$= -0.30 + 2$$

$$= 1.7$$

We determine the pH as follows:

The water constant is

$$[H^+][OH^-] = 1.0 \times 10^{-14}$$

We arrange the preceding equation and substitute values into it:

$$[H^+] = \frac{1.0 \times 10^{-14}}{[OH^-]}$$

$$= \frac{1.0 \times 10^{-14}}{2.0 \times 10^{-2}}$$

$$= 5.0 \times 10^{-13} M$$

We calculate pH in either of two ways:

$$pH = - \log [H^+]$$

$$= - \log (5.0 \times 10^{-13})$$

$$= -0.70 + 13$$

$$= 12.3$$

or

$$pH + pOH = 14$$
$$pH = 14 - pOH$$
$$= 14 - 1.7$$
$$= 12.3$$

5. pH = 2.4
   [16.1, 16.4]

The expression for the ionization constant is

$$K = 1.77 \times 10^{-4} = \frac{[H^+][HCOO^-]}{[HCOOH]}$$

We make the following assumptions:

| SPECIES | EQUILIBRIUM CONCENTRATION |
|---------|---------------------------|
| $H^+$ | $(x)M$ |
| $HCOO^-$ | $(x)M$ |
| $HCOOH$ | $\left((1.0 \times 10^{-1}) - x \cong 1.0 \times 10^{-1}\right)M$ |

We substitute concentrations into the expression for the ionization constant:

$$K = 1.77 \times 10^{-4} = \frac{[H^+][HCOO^-]}{[HCOOH]}$$

$$1.77 \times 10^{-4} = \frac{x^2}{1.0 \times 10^{-1}}$$

$$x^2 = 1.77 \times 10^{-5}$$

$$= 4.2 \times 10^{-3}$$

Thus, $[H^+] = 4.2 \times 10^{-3} M$ and the pH is

$$pH = -\log[H^+]$$

$$= -\log(4.2 \times 10^{-3})$$

$$= -0.62 + 3$$

$$= 2.4$$

6. $[H^+] = 5.7 \times 10^{-5} M$   The expression for the ionization constant is
$[Clo^-] = 5.7 \times 10^{-5} M$
$[HClO] = 1.0 \times 10^{-1} M$   $\quad K = 3.2 \times 10^{-8} = \dfrac{[H^+][Clo^-]}{[HClO]}$
[16.1]

We make the following assumptions:

| SPECIES | ORIGINAL CONCENTRATION | EQUILIBRIUM CONCENTRATION |
|---------|------------------------|---------------------------|
| $H^+$ | $10^{-7} M$ | $(10^{-7} + x \cong x)M$ |
| $Clo^-$ | $O$ | $(x)M$ |
| $HClO$ | $1.0 \times 10^{-1} M$ | $\left((1.0 \times 10^{-1}) - x \cong 1.0 \times 10^{-1}\right)M$ |

We substitute concentrations into the expression for the ionization constant:

$$3.2 \times 10^{-8} = \frac{[H^+][ClO^-]}{[HClO]}$$

$$= \frac{x^2}{1.0 \times 10^{-1}}$$

$$x^2 = 3.2 \times 10^{-9}$$

$$x = 5.7 \times 10^{-5}$$

Thus,

$$[H^+] = 5.7 \times 10^{-5} M$$

7. $3.2 \times 10^{-3}$ mol HNO$_2$

[16.1, 16.3]

The equation for the ionization of nitrous acid is

$$HNO_2 \rightleftharpoons H^+ + NO_2^-$$

A pH of 3.0 corresponds to a $[H^+]$ of $1.0 \times 10^{-3} M$:

$$[H^+] = 10^{-pH}$$
$$= 10^{-3.0}$$
$$= 1.0 \times 10^{-3} M$$

We make the following assumptions:

| SPECIES | ORIGINAL CONCENTRATION | EQUILIBRIUM CONCENTRATION |
|---------|------------------------|----------------------------|
| $H^+$ | $10^{-7} M$ | $\left(10^{-7} + (1.0 \times 10^{-3}) = 1.0 \times 10^{-3}\right) M$ |
| $NO_2^-$ | 0 | $1.0 \times 10^{-3} M$ |
| $HNO_2$ | $(x) M$ | $\left(x - (1.0 \times 10^{-3})\right) M$ |

We substitute concentrations into the expression of the ionization constant:

$$K = 4.5 \times 10^{-4} = \frac{[H^+][NO_2^-]}{[HNO_2]}$$

$$4.5 \times 10^{-4} = \frac{(1.0 \times 10^{-3})^2}{x - (1.0 \times 10^{-3})}$$

$$x = 3.2 \times 10^{-3}$$

Thus, $3.2 \times 10^{-3}$ mol of $HNO_2$ must be added.

8. [HX] = 5.0$M$      The ionization of HX is
   [16.1]

$$HX \rightleftharpoons H^+ + X^-$$

We make the following assumptions:

| SPECIES | ORIGINAL CONCENTRATION | EQUILIBRIUM CONCENTRATION |
|---------|------------------------|----------------------------|
| $H^+$ | $10^{-7}M$ | $(10^{-7} + 0.01x \cong 0.01x)M$ |
| $X^-$ | 0 | |
| HX | $(x)M$ | $(x - 0.01x \cong x)M$ |

We substitute concentrations into the expression for the ionization constant:

$$K = 5.0 \times 10^{-4} = \frac{[H^+][X^-]}{[HX]}$$

$$= \frac{(0.01x)^2}{x}$$

$$(5.0 \times 10^{-4})x = (1.0 \times 10^{-4})x^2$$

$$5.0 \times 10^{-4} = (1.0 \times 10^{-4})x$$

$$x = \frac{5.0 \times 10^{-4}}{1.0 \times 10^{-4}}$$

$$= 5.0$$

Thus,

$$[HX] = 5.0M$$

9. $pK = 6.1$
   [16.8]

The ionization of carbonic acid is

$$CO_2 + H_2O \rightleftharpoons H^+ + HCO_3^-$$

The expression for the ionization constant is

$$K = \frac{[H^+][HCO_3^-]}{[CO_2]}$$

A pH of 7.3 corresponds to a $[H^+]$ of $5.0 \times 10^{-8}\,M$:

$$-\log [H^+] = pH$$

$$= 7.3$$

$$[H^+] = 10^{-7.3}$$

$$= (10^{0.7})(10^{-8.0})$$

$$= 5.0 \times 10^{-8}\,M$$

Substituting concentrations into the expression for the ionization constant, we find

$$K = \frac{[H^+][HCO_3^-]}{[CO_2]}$$

$$= (5.0 \times 10^{-8})(16)$$

$$= 8.0 \times 10^{-7}$$

and

$$pK = -\log K$$

$$= -\log(8.0 \times 10^{-7})$$

$$= -0.90 + 7$$

$$= 6.1$$

We could also determine $pK$ as follows:

$$pH = pK - \log \frac{[CO_2]}{[HCO_3^-]}$$

$$pK = pH + \log \frac{[CO_2]}{[HCO_3^-]}$$

$$= 7.3 + \log\left(\frac{1}{16}\right)$$

$$= 7.3 + \log\left(6.25 \times 10^{-2}\right)$$

$$= 7.3 + 0.796 - 2$$

$$= 6.1$$

10. pH = 7.7
    [16.7, 16.8]

Substituting concentrations into the expression for the ionization constant, we find

$$K = \frac{[H^+][HCO_3^-]}{[CO_2]}$$

$$8.0 \times 10^{-7} = [H^+]\ 40$$

$$[H^+] = \left(\frac{1}{40}\right)(8.0 \times 10^{-7})$$

$$= 2.0 \times 10^{-8} M$$

and

$$pH = -\log [H^+]$$

$$= -\log(2.0 \times 10^{-8})$$

$$= -0.30 + 8$$

$$= 7.7$$

We could also determine pH as follows:

$$pH = pK - \log \frac{[CO_2]}{[HCO_3^-]}$$

$$= 6.1 - \log\left(\frac{1}{40}\right)$$

$$= 6.1 - \log(2.5 \times 10^{-2})$$

$$= 6.1 - 0.40 + 2$$

$$= 7.7$$

11. [16.4]
    a. pH = 9.30

We substitute concentrations into the equation relating pH and $\mathscr{E}$ at 25°C:

$$pH = \frac{\mathscr{E} - 0.2825}{0.05916}$$

$$= \frac{0.833 - 0.2825}{0.05916}$$

$$= 9.30$$

    b. $[H^+] =$
       $4.96 \times 10^{-10} M$

We calculate $[H^+]$:

$$pH = -\log [H^+]$$

$$[H^+] = 10^{-pH}$$
$$= 10^{-9.30}$$

$$= (10^{0.30})(10^{-10})$$

$$= 4.96 \times 10^{-10} M$$

12. pH range
    = 6.0 to 7.6
    [16.5]

The ionization of HIn is

$$HIn \rightleftharpoons H^+ + In^-$$

and the expression for the ionization constant is

$$K = 5.0 \times 10^{-8} = \frac{[H^+][In^-]}{[HIn]}$$

The yellow form is HIn and the blue form is In⁻. Thus, the ratio of yellow to blue is the ratio of HIn to In⁻, which is equal to 20 when the yellow color is visible. We calculate the pH corresponding to the ratio of HIn to In⁻ equal to 20:

$$K = 5.0 \times 10^{-8} = \frac{[H^+][In^-]}{[HIn]}$$

$$[H^+] = (5.0 \times 10^{-8})(2.0 \times 10^1)$$
$$= 1.0 \times 10^{-6} M$$

and

$$pH = -\log [H^+]$$

$$= -\log (1.0 \times 10^{-6})$$

$$= 6.0$$

The blue color is visible when the ratio of blue to yellow is 2, or when the ratio of $In^-$ to HIn is 2. We calculate the pH corresponding to the ratio of HIn to $In^-$ equal to 1/2:

$$K = 5.0 \times 10^{-8} = \frac{[H^+][In^-]}{[HIn]}$$

$$[H^+] = (5.0 \times 10^{-8})\left(\frac{1}{2}\right)$$

$$= 2.5 \times 10^{-8}M$$

and

$$pH = -\log(2.5 \times 10^{-8})$$

$$= -0.40 + 8$$

$$= 7.6$$

13. pH = 4.7
    [16.6]

The ionization of HCNO is

$$HCNO \rightleftharpoons H^+ + CNO^-$$

Since NaCNO is a strong electrolyte,

$$NaCNO \rightarrow Na^+ + CNO^-$$

We make the following assumptions:

| SPECIES | ORIGINAL CONCENTRATION | EQUILIBRIUM CONCENTRATION |
|---------|------------------------|---------------------------|
| $H^+$ | $10^{-7}M$ | $(10^{-7} + x \cong x)M$ |
| $CNO^-$ | $\dfrac{0.040 \text{ mol}}{0.250 \text{ liter}} = 0.16M$ | $(0.16 + x \cong 0.16)M$ |
| HCNO | $0.025M$ | $(0.025 - x \cong 0.025)M$ |

Substituting concentrations into the expression for the ionization constant of HCNO, we find

$$K = 1.2 \times 10^{-4} = \frac{[H^+][CNO^-]}{[HCNO]}$$

$$1.2 \times 10^{-4} = \frac{x(0.16)}{0.025}$$

$$x = 1.9 \times 10^{-5}$$

and

$$pH = -\log [H^+]$$

$$= -\log (1.9 \times 10^{-5})$$

$$= -0.28 + 5$$

$$= 4.7$$

We could also determine pH as follows:

$$pH = pK - \log \frac{[HCNO]}{[CNO^-]}$$

$$= -\log (1.2 \times 10^{-4}) - \log \left( \frac{2.5 \times 10^{-2}}{1.6 \times 10^{-1}} \right)$$

$$= -\log (1.2 \times 10^{-4}) - \log (1.6 \times 10^{-1})$$

$$= -0.079 + 4 - 0.20 + 1$$

$$= 4.7$$

14. $K = 1.0 \times 10^{-6}$
   [16.1, 16.3]

The expression for the ionization constant of strychnine is

$$K = \frac{[SH^+][OH^-]}{[S]}$$

We calculate $[OH^-]$ as follows:

$$[H^+] = 10^{-pH}$$

$$= 10^{-11}$$

$$[H^+][OH^-] = 1.0 \times 10^{-14}$$

$$[OH^-] = \frac{1.0 \times 10^{-14}}{[H^+]}$$

$$= \frac{1.0 \times 10^{-14}}{1.0 \times 10^{-11}}$$

$$= 1.0 \times 10^{-3}$$

We make the following assumptions:

| SPECIES | EQUILIBRIUM CONCENTRATION |
|---------|---------------------------|
| S | $\left(1.0 - (1.0 \times 10^{-3}) \cong 1.0\right) M$ |
| $SH^+$ | $1.0 \times 10^{-3} M$ (since the only source is S in the ionization) |
| $OH^-$ | $1.0 \times 10^{-3} M$ |

Substituting concentrations into the expression for the ionization constant, we find

$$K = \frac{(1.0 \times 10^{-3})^2}{1.0}$$

$$K = 1.0 \times 10^{-6}$$

15. pH = 5.0
   [16.7]

The ionization of HZ is

$$HZ \rightleftharpoons H^+ + Z^-$$

Since NaZ is a strong electrolyte,

$$NaZ \rightarrow Na^+ + Z^-$$

We make the following assumptions:

| SPECIES | ORIGINAL CONCENTRATION | EQUILIBRIUM CONCENTRATION |
|---------|------------------------|----------------------------|
| $H^+$ | $10^{-7} M$ | $(10^{-7} + x \cong x) M$ |
| HZ | $3.0 \times 10^{-1} M$ | $\left((3.0 \times 10^{-1}) - x \cong 3.0 \times 10^{-1}\right) M$ |
| $Z^-$ | $1.5 \times 10^{-1} M$ | $\left((1.5 \times 10^{-1}) + x \cong 1.5 \times 10^{-1}\right) M$ |

We substitute concentrations into the expression for the ionization constant:

$$K = 5.0 \times 10^{-6} = \frac{x(1.5 \times 10^{-1})}{3.0 \times 10^{-1}}$$

$$x = \frac{(5.0 \times 10^{-6})(3.0 \times 10^{-1})}{1.5 \times 10^{-1}}$$

$$= 1.0 \times 10^{-5} \quad \text{(Note the assumption that } x \ll 10^{-1} \text{ is valid.)}$$

Thus,

$$[H^+] = 1.0 \times 10^{-5} M$$

and

$$pH = -\log [H^+]$$
$$= -\log (1.0 \times 10^{-5})$$
$$= 5.0$$

We can also determine pH as follows:

$$pH = pK - \log \frac{[HZ]}{[Z^-]}$$

$$= -\log (5.0 \times 10^{-6}) - \log \left(\frac{0.30}{0.15}\right)$$

$$= -0.70 + 6 - 0.30$$

$$= 5.0$$

16. 0.022 mol $NH_4Cl$     The ionization of $NH_3$ is
    [16.6]

$$NH_3 + H_2O \rightleftharpoons NH_4^+ + OH^-$$

The expression for the ionization constant is

$$K = 1.8 \times 10^{-5} = \frac{[NH_4^+][OH^-]}{[NH_3]}$$

We determine $[OH^-]$ as follows:

pH = 10.0

$[H^+] = 10^{-pH}$

$\quad = 1.0 \times 10^{-10} M$

$[OH^-] = \dfrac{1.0 \times 10^{-14}}{[H^+]}$

$\quad = \dfrac{1.0 \times 10^{-14}}{1.0 \times 10^{-10}}$

$\quad = 1.0 \times 10^{-4} M$

We make the following assumptions:

| SPECIES | EQUILIBRIUM CONCENTRATION |
|---------|---------------------------|
| $NH_3$ | $\left(0.250 - (1.0 \times 10^{-4}) \cong 0.250\right) M$ |
| $NH_4^+$ | $(x) M$ |
| $OH^-$ | $1.0 \times 10^{-4} M$ |

We substitute concentrations into the expression for the ionization constant:

$$K = 1.8 \times 10^{-5} = \frac{[NH_4^+][OH^-]}{[NH_3]}$$

$$1.8 \times 10^{-5} = \frac{x(1.0 \times 10^{-4})}{2.50 \times 10^{-1}}$$

$$x = \frac{(1.8 \times 10^{-5})(2.50 \times 10^{-1})}{1.0 \times 10^{-14}}$$

$$= 4.5 \times 10^{-2}$$

Thus,

$$[NH_4^+] = 0.045M$$

and

$$? \text{ mol } NH_4^+ = (0.045M \text{ } NH_4^+)(0.500 \text{ liter})$$

$$= 0.022 \text{ mol } NH_4^+$$

17. pH = 4.7
[16.7]

The expression for the ionization constant of $HC_2H_3O_2$ is

$$K = 1.8 \times 10^{-5} = \frac{[H^+][C_2H_3O_2^-]}{[HC_2H_3O_2]}$$

We make the following assumptions:

| SPECIES | ORIGINAL CONCENTRATION | EQUILIBRIUM CONCENTRATION |
|---|---|---|
| $H^+$ | $10^{-7}M$ | $(10^{-7} + x \approx x)M$ |
| $C_2H_3O_2^-$ | $1.0M$ | $\left(\frac{(50\text{ml})(1.0)}{(100\text{ml})} + x = 0.50 + x\right)M \cong 0.50M$ |
| $HC_2H_3O_2$ | $1.0M$ | $\left(\frac{(50\text{ml})(1.0)}{(100\text{ml})} - x = 0.50 - x\right)M \cong 0.50M$ |

We substitute concentrations into the expression for the ionization constant:

$$K = 1.8 \times 10^{-5} = \frac{[H^+][C_2H_3O_2^-]}{[HC_2H_3O_2]}$$

$$1.8 \times 10^{-5} = \frac{x(0.50)}{0.50}$$

$$x = 1.8 \times 10^{-5}$$

Thus,

$$[H^+] = 1.8 \times 10^{-5} M$$

and

$$pH = -\log [H^+]$$

$$= -\log(1.8 \times 10^{-5})$$

$$= -0.26 + 5$$

$$= 4.7$$

18. $[H^+] = 6.3 \times 10^{-5} M$   The expression for the primary ionization constant
    [16.8]   of $H_2S$ is

$$K_1 = 1.1 \times 10^{-7} = \frac{[H^+][HS^-]}{[H_2S]}$$

We make the following assumptions

| SPECIES | EQUILIBRIUM CONCENTRATION |
|---------|---------------------------|
| $H^+$ | $(x) M$ |
| $HS^-$ | $(x) M$ |
| $H_2S$ | $\left((3.6 \times 10^{-2}) - x \cong 3.6 \times 10^{-2}\right) M$ |

We substitute concentrations into the expression
for the ionization constant:

$$K = 1.1 \times 10^{-7} = \frac{[H^+][HS^-]}{[H_2S]}$$

$$1.1 \times 10^{-7} = \frac{x^2}{3.6 \times 10^{-2}}$$

$$x^2 = 40 \times 10^{-10}$$

$$x = 6.3 \times 10^{-5}$$

Thus,

$$[H^+] = 6.3 \times 10^{-5} M$$

19. $[H^+] = 2.8 \times 10^{-3} M$    The ionizations of ascorbic acid are

$[C_6H_7O_6^-] =$

    $2.8 \times 10^{-3} M$

$[C_6H_6O_6^{2-}] =$

    $1.6 \times 10^{-12} M$

[16.8]

$$C_6H_8O_6 \rightleftharpoons H^+ + C_6H_7O_6^-$$

$$C_6H_7O_6^- \rightleftharpoons H^+ + C_6H_6O_6^{2-}$$

For the first ionization we make the following assumptions:

| SPECIES | EQUILIBRIUM CONCENTRATION |
|---|---|
| $C_6H_8O_6$ | $\left((1.0 \times 10^{-1}) - x = 1.0 \times 10^{-1}\right) M$ |
| $C_6H_7O_6^-$ | $(x) M$ |
| $H^+$ | $(x) M$ |

We substitute concentrations into the expression for the ionization constant of ascorbic acid $K_1$:

$$K_1 = 7.9 \times 10^{-5} = \frac{[H^+]\,[C_6H_7O_6^-]}{[C_6H_8O_6]}$$

$$7.9 \times 10^{-5} = \frac{x^2}{1.0 \times 10^{-1}}$$

$$x^2 = 7.9 \times 10^{-6}$$

$$x = 2.8 \times 10^{-3}$$

Thus,

$$[H^+] = [C_6H_7O_6^-] = 2.8 \times 10^{-3} M$$

For the second ionization of ascorbic acid we make the following assumptions:

| SPECIES | EQUILIBRIUM CONCENTRATION |
|---|---|
| $H^+$ | $2.8 \times 10^{-3} M$ |
| $C_6H_7O_6^-$ | $2.8 \times 10^{-3} M$ |
| $C_6H_6O_6^{2-}$ | $(y) M$ |

We substitute concentrations into the expression for the ionization constant of ascorbic acid $K_2$:

$$K_2 = 1.6 \times 10^{-12} = \frac{[H^+][C_6H_6O_6^{2-}]}{[C_6H_7O_6^-]}$$

$$1.6 \times 10^{-12} = \frac{(2.8 \times 10^{-3})(y)}{2.8 \times 10^{-3}}$$

$$y = 1.6 \times 10^{-12}$$

Thus,

$$[C_6H_6O_6^{2-}] = 1.6 \times 10^{-12} M$$

20. [16.7]
   a. pH = 4.7

The ionization of $HC_2H_3O_2$ is

$$HC_2H_3O_2 \rightleftharpoons H^+ + C_2H_3O_2^-$$

The expression for the ionization constant is

$$K = 1.8 \times 10^{-5} = \frac{[H^+][C_2H_3O_2^-]}{[HC_2H_3O_2]}$$

We make the following assumptions:

| SPECIES | EQUILIBRIUM CONCENTRATION |
|---|---|
| $H^+$ | $(x)\,M$ |
| $C_2H_3O_2^-$ | $\left(\dfrac{(500\ ml)(1.0)}{1000\ ml} + x \cong 0.50\right)M$ |
| $HC_2H_3O_2$ | $\left(\dfrac{(500ml)(1.0)}{1000\ ml} - x \cong 0.50\right)M$ |

We substitute concentrations into the expression for the ionization constant:

$$K = 1.8 \times 10^{-5} = \frac{[H^+]\,[C_2H_3O_2^-]}{[HC_2H_3O_2]}$$

$$1.8 \times 10^{-5} = \frac{x(0.50)}{(0.50)}$$

$$x = 1.8 \times 10^{-5} \quad \left(\begin{array}{l}\text{The assumption that}\\ x \ll 0.50 \text{ is valid.}\end{array}\right)$$

Thus,

$$[H^+] = 1.8 \times 10^{-5}M$$

and

$$pH = -\log\,[H^+]$$

$$= -\log\,(1.8 \times 10^5)$$

$$= -0.26 + 5$$

$$= 4.7$$

We could also calculate pH as follows:

$$pH = pK - \log \frac{[HC_2H_3O_2]}{[C_2H_3O_2^-]}$$

$$= 4.7 - \log\left(\frac{0.5}{0.5}\right)$$

$$= 4.7 - 0$$

$$= 4.7$$

b. pH = 4.7   The $H^+$ from the completely dissociated HCl will react with $C_2H_3O_2^-$ to form $HC_2H_3O_2$:

$$H^+ + C_2H_3O_2^- \rightleftharpoons HC_2H_3O_2$$

The protonation of $C_2H_3O_2^-$ has a $K$ equal to $1/(1.8 \times 10^{-5})$, or $5.6 \times 10^4$   Since this constant is large, we assume that all the $H^+$ reacts to form $HC_2H_3O_2$. Thus, we make the following assumptions:

| SPECIES | EQUILIBRIUM CONCENTRATION |
|---------|---------------------------|
| $H^+$ | $(y)M$ |
| $C_2H_3O_2^-$ | $(0.50 - 0.005 = 0.495)M$ |
| $HC_2H_3O_2$ | $(0.50 + 0.005 = 0.505)M$ |

We substitute concentrations into the equation relating pH and p$K$:

$$pH = pK - \log \frac{[HC_2H_3O_2]}{[C_2H_3O_2^-]}$$

$$= 4.7 - \log \left( \frac{0.505}{0.495} \right)$$

$$= 4.7 - \log 1.02$$

$$= 4.7 - 0.009$$

$$= 4.7$$

The buffer solution maintains its pH in spite of the addition of the hydrochloric acid.

SELF-TEST                   Complete the test in 30 minutes:

1. What is the pH of a solution that has an $OH^-$
   concentration of 0.075$M$?

2. What is the $H^+$ concentration of a solution with
   a pH of 9.6?

3. A weak acid, HX, is 0.20% ionized in 0.50$M$
   solution.  What is the ionization constant of
   the acid?

4. The ionization constant of a weak acid HX is
   $6.0 \times 10^{-6}$.  What concentration of NaX must be
   present in a 0.30$M$ solution of HX to prepare
   a buffer with a pH of 5.00?

# 17

# Ionic Equilibria, Part II

OBJECTIVES

(a) You should be able to demonstrate your knowledge of the following terms by defining them, describing them, or giving specific examples of them:

amphoteric hydroxide [17.5]
complex ion [17.4]
formation constant (also called stability constant) [17.4]
hydrolysis [17.6]
ion product [17.2]
ligand [17.4]
precipitation indicator [17.2]
salt effect [17.1]
solubility product [17.1]

(b) You should be able to determine the value of a $K_{SP}$ from experimental data and to calculate solubility from the value of the $K_{SP}$.

(c) You should be able to determine the concentrations of species in a solution that contains complex ions.

(d) You should be able to calculate the pH of a solution containing the conjugate base of a weak acid.

(e) You should be able to calculate the equilibrium constant for a oxidation-reduction reaction.

(f) You should be able to determine titration curves.

EXERCISES

I.  Answer each of the following questions:  The questions relate to Chapters 16 and 17 of your text and are designed to enhance your qualitative understanding of ionic equilibria.  Answer each question without performing any calculations.

_____

1. The pH of pure water at 25°C is closest to

    a. 6.0  b. 6.5  c. 7.0  d. 7.5  e. 8.0

_____

2. The pH of a $10^{-6}M$ HCl solution is closest to

    a. 6.0  b. 6.5  c. 7.0  d. 7.5  e. 8.0

_____

3. The pH of a $10^{-8}M$ HCl solution is closest to

    a. 6.0  b. 6.5  c. 7.0  d. 7.5  e. 8.0

_____

4. The pH of a $10^{-3}M$ NaOH solution is closest to

    a. 3.0  b. 5.0  c. 9.0  d. 11.0

_____

5. The pH of a $10^{-3}M$ NaCl solution is closest to

    a. 3.0  b. 6.0  c. 7.0  d. 8.0  e. 11.0

_____

6. The pH of $10^{-3}M$ acetic acid ($K = 1.8 \times 10^{-5}$) is closest to

    a. 1.0  b. 4.0  c. 7.0  d. 10.0  e. 13.0

7. The species in a $10^{-3}M$ acetic acid solution that has the highest concentration is

    a. $H^+$  b. $OH^-$  c. $Na^+$  d. $C_2H_3O_2^-$  e. $HC_2H_3O_2$

8. The species in a $10^{-5}M$ HCl solution that has the highest concentration is

    a. $H^+$  b. $OH^-$  c. $Na^+$  d. $Cl^-$  e. HCl

9. Five solutions are prepared. Each solution is $10^{-2}M$ in one of the following. Which solution is the most acidic?

    a. $HClO_2$, $K = 1.1 \times 10^{-2}$

    b. $HC_2H_3O_2$, $K = 1.8 \times 10^{-5}$  3

    c. HCN, $K = 4.0 \times 10^{-10}$

    d. HF, $K = 6.7 \times 10^{-4}$

    e. $NH_3$, $K = 1.8 \times 10^{-5}$

10. A solution that is $10^{-2}M$ in acetic acid and $10^{-2}M$ in sodium acetate is prepared. The pK of acetic acid is 4.7. The pH of the solution is closest to

    a. 2.0  b. 4.7  c. 5.0  d. 5.7  e. 7.0

11. The most soluble of the following carbonates is

    a. $BaCO_3$, $K_{SP} = 1.6 \times 10^{-9}$.

    b. $CdCO_3$, $K_{SP} = 5.2 \times 10^{-12}$.

    c. $CaCO_3$, $K_{SP} = 4.7 \times 10^{-9}$.

    d. $PbCO_3$, $K_{SP} = 1.5 \times 10^{-15}$.

    e. $MgCO_3$, $K_{SP} = 1 \times 10^{-15}$

12. The $K_{SP}$ of $CdCO_3$ is $5.2 \times 10^{-12}$, and the solubility of $CdCO_3$ calculated with that value is $2.6 \times 10^{-6}$ mol/liter. The experimentally determined solubility is

a. much lower because of temperature effects.

b. much higher because of carbonate hydrolysis.

c. much higher because of temperature effects.

d. much higher because of supersaturation.

e. the same.

13. Which of the following indicators would be best for the titration of $0.1M$ $HNO_2$ with $0.1M$ NaOH?

a. methyl orange, which has a pH range for color change of 3.1 to 4.5

b. methyl red, which has a pH range for color change of 4.2 to 6.3

c. litmus, which has a pH range for color change of 5.0 to 8.0

d. thymol blue, which has a pH range for color change of 8.0 to 9.6

e. all of the preceding

14. The value of the solubility constant of silver(I) sulfate is approximately the same as that of calcium sulfate. Which of the following statements is true?

a. Calcium sulfate is more soluble than silver sulfate.

b. Silver sulfate is more soluble than calcium sulfate.

c. Silver sulfate is as soluble as calcium sulfate.

d. Nothing can be said about the relative solubilities of silver sulfate and calcium sulfate.

e. None of the preceding are true.

15. The pH of a solution is 4.50.  What is the pOH
    of the solution?

    a. 4.50  b. 5.50  c. 6.50  d. 8.50  e. 9.50

16. How many milliliters of 0.05$M$ NaOH are needed to
    titrate 50 ml of 0.1$M$ acetic acid?

    a. 25  **b.** 50  c. 100  d. 200  e. 500


II.  Answer the following:

1. What is the pH of a 0.10$M$ $NaNO_2$ solution?  The
   ionization constant of $HNO_2$, a weak acid, is
   $4.5 \times 10^{-4}$.

2. At $25^{\circ}C$ a saturated solution of $Ca(OH)_2$ has a pH
   of 12.45.   Answer the following:

   a. What is the molar solubility of $Ca(OH)_2$?

   b. What is the $K_{SP}$ of $Ca(OH)_2$?

3. A 0.10$M$ solution of KCN has a pH of 11.0.   What
   is the ionization constant, $K$, of HCN?

4. What is the pH of a 0.10$M$ $Na_2S$ solution?  The
   ionization constant for the primary ionization
   of $H_2S$ is $1.1 \times 10^{-7}$ and that for the secondary
   ionization is $1.0 \times 10^{-14}$.

5. Calculate the $K_{SP}$ of lead iodide from the following standard electrode potentials:

   $2e^- + PbI_2(s) \rightleftharpoons Pb(s) + 2I^-(aq)$   $\mathscr{E}^{\circ} = -0.365$ V

   $2e^- + Pb^{2+}(aq) \rightleftharpoons Pb(s)$           $\mathscr{E}^{\circ} = -0.126$ V

6. Will a precipitate form in a solution that is
   $1.00 \times 10^{-3}M$ in $Ba^{2+}$ and $1.00 \times 10^{-3}M$ in $SO_4^{2-}$.
   The $K_{SP}$ of $BaSO_4$ is $1.5 \times 10^{-9}$.

7. How many moles of $Ag_2C_2O_4$ will dissolve in 100
   ml of 0.050$M$ $Na_2C_2O_4$?  Neglect the hydrolysis
   of ions.   The $K_{SP}$ of $Ag_2C_2O_4$ is $1.1 \times 10^{-11}$.

8. Which is the least soluble:  AgCl or $Ag_2CrO_4$
The $K_{sp}$ of AgCl is $1.7 \times 10^{-10}$ and that of
$Ag_2CrO_4$ is $1.9 \times 10^{-12}$.

9. A solution is $0.20M$ in $Ca^{2+}$ and $0.10M$ in $Sr^{2+}$.
Answer the following:

    a. If solid $Na_2SO_4$ is added very slowly to the
solution, which will precipitate first:
$CaSO_4$ or $SrSO_4$?  Neglect volume changes.
The $K_{sp}$ of $CaSO_4$ is $2.4 \times 10^{-5}$ and that of
$SrSO_4$ is $7.6 \times 10^{-7}$.

    b. The addition of $Na_2SO_4$ is continued until
the second cation starts to precipitate as
the sulfate.  What is the concentration of
the first cation at that point?

10. What are the final $[Fe^{2+}]$ and $[Cd^{2+}]$ in a solution originally $0.10M$ in $Fe^{2+}$ and $0.10M$ in $Cd^{2+}$
which is buffered with $1.0M$ $SO_4^{2-}$ and $1.0M$ $HSO_4^-$
and saturated with $H_2S$?  For a saturated solution
of $H_2S$

$$[H^+]^2\ [S^{2-}] = 1.1 \times 10^{-22}$$

The ionization constant for the secondary ion-
ization of $H_2SO_4$ is $1.3 \times 10^{-2}$.  The $K_{sp}$ of FeS
is $4.0 \times 10^{-19}$ and that of CdS is $1.0 \times 10^{-28}$.

11. A solution is prepared by mixing 50 ml of $0.020M$
$Ag^+$ with 50 ml of $0.20M$ $NH_3$.  The principal
equilibrium is

$$Ag^+ + 2NH_3 \rightleftharpoons Ag(NH_3)_2^+$$

and the equilibrium expression is

$$K = \frac{[Ag(NH_3)_2^+]}{[Ag^+][NH_3]^2} = 1.67 \times 10^7$$

What are the final concentrations of $Ag^+$, $NH_3$,
$Ag(NH_3)_2^+$, $H^+$, $OH^-$, $NH_3$, and $NH_4^+$ in the solution?
The equilibrium constant for the hydrolysis of
$NH_3$ is $1.8 \times 10^{-5}$.

12. What is the equilibrium constant for the reduction of $Cu^{2+}$ by Zn. The necessary standard electrode potentials are

$$Cu^{2+} + 2e^- \rightleftharpoons Cu \quad \mathscr{E}° = +0.34 \text{ V}$$

$$Zn^{2+} + 2e^- \rightleftharpoons Zn \quad \mathscr{E}° = -0.76 \text{ V}$$

13. Glutaramic acid has an ionization constant of $4.0 \times 10^{-5}$. Draw the titration curve for the titration of 50 ml of $0.200N$ glutaramic acid with $0.200N$ NaOH. The $0.200N$ glutaramic acid solution is diluted to 100 ml before any base is added.

## ANSWERS TO EXERCISES

I. Concepts of ionic equilibria

1. 7.0
   [16.3]

Since

$$K = 1.0 \times 10^{-14} = [H^+][OH^-]$$

and in pure water

$$[H^+] = [OH^-]$$

we determine the pH to be equal to 7:

$$[H^+][OH^-] = 1.0 \times 10^{-14}$$

$$[H^+]^2 = 1.0 \times 10^{-14}$$

$$[H^+] = 1.0 \times 10^{-7}$$

$$pH = -\log [H^+]$$

$$= -\log (1.0 \times 10^{-7})$$

$$= 0 + 7$$

$$= 7$$

2. 6.0
   [16.3]

Since HCl completely dissociates,

$$[H^+] \cong 10^{-6} M$$

Therefore,

   pH = 6

3. 7.0
   [16.3]

The $[H^+]$ from HCl is $10^{-8}M$, but that from the dissociation of water is much greater, $\sim 10^{-7}M$. Thus, pH = 7. The pH is slightly less than 7 because of the presence of $[H^+]$ from HCl.

4. 11.0
   [16.3]

Since NaOH completely dissociates

   $NaOH \rightarrow Na^+ + OH^-$

We determine the pH as follows:

$$pOH = -log\ [OH^-]$$

$$= -log\ (10^{-3})$$

$$= 3$$

$$pOH + pH = 14$$

$$pH = 14\ -pOH$$

$$pH = 14\ -3$$

$$= 11$$

5. 7.0
   [16.3]

A $10^{-3}M$ NaCl solution is neutral.

6. 4.0
   [16.7]

If acetic acid were a strong acid, i.e., if it were completely dissociated in water, the pH would be 3. If it were not dissociated at all, the pH would be 7. Acetic acid actually lies somewhere between the limits; thus, 4.0 is the best choice.

7. $HC_2H_3O_2$
   [16.1]

In solutions of weak acids the most concentrated species is usually the acid.

8. $H^+$
   [16.3]

In water HCl completely dissociates, giving $10^{-5}M\ H^+$ and $10^{-5}M\ Cl^-$. There is a small contribution of $H^+$ from the ionization of water; thus, $[H^+]$ is slightly greater than $[Cl^-]$.

9. $HClO_2$
   [16.1]

A larger dissociation constant corresponds to a greater degree of dissociation.

10. 4.7
    [16.6]

The pH equals the p$K$ since $[HC_2H_3O_2]$ equals $[C_2H_3O_2^-]$:

$$pH = pK + \log \frac{[C_2H_3O_2^-]}{[HC_2H_3O_2]}$$

$$= 4.7 + 0$$

$$= 4.7$$

11. $CaCO_3$
    [17.1]

The most soluble of the carbonates is the one with the largest value of $K_{SP}$:

$$K_{SP} = [M^{2+}][CO_3^{2-}]$$

The largest value of $K_{SP}$ corresponds to the compound that gives the most ions in solution.

12. much higher because of carbonate hydrolysis [17.6]

13. thymol blue
    [17.7]

At the equivalence point the number of moles of acid is equal to the number of moles of base. Since $HNO_2$ is a weak acid, the conjugate base $NO_2^-$ will hydrolyze, thus producing a basic solution. The indicator must therefore change color in the basic region.

14. Calcium sulfate is more soluble than silver sulfate. [17.1]

If we indicate molar solubility by $S$, we find that for $Ag_2SO_4$

$$K_{SP} = S(2S)^2$$

$$= 4S^3$$

and for $CaSO_4$

$$K_{SP} = S^2$$

If the value of $K_{SP}$ is the same for both compounds, we find that for $Ag_2SO_4$

$$S = \sqrt[3]{\tfrac{1}{4}K_{SP}}$$

and for $CaSO_4$

$$S = \sqrt{K_{SP}}$$

Thus, the solubility of $CaSO_4$ is larger.

15. 9.50
   [16.3]

We determine pOH as follows:

$$pH + pOH = 14.00$$
$$pOH = 14.00 - pH$$
$$= 14.00 - 4.50$$
$$= 9.50$$

16. 100 ml
   [17.7]

The number of moles of acid equals the number of moles of base:

$$\text{moles of acid} = \text{moles of base}$$

$$(VM)_{acid} = (VM)_{base}$$

$$(0.050 \text{ liter})\left(\frac{0.1 \text{ mol}}{1 \text{ liter}}\right) = V_{base}\left(\frac{0.05 \text{ mol}}{1 \text{ liter}}\right)$$

$$V_{base} = 100 \text{ ml}$$

## II. Ionic equilibria calculations

1. pH = 7.2
   [17.6]

Sodium nitrite is a strong electrolyte. Thus,

$$NaNO_2(s) \rightarrow Na^+(aq) + NO_2^-(aq)$$

The hydrolysis of $NO_2^-$ is

$$NO_2^-(aq) + H_2O \rightleftharpoons HNO_2(aq) + OH^-(aq)$$

We obtain the equilibrium constant for the hydrolysis as follows:

$$K_H = \frac{[HNO_2][OH^-]}{[NO_2^-]} = \frac{K_W}{K_{HNO_2}}$$

$$= \frac{1.0 \times 10^{-14}}{4.5 \times 10^{-4}}$$

$$= 2.2 \times 10^{-9}$$

We make the following assumptions:

| SPECIES | EQUILIBRIUM CONCENTRATION |
|---------|---------------------------|
| $HNO_2$ | $(x)M$ |
| $OH^-$ | $(x)M$ |
| $NO_2^-$ | $(0.10 - x \cong 0.10)M$ |

We substitute concentrations into the equilibrium expression for the hydrolysis of $NO_2^-$:

$$K_H = 2.2 \times 10^{-9} = \frac{[HNO_2][OH^-]}{[NO_2^-]}$$

$$= \frac{x^2}{0.10}$$

$$x^2 = 2.2 \times 10^{-10}$$

$$x = 1.5 \times 10^{-5}$$

Thus,

$$[OH^-] = 1.5 \times 10^{-5} M$$

We calculate pH as follows:

$$[H^+][OH^-] = 1.0 \times 10^{-14}$$

$$[H^+] = \frac{1.0 \times 10^{-14}}{[OH^-]}$$

$$= \frac{1.0 \times 10^{-14}}{1.5 \times 10^{-5}}$$

$$= 6.7 \times 10^{-8}$$

$$pH = -\log(6.7 \times 10^{-8})$$

$$= 0.8 + 8$$

$$= 7.2$$

2. [17.1]
  a. $K_{SP} =$
     $2.3 \times 10^{-5}$

The equilibrium between solid $Ca(OH)_2$ and a saturated solution of $Ca(OH)_2$ is

$$Ca(OH)_2(s) \rightleftharpoons Ca^{2+}(aq) + 2OH^-(aq)$$

and the $K_{SP}$ is

$$K_{SP} = [Ca^{2+}][OH^-]^2$$

We calculate the $[OH^-]$ as follows:

$$pH + pOH = 14$$

$$pOH = 14 - pH$$

$$= 14 - 12.45$$

$$= 1.55$$

$$[OH^-] = 10^{-pOH}$$

$$= 10^{-1.55}$$

$$= (10^{0.55})(10^{-2})$$

$$= 3.6 \times 10^{-2}$$

According to the balanced equation two moles of $OH^-$ are produced for every mole of $Ca^{2+}$ that reacts. Thus,

$$[Ca^{2+}] = \frac{[OH^-]}{2}$$

$$= \frac{(3.6 \times 10^{-2})}{2}$$

$$= 1.8 \times 10^{-2}$$

We calculate the $K_{SP}$:

$$K_{SP} = [Ca^{2+}][OH^-]^2$$

$$= (1.8 \times 10^{-2})(3.6 \times 10^{-2})^2$$

$$= 2.3 \times 10^{-5}$$

b. $1.8 \times 10^{-2}$

The molar solubility of $Ca(OH)_2$ at $25^{\circ}C$ is the number of moles of $Ca(OH)_2$ dissolved per liter of solution at that temperature.  Since one mole of $Ca^{2+}$ is present per mole of $Ca(OH)_2$ dissolved,

$$\text{molar solubility of } Ca(OH)_2 = [Ca^{2+}]$$

$$= 1.8 \times 10^{-2}$$

3. $K = 1.0 \times 10^{-9}$
   [17.6]

Potassium cyanide is a strong electrolyte:

$$KCN \rightarrow K^+ + CN^-$$

The cyanide ion is hydrolyzed:

$$CN^- + H_2O \rightleftharpoons HCN + OH^-$$

The equilibrium expression for the hydrolysis is

$$K_H = \frac{K_W}{K_{HCN}} = \frac{[HCN][OH^-]}{[CN^-]}$$

We calculate the $[OH^-]$:

$$pOH = 14 - pH$$

$$= 14 - 11.0$$

$$= 3.0$$

$$[OH^-] = 10^{-pOH}$$

$$= 1 \times 10^{-3}$$

We make the following assumptions:

| SPECIES | EQUILIBRIUM CONCENTRATION |
|---------|---------------------------|
| $OH^-$ | $1 \times 10^{-3}M$ |
| $HCN$ | $1.0 \times 10^{-3}M$ |
| $CN^-$ | $(1.0 \times 10^{-1}M) - (1.0 \times 10^{-3}M) \cong 1.0 \times 10^{-1}M$ |

We substitute concentrations into the expression for the $K_H$:

$$K_H = \frac{K_W}{K_{HCN}} = \frac{[HCN][OH^-]}{[CN^-]}$$

$$\frac{1.0 \times 10^{-14}}{K_{HCN}} = \frac{(1.0 \times 10^{-3})(1.0 \times 10^{-3})}{1.0 \times 10^{-1}}$$

$$K_{HCN} = \frac{(1.0 \times 10^{-14})(1.0 \times 10^{-1})}{(1.0 \times 10^{-3})^2}$$

$$= 1.0 \times 10^{-9}$$

4. pH = 13
   [17.6]

We consider the following equilibria:

$$S^{2-} + H_2O \rightleftharpoons HS^- + OH^- \qquad K_H = \frac{K_W}{K_{2_{H_2S}}} = \frac{1.0 \times 10^{-14}}{1.0 \times 10^{-14}} = 1.0$$

$$HS^- + H_2O \rightleftharpoons H_2S + OH^- \qquad K_H = \frac{K_W}{K_{1_{H_2S}}} = \frac{1.0 \times 10^{-14}}{1.1 \times 10^{-7}} = 9.1 \times 10^{-8}$$

$$H_2O \rightleftharpoons H^+ + OH^- \qquad K_W = 1.0 \times 10^{-14}$$

Since the $K$ of the first equilibrium is much larger than either of the other two, the first equilibrium is the most important and the only one we must consider. In addition, the $OH^-$ produced in the hydrolysis of $S^{2-}$ would repress the other two equilibria as written.

The expression for hydrolysis of $S^{2-}$ is

$$K_H = 1.0 = \frac{[HS^-][OH^-]}{[S^{2-}]}$$

We make the following assumptions:

| SPECIES | EQUILIBRIUM CONCENTRATION |
|---------|---------------------------|
| $OH^-$ | $(x)M$ |
| $HS^-$ | $(x)M$ |
| $S^{2-}$ | $(0.10 - x)M$ |

We substitute concentrations into the expression for the hydrolysis constant of $S^{2-}$:

$$K_H = 1.0 = \frac{[HS^-][OH^-]}{[S]^{2-}}$$

$$1.0 = \frac{x^2}{(0.10 - x)}$$

$$x^2 = 0.10 - x$$

$$x^2 + x - 0.10 = 0$$

We use the quadratic formula to solve for $x$:

$$x = \frac{-1.0 \pm \sqrt{(1.0)^2 - 4(1.0)(-0.10)}}{2(1.0)}$$

$$x = \frac{-1.0 \pm 1.2}{2}$$

$$x = 1 \times 10^{-1}$$

Thus,

$$[OH^-] = 1.0 \times 10^{-1}M$$

If we had assumed that $x \ll 0.10$, the solution would have been $x = 3.2 \times 10^{-1}$, a valve greater than 0.10. Thus, the assumption that $x$ is not negligible is valid.

We calculate the pH as follows:

$$[H^+][OH^-] = 1.0 \times 10^{-14}$$

$$[H^+] = \frac{1.0 \times 10^{-14}}{[OH^-]}$$

$$= \frac{1.0 \times 10^{-14}}{1.0 \times 10^{-1}}$$

$$= 1.0 \times 10^{-13}$$

$$pH = -\log [H^+]$$

$$= -\log (1.0 \times 10^{-13})$$

$$= 13$$

5. $K_{SP} = 8.5 \times 10^{-9}$
   [17.1]

The equilibrium between solid lead iodide and a saturated solution of lead iodide is

$$PbI_2(s) \rightleftharpoons Pb^{2+}(aq) + 2I^-(aq)$$

We calculate the $\mathscr{E}°$ of the cell from the half reactions:

$$2e^- + PbI_2(s) \rightleftharpoons Pb(s) + 2I^-(aq) \qquad \mathscr{E}° = -0.365 \text{ V}$$

$$Pb(s) \rightleftharpoons Pb^{2+}(aq) + 2e^- \qquad \mathscr{E}° = +0.126 \text{ V}$$

$$PbI_2(s) \rightleftharpoons Pb^{2+}(aq) + 2I^-(aq) \qquad \mathscr{E}° = -0.239 \text{ V}$$

We calculate $K_{SP}$ as follows:

$$\Delta G° = -nF\mathscr{E}°$$

$$= -(2)(23.06 \text{ kcal/V})(-0.239 \text{ V})$$

$$= 11.0 \text{ kcal}$$

$$\Delta G^0 = -2.303 \, RT \, \log K$$

$$\log K = -\frac{\Delta G^0}{2.303 \, RT}$$

$$\log K = -\frac{11.0 \text{ kcal}}{(2.303)(1.987 \text{ cal}/^\circ K \text{ mol})(298^\circ K)}$$

$$\log K = -8.07$$

$$K = 10^{-8.07}$$

$$= (10^{+0.93})(10^{-9})$$

$$= 8.5 \times 10^{-9}$$

6. yes
   [17.2]

The $K_{SP}$ of $BaSO_4$ is

$$K_{SP} = 1.5 \times 10^{-9} = [Ba^{2+}][SO_4^{2-}]$$

In a solution that is $0.001M$ in $Ba^{2+}$ and $0.001M$ in $SO_4^{2-}$, the ion product is

$$\text{ion product} = [Ba^{2+}][SO_4^{2-}]$$

$$= (1 \times 10^{-3})(1 \times 10^{-3})$$

$$= 1 \times 10^{-6}$$

If the ion product is greater than $K_{SP}$, precipitation will occur until the ion product equals the $K_{SP}$. Since the ion product of $BaSO_4$ is greater than the $K_{SP}$, precipitation will occur.

7. $7.4 \times 10^{-7}$ mol
   [17.2]

Sodium oxalate is a strong electrolyte:

$$Na_2C_2O_4 \rightarrow 2Na^+ + C_2O_4^{2-}$$

In a saturated solution the ion product is equal to the $K_{SP}$ and no more solid will dissolve. The equilibrium between solid $Ag_2C_2O_4$ and a saturated solution of $Ag_2C_2O_4$ is

$$Ag_2C_2O_4 \rightleftharpoons 2 \, Ag^+ + C_2O_4^{2-}$$

and the $K_{SP}$ for $Ag_2C_2O_4$ is

$$K_{SP} = 1.1 \times 10^{-11} = [Ag^+]^2[C_2O_4^{2-}]$$

We make the following assumptions:

| SPECIES | EQUILIBRIUM CONCENTRATION |
|---|---|
| $Ag^+$ | $(2x)M$ (where $x$ = molar solubility of $Ag_2C_2O_4$) |
| $C_2O_4^{2-}$ | $\left((5.0 \times 10^{-2}) + x \cong 5.0 \times 10^{-2}\right)M$ |

We substitute concentrations into the expression for the $K_{SP}$ :

$$K_{SP} = 1.1 \times 10^{-11} = [Ag]^2[C_2O_4^{2-}]$$

$$= (2x)^2 (5.0 \times 10^{-2})$$

$$= (2.0 \times 10^{-1})(x^2)$$

$$x^2 = \frac{1.1 \times 10^{-11}}{2.0 \times 10^{-1}}$$

$$x^2 = 5.5 \times 10^{-11}$$

$$x = 7.4 \times 10^{-6}$$

Thus, $7.4 \times 10^{-6}$ mol of $Ag_2C_2O_4$ will dissolve per liter of solution. We calculate the number of moles that will dissolve in 100 ml of solution:

$$? \text{ mol} = \left(\frac{7.4 \times 10^{-6} \text{ mol}}{1000 \text{ ml}}\right) 100 \text{ ml}$$

$$= 7.4 \times 10^{-7} \text{ mol}$$

8. AgCl
   [17.2]

The $K_{SP}$ of AgCl is $1.7 \times 10^{-10}$ and that of $Ag_2CrO_4$ is $1.9 \times 10^{-12}$. The compound with the smaller molar solubility is the less soluble. We calculate the molar solubility of each compound:

AgCl:

The equilibrium between solid silver chloride and a saturated solution of silver chloride is

$$AgCl \rightleftharpoons Ag^+ + Cl^-$$

Since 1 mol of $Ag^+$ and 1 mol of $Cl^-$ are produced per mole of AgCl dissolved, the molar solubility of AgCl, $x$, is

$$x = [Ag^+] = [Cl^-]$$

We calculate $x$ from the $K_{SP}$ of AgCl:

$$K_{SP} = 1.7 \times 10^{-10} = [Ag^+][Cl^-]$$

$$1.7 \times 10^{-10} = x^2$$

$$x = 1.3 \times 10^{-5}$$

$Ag_2CrO_4$:

The equilibrium between solid silver chromate and a saturated solution of silver chromate is

$$Ag_2CrO_4 \rightleftharpoons 2\,Ag^+ + CrO_4^{2-}$$

Since 2 mol of $Ag^+$ and 1 mol of $CrO_4^{2-}$ are produced per 1 mol of $Ag_2CrO_4$ dissolved, the molar solubility of $Ag_2CrO_4$, $y$, is

$$y = \frac{[Ag^+]}{2} = [CrO_4^{2-}]$$

We calculate $y$ from the $K_{SP}$ of $Ag_2CrO_4$

$$K_{SP} = 1.9 \times 10^{-12} = [Ag^+]^2[CrO_4]$$

$$= [2y]^2[y]$$

$$= 4y^3$$

$$y = 7.8 \times 10^{-5}$$

Thus, AgCl is the less soluble.

9. [17.2]
   a. $SrSO_4$

The $K_{SP}$ of $CaSO_4$ is $2.4 \times 10^{-5}$ and that of $SrSO_4$ is $7.6 \times 10^{-7}$:

$$K_{SP} = 2.4 \times 10^{-5} = [Ca^{2+}][SO_4^{2-}]$$

$$K_{SP} = 7.6 \times 10^{-7} = [Sr^{2+}][SO_4^{2-}]$$

Precipitation begins when the ion product just exceeds the $K_{SP}$. A greater concentration of $SO_4^{2-}$ is necessary to exceed the $K_{SP}$ of $CaSO_4$ than to exceed that of $SrSO_4$:

$$[Ca^{2+}][SO_4^{2-}] = 2.4 \times 10^{-5}$$

$$[SO_4^{2-}] = \frac{2.4 \times 10^{-5}}{[Ca^{2+}]}$$

$$= \frac{2.4 \times 10^{-5}}{2.0 \times 10^{-1}}$$

$$= 1.2 \times 10^{-4}$$

$$[Sr^{2+}][SO_4^{2-}] = 7.6 \times 10^{-7}$$

$$[SO_4^{2-}] = \frac{7.6 \times 10^{-7}}{1.0 \times 10^{-1}}$$

$$= 7.6 \times 10^{-6}$$

Thus, $SrSO_4$ will precipitate first.

   b. $[Sr^{2+}] =$
   $6.3 \times 10^{-3} M$

We know that $[SO_4^{2-}] = 1.2 \times 10^{-4} M$ when $Ca^{2+}$ begins to precipitate as $CaSO_4$ [see answer to part (a) of this problem]. We can determine the $[Sr^{2+}]$ when $Ca^{2+}$ begins to precipitate as follows:

$$K_{SP} = 7.6 \times 10^{-7} = [Sr^{2+}][SO_4^{2-}]$$

$$[Sr^{2+}] = \frac{7.6 \times 10^{-7}}{1.2 \times 10^{-4}}$$

$$= 6.3 \times 10^{-3}$$

10. $[Fe^{2+}] = 0.10M$

$[Cd^{2+}] = 1.3 \times 10^{-10}M$

[17.2]

To determine whether a metal sulfide precipitates, we calculate the ion product for each compound. First we determine $[S^{2-}]$ in a saturated $H_2S$ solution:

$$[H^+]^2[S^{2-}] = 1.1 \times 10^{-22}$$

The $[H^+]$ is controlled by the $HSO_4^-/SO_4^{2-}$ buffer:

$$HSO_4^- \rightleftharpoons H^+ + SO_4^{2-}$$

Thus, we calculate the $[H^+]$ from the ionization expression for $HSO_4^-$:

$$K = 1.3 \times 10^{-2} = \frac{[H^+][SO_4^{2-}]}{[HSO_4^-]}$$

We make the following assumptions:

| SPECIES | EQUILIBRIUM CONCENTRATION |
|---------|---------------------------|
| $H^+$ | $(x)M$ |
| $SO_4^{2-}$ | $(1.0 + x \cong 1.0)M$ |
| $HSO_4^-$ | $(1.0 - x \cong 1.0)M$ |

We substitute concentrations into the expression for the ionization constant of $HSO_4^-$:

$$K = 1.3 \times 10^{-2} = \frac{[H^+][SO_4^{2-}]}{[HSO_4^-]}$$

$$= \frac{(x)(1.0)}{(1.0)}$$

$$x = 1.3 \times 10^{-2}$$

Thus, $[H^+] = 1.3 \times 10^{-2}M$

We determine the $[S^{2-}]$:

$$[H^+]^2[S^{2-}] = 1.1 \times 10^{-22}$$

$$[S^{2-}] = \frac{1.1 \times 10^{-22}}{[H^+]^2}$$

$$= \frac{1.1 \times 10^{-22}}{(1.3 \times 10^{-2})^2}$$

$$= 6.5 \times 10^{-19}$$

The ion product just exceeds $K_{SP}$ when the metal sulfide begins to precipitate. Thus, for FeS the ion product is

$$[Fe^{2+}][S^{2-}] = (1.0 \times 10^{-1})(6.5 \times 10^{-19})$$

$$= 6.5 \times 10^{-20}$$

and the $K_{SP}$ is $4.0 \times 10^{-19}$. Since the $K_{SP}$ of FeS is greater than the ion product, FeS will not precipitate and the final $[Fe^{2+}]$ is $0.10M$.

For CdS the ion product is

$$[Cd^{2+}][S^{2-}] = (1.0 \times 10^{-1})(6.5 \times 10^{-19})$$

$$= 6.5 \times 10^{-20}$$

and the $K_{SP}$ is $1 \times 10^{-28}$.

Since the $K_{SP}$ of CdS is less than the ion product, CdS will precipitate:

$$Cd^{2+} + H_2S \rightarrow 2H^+ + CdS$$

We calculate the concentration of $Cd^{2+}$ after precipitation of CdS:

$$K_{SP} = 1.0 \times 10^{-28} = [Cd^{2+}][S^{2-}]$$

$$[Cd^{2+}] = \frac{1.0 \times 10^{-28}}{[S^{2-}]}$$

$$[Cd^{2+}] = \frac{1.0 \times 10^{-28}}{6.5 \times 10^{-19}}$$

$$= 1.5 \times 10^{-10}$$

11. $[Ag^+] =$    $6.0 \times 10^{-8}M$

$[NH_3] =$    $1.0 \times 10^{-1}M$

$[Ag(NH_3)_2^+] =$    $1.0 \times 10^{-2}M$

$[NH_4^+] =$    $1.3 \times 10^{-3}M$

$[OH^-] =$    $1.3 \times 10^{-3}M$

$[H^+] =$    $7.7 \times 10^{-12}M$

[17.4]

We make the following assumptions:

| SPECIES | EQUILIBRIUM CONCENTRATION |
|---------|---------------------------|
| $Ag^+$ | $(x)M$ |
| $NH_3$ | $(0.10 - 2x \cong 0.10)M$ |
| $Ag(NH_3)_2^+$ | $(0.010 - x \cong 0.010)M$ |

We substitute concentrations into the expression for the formation constant of $Ag(NH_3)_2^+$:

$$K = 1.67 \times 10^7 = \frac{[Ag(NH_3)_2^+]}{[Ag^+][NH_3]^2}$$

$$= \frac{0.01}{(x)(0.10)^2}$$

$$x = \frac{0.010}{(0.10)^2(1.67 \times 10^7)}$$

$$= 6.0 \times 10^{-8}$$

Thus,

$$[NH_3] = 0.10M - 2(6.0 \times 10^{-8}M)$$

$$= 0.10M$$

$$[Ag(NH_3)_2^+] = (0.010M) - (6.0 \times 10^{-8}M)$$

$$= 0.010M$$

$$[Ag^+] = 6.0 \times 10^{-8} M$$

We use the equilibrium expression for the hydrolysis of $NH_3$ to determine $[OH^-]$ and $[NH_4^+]$:

$$NH_3(aq) + H_2O \rightleftharpoons NH_4^+(aq) + OH^-(aq)$$

$$K_H = 1.8 \times 10^{-5} = \frac{[NH_4^+][OH^-]}{[NH_3]}$$

We make the following assumptions:

| SPECIES | EQUILIBRIUM CONCENTRATION |
|---------|---------------------------|
| $OH^-$ | $(x)M$ |
| $NH_4^+$ | $(x)M$ |
| $NH_3$ | $(0.10 - x \cong 0.10)M$ |

We substitute concentrations into the expression for the hydrolysis constant of $NH_3$:

$$K_H = 1.8 \times 10^{-5} = \frac{[NH_4^+][OH^-]}{[NH_3]}$$

$$= \frac{x^2}{0.10}$$

$$x^2 = 1.8 \times 10^{-6}$$

$$x = 1.3 \times 10^{-3}$$

Thus,

$$[NH_3] = 0.10 M$$

$$[NH_4^+] = 1.3 \times 10^{-3} M$$

$$[OH^-] = 1.3 \times 10^{-3} M$$

We determine $[H^+]$:

$$[H^+][OH^-] = 1.0 \times 10^{-14}$$

$$[H^+] = \frac{1.0 \times 10^{-14}}{1.3 \times 10^{-3}}$$

$$= 7.7 \times 10^{-12}$$

Thus,

$$[H^+] = 7.7 \times 10^{-12} M$$

12. $K = 1.6 \times 10^{37}$

[17.8]

We determine the value of $\mathscr{E}°$ for the reduction of $Cu^{2+}$ by Zn:

| | |
|---|---|
| $Cu^{2+} + 2e^- \rightarrow Cu$ | $\mathscr{E}° = +0.34$ V |
| $Zn \rightarrow Zn^{2+} + 2e^-$ | $\mathscr{E}° = +0.76$ V |
| $Cu^{2+} + Zn \rightarrow Cu + Zn^{2+}$ | $\mathscr{E}° = +1.10$ V |

We calculate the equilibrium constant for the reaction:

$$\log K = \frac{n\mathscr{E}°}{0.05916}$$

$$= \frac{2(1.10)}{0.05916}$$

$$= 37.2$$

$$K = \frac{[Zn^{2+}]}{[Cu^{2+}]}$$

$$= 10^{37.2}$$

$$= 1.6 \times 10^{37}$$

13. [17.7]               The titration curve is

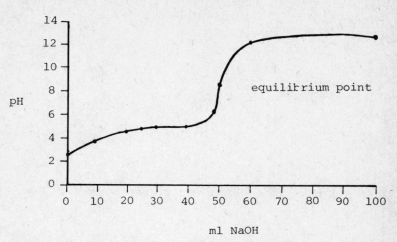

ml NaOH

We determine the titration curve for the titration
of glutaramic acid, a weak acid, with sodium hydroxide, a strong base, as follows:

For simplicity we will use HA to designate glutaramic
acid.   Since HA is a weak acid,

$$[HA] \neq [H^+]$$

Thus, at the initial point of the titration curve
(i.e., the point at which no base has been added)
the solution contains only HA, $H^+$, and $A^-$.   We
make the following assumptions:

| SPECIES | EQUILIBRIUM CONCENTRATION |
|---------|---------------------------|
| $H^+$ | $(x)M$ |
| $A^-$ | $(x)M$ |
| HA | $\left(\dfrac{(50.0 \text{ ml}) \ (0.200M)}{100 \text{ ml}} - x\right)M \cong 0.100M$ |

We substitute concentrations into the expression for the ionization constant of HX:

$$K = 4.0 \times 10^{-5} = \frac{[H^+][X^-]}{[HX]}$$

$$= \frac{x^2}{0.100}$$

$$x^2 = 4.0 \times 10^{-6}$$

$$x = 2.0 \times 10^{-3}$$

Thus,

$$[H^+] = 2.0 \times 10^{-3}M$$

and

$$pH = -\log [H^+]$$

$$= -\log(2.0 \times 10^{-3})$$

$$= -0.30 + 3$$

$$= 2.7$$

Thus, the pH of the solution is 2.7 before base is added. We now determine the pH of the solution as specified amounts of NaOH are added:

## Addition of 10.0 ml of NaOH

$$\text{mmol HA before addition of any base} = (100 \text{ ml}) (0.100M \text{ HA})$$

$$= 10.0 \text{ mmol HA}$$

$$\text{mmol NaOH added} = (10.0 \text{ ml}) (0.200M \text{ NaOH})$$

$$= 2.00 \text{ mmol NaOH}$$

$$\text{mmol NaA formed} = \text{mmol NaOH added}$$

$$= 2.00 \text{ mmol}$$

$$\text{mmol HA remaining} = (10.0 - 2.00) \text{ mmol}$$

$$= 8.0 \text{ mmol}$$

$$\text{total volume of solution} = 100 \text{ ml} + 10 \text{ ml}$$

$$= 110 \text{ ml}$$

Because of the NaA formed by the neutralization, $[H^+] \neq [A^-]$. We substitute concentrations into the expression for the ionization constant of HA and solve for $[H^+]$:

$$K = 4.0 \times 10^{-5} = \frac{[H^+][A^-]}{[HA]}$$

$$= \frac{(x)(2.0 \text{ mmol}/110 \text{ ml})}{(8.0 \text{ mmol}/110 \text{ ml})}$$

$$x = 1.6 \times 10^{-4}$$

Thus,

$$[H^+] = 1.6 \times 10^{-4} M$$

and

$$pH = -\log [H^+]$$

$$= -\log (1.6 \times 10^{-4})$$

$$= -0.2 + 4$$

$$= 3.8$$

Addition of 20.0 ml of NaOH

$$\text{mmol HA before addition of any base} = (100 \text{ ml})(0.100M \text{ HA})$$

$$= 10.0 \text{ mmol HA}$$

$$\text{mmol NaOH added} = \left(20.0 \text{ ml}\right)\left(0.200M \text{ NaOH}\right)$$

$$= 4.00 \text{ mmol NaOH}$$

$$\text{mmol NaA formed} = \text{mmol NaOH added}$$

$$= 4.00 \text{ mmol}$$

mmol HA remaining = (10.0 - 4.00)mmol

$$= 6.0 \text{ mmol}$$

total volume of solution = 100 ml + 20 ml

$$= 120 \text{ ml}$$

We substitute concentrations into the expression for the ionization constant of HA and solve for $[H^+]$

$$K = 4.0 \times 10^{-5} = \frac{[H^+][A^-]}{[HA]}$$

$$= \frac{(x)(4.0 \text{ mmol}/120 \text{ ml})}{(6.0 \text{ mmol}/120 \text{ ml})}$$

$$x = 6.0 \times 10^{-5}$$

Thus,

$$[H^+] = 6.0 \times 10^{-5} M$$

and

$$pH = -\log [H^+]$$

$$= -\log (6.0 \times 10^{-5})$$

$$= -0.78 + 5$$

$$= 4.2$$

### Addition of 25.0 ml of NaOH

mmol HA before addition of any base = (100 ml) (0.100$M$ HA)

$$= 10.0 \text{ mmol HA}$$

mmol NaOH added = (25.0 ml) (0.200$M$ NaOH)

$$= 5.00 \text{ mmol NaOH}$$

mmol NaA formed = mmol NaOH added

$$= 5.00 \text{ mmol}$$

mmol HA remaining = (10.0 − 5.00) mmol

$$= 5.0 \text{ mmol}$$

total volume of solution = 100 ml + 25 ml

$$= 125 \text{ ml}$$

We substitute concentrations into the expression for the ionization constant of HA and solve for $[H^+]$:

$$K = 4.0 \times 10^{-5} = \frac{[H^+][A^-]}{[HA]}$$

$$= \frac{(x)(5.0 \text{ mmol}/125 \text{ ml})}{(5.0 \text{ mmol}/125 \text{ ml})}$$

$$x = 4.0 \times 10^{-5}$$

Thus,

$$[H^+] = 4.0 \times 10^{-5} M$$

and

$$pH = -\log [H^+]$$

$$= -\log (4.0 \times 10^{-5})$$

$$= -0.60 + 5$$

$$= 4.4$$

### Addition of 30.0 ml of NaOH

mmol HA before addition of any base = (100 ml) (0.100$M$ HA)

$$= 10.0 \text{ mmol} \quad HA$$

mmol NaOH added = (30.0 ml) (0.200$M$ NaOH)

$$= 6.00 \text{ mmol NaOH}$$

mmol NaA formed = mmol NaOH added

$$= 6.00 \text{ mmol}$$

mmol HA remaining = (10.0 - 6.00) mmol

$$= 4.0 \text{ mmol}$$

total volume of solution = 100 ml + 30.0 ml

$$= 130 \text{ ml}$$

We substitute concentrations into the expression for the ionization constant of HA and solve for $[H^+]$:

$$K = 4.0 \times 10^{-5} = \frac{[H^+][A^-]}{[HA]}$$

$$= \frac{(x)(6.0 \text{ mmol}/130 \text{ ml})}{(4.0 \text{ mmol}/130 \text{ ml})}$$

$$x = 2.7 \times 10^{-5}$$

Thus,

$$[H^+] = 2.7 \times 10^{-5}M$$

and

$$pH = -\log [H^+]$$

$$= -\log (2.7 \times 10^{-5})$$

$$= -0.43 + 5$$

$$= 4.6$$

Addition of 40.0 ml of NaOH

mmol HA before addition of any base = (100 ml) (0.100M HA)

$$= 10.0 \text{ mmol HA}$$

mmol NaOH added = (40.0 ml) (0.200M NaOH)

$$= 8.00 \text{ mmol NaOH}$$

mmol NaA formed = mmol NaOH added

$$= 8.00 \text{ mmol}$$

$$\text{mmol HA remaining} = (10.0 - 8.00)\text{mmol}$$

$$= 2.00 \text{ mmol}$$

$$\text{total volume of solution} = 100 \text{ ml} + 40.0 \text{ ml}$$

$$= 140 \text{ ml}$$

We substitute concentrations into the expression for the ionization constant of HA and solve for $[H^+]$:

$$K = 4.0 \times 10^{-5} = \frac{[H^+][A^-]}{[HA]}$$

$$= \frac{(x)(8.00 \text{ mmol}/140 \text{ ml})}{(2.00 \text{ mmol}/140 \text{ ml})}$$

$$x = 1.0 \times 10^{-5}$$

Thus,

$$[H^+] = 1.0 \times 10^{-5}M$$

and

$$pH = -\log[H^+]$$

$$= -\log(1.0 \times 10^{-5})$$

$$= 0 + 5$$

$$= 5.0$$

### Addition of 49.0 ml of NaOH

$$\text{mmol HA before addition of any base} = (100 \text{ ml})(0.100M \text{ HA})$$

$$= 10.0 \text{ mmol HA}$$

$$\text{mmol NaOH added} = (49.0 \text{ ml})(0.200M \text{ NaOH})$$

$$= 9.80 \text{ mmol NaOH}$$

$$\text{mmol NaA formed} = \text{mmol NaOH added}$$

$$= 9.80 \text{ mmol}$$

mmol HA remaining = (10.0 - 9.80) mmol

$$= 0.2 \text{ mmol}$$

total volume of solution = 100 ml + 49 ml

$$= 149 \text{ ml}$$

We substitute concentrations into the expression for the ionization constant of HA and solve for $[H^+]$:

$$K = 4.0 \times 10^{-5} = \frac{[H^+][A^-]}{[HA]}$$

$$= \frac{(x)(9.80 \text{ mmol}/149 \text{ ml})}{(0.2 \text{ mmol}/149 \text{ ml})}$$

$$x = 8.2 \times 10^{-7}$$

Thus,

$$[H^+] = 8.2 \times 10^{-7} M$$

and

$$pH = -\log [H^+]$$

$$= -\log (8.2 \times 10^{-7})$$

$$= -0.91 + 7$$

$$= 6.1$$

### Addition of 50.0 ml of NaOH

mmol HA before addition of any base = (100 ml) (0.100M HA)

$$= 10.0 \text{ mmol HA}$$

mmol NaOH added = (50.0 ml) (0.200M NaOH)
$$= 10.0 \text{ mmol NaOH}$$

mmol NaA formed = mmol  NaOH added

$$= 10.0 \text{ mmol}$$

mmol HA remaining = (10.0 - 10.0) mmol

$$= 0 \text{ mmol}.$$

The solution contains only the salt of a weak acid and a strong base.  We determine the pH from the expression for the hydrolysis of the salt:

$$A^- + H_2O \rightleftharpoons HA + OH^-$$

$$K_H = \frac{K_W}{K_{HA}} = \frac{[HA][OH^-]}{[A^-]}$$

We substitute concentrations into the expression for the hydrolysis constant and solve for $[OH^-]$. Note that $[OH^-] = [HA]$.

$$K_H = \frac{K_W}{K_{HA}} = \frac{[HA][OH^-]}{[A^-]}$$

$$[OH^-]^2 = \frac{K_W[A^-]}{K_{HA}}$$

$$[OH^-]^2 = \frac{(1.0 \times 10^{-14})(10.0 \text{ mmol}/150 \text{ ml})}{(4.0 \times 10^{-5})}$$

$$[OH^-]^2 = 1.7 \times 10^{-11}$$

$$[OH^-] = 4.1 \times 10^{-6}$$

$$pOH = -\log[OH^-]$$

$$= -\log(4.1 \times 10^{-6})$$

$$= -0.61 + 6$$

$$= 5.4$$

$$pH + pOH = 14$$

$$pH = 14 - pOH$$

$$= 14 - 5.4$$

$$= 8.6$$

After the equivalence point there is an excess of strong base, and the salt of the weak acid and strong base does not hydrolyze appreciably because of the common-ion effect.  We can calculate the pH from the amount of base in excess of that needed to neutralize the acid:

## Addition of 60.0 ml NaOH

$$[OH^-] = \frac{(\text{mmol of NaOH added}) - (\text{mmol HA originally present})}{\text{total volume of solution}}$$

$$= \frac{(60.0 \text{ ml}) \ (0.200M) \ - \ (100 \text{ ml}) \ (0.100M)}{160 \text{ ml}}$$

$$= 0.0125M$$

$$pOH = -\log [OH^-]$$

$$= -\log (1.25 \times 10^{-2})$$

$$= -0.0969 + 2$$

$$= 1.90$$

$$pH + pOH = 14$$

$$pH = 14 - 1.90$$

$$pH = 12.1$$

## Addition of 100.0 ml NaOH

$$[OH^-] = \frac{(\text{mmol of NaOH added}) - (\text{mmol HA originally present})}{\text{total volume of solution}}$$

$$[OH^-] = \frac{(100 \text{ ml}) \ (0.200M) \ - \ (100 \text{ ml}) \ (0.100M)}{200 \text{ ml}}$$

$$= 0.0500M$$

$$pOH = -\log[OH^-]$$

$$= -\log (5.00 \times 10^{-2})$$

$$= 0.699 + 2$$

$$= 1.30$$

$$pH + pOH = 14$$

$$pH = 14 - 1.30$$

$$pH = 12.7$$

SELF-TEST

Complete the test in 40 minutes.   The test covers Chapters 16 and 17.

1. How many grams of $Ag^+$ are in a 250 ml of a 0.100$M$ $K_2CrO_4$ solution that is saturated with $Ag_2CrO_4$? The $K_{SP}$ of $Ag_2CrO_4$ is $1.9 \times 10^{-12}$.

2. What is the pH at the equivalence point of a titration of 50.0 ml of 0.10$M$ trimethylacetic acid with 0.10$M$ NaOH?   The p$K$ of the ionization of trimethylacetic acid is 5.0.

3. What is the pH of the solution described in problem 3 of the Self-test after 30.0 ml of 0.10$M$ NaOH is added?

4. What is the % hydrolysis of a 0.100$M$ solution of sodium acetate?   The ionization constant of acetic acid is $1.8 \times 10^{-5}$.

# 18
## Metals

Kroll process [18.4]
lanthanide [18.9]
lanthanide contraction [18.8]
matte [18.4]
metallurgy [18.3]
Mond process [18.5]
open hearth process [18.5]
ore [18.2]
Parkes process [18.5]
slag [18.4]
smelting [18.4]
thermite process [18.4]
transition metals [18.8]
Van Arkel process [18.5]
zone refining [18.5]

(b) You should be familiar with the chemical and physical properties of the metals shown in the shaded area of the periodic table in Figure 18.1 of the study guide.

(c) You should be familiar with the natural sources of the most important metals and the methods of refining or preparing them.

(d) You should be able to apply the basic information of all previous chapters of your text toward an understanding of the chemistry of metals.

EXERCISES

I.  Answer each of the following:

    1. Which of the following metals does not have an

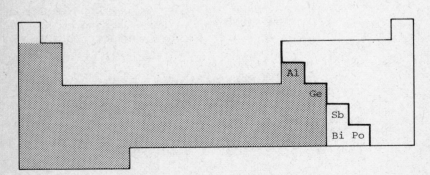

Figure 18.1 Periodic table

important oxidation state of 1+?
a. Cu   b. Na   c. Au   d. Fe   e. Tl

2. Which of the following is not used as a
   dessicant?
   a. NaCl                    d. $CaSO_4$
   b. $CaCl_2$                e. $Ba(ClO_4)_2$
   c. $Mg(ClO_4)_2$

3. Which of the following is the softest metal?
   a. Li   b. Cs   c. Be   d. Ba   e. Na

4. Aluminum is produced by
   a. the Hall process
   b. zone refining
   c. hydrogen reduction
   d. the Bessemer process
   e. the Van Arkel process

5. The most stable oxidation state of the
   lanthanide elements is
   a. 1+   b. 2+   c. 3+   d. 5+   e. 7+

6. Which of the following statements is not true
   of Be with respect to other elements of
   group II A?
   a. It has the highest melting point.
   b. It is the most dense.
   c. It has the highest first ionization potential.
   d. It is the hardest.
   e. It has the smallest atomic radius.

7. Which of the following transition metals has
   the greatest number of possible oxidation states?
   a. Y,yttrium              d. Zn,zinc
   b. Sc,scandium            e. Au,gold
   c. Ru,ruthenium

8. Which of the following has the smallest atomic
   radius?
   a. Mn,manganese           d. Sc,scandium
   b. Tc,technetium          e. Zn,zinc
   c. Re,rhenium

_____  9. Common impurities in pig iron obtained from
a blast furnace include all but one of the
following. Identify the element that is not a
contaminant.
   a. carbon              d. sulfur
   b. silicon             e. titanium
   c. phosphorus

_____  10. Which of the following is commonly found free
in nature?
   a. Fe   b. W   c. Ba   d. Li   e. Au

_____  11. Which of the following metals has the lowest
melting point?
   a. Ag   b. Cd   c. In   d. Pd   e. Rh

_____  12. Which of the following does not contain an
unpaired $s$ electron in the valence shell of
the ground state atom?
   a. Rb,rubidium         d. Sn,tin
   b. Cu,copper           e. Au,gold
   c. Mo,molybdenum

_____  13. Which of the following compounds is the least
soluble in water?
   a. $Be(OH)_2$           d. $Ba(OH)_2$
   b. $Ca(OH)_2$           e. $Mg(OH)_2$
   c. $Sr(OH)_2$

_____  14. Which of the following elements has common
oxidation states of 2+, 4+, and 7+?
   a. Mn   b. Ni   c. Pt   d. La   e. Cs

_____  15. Which of the following elements has common
oxidation states of 1+ and 3+?
   a. Au   b. Cd   c. Sc   d. K   e. Hg

II. Answer the following:

1. Complete the following equations.  If no reaction occurs, write *NA* in the space provided for products.

   a. $Al(s) + Cl_2(g) \longrightarrow$

   b. $CaO(s) + H_2O(l) \longrightarrow$

   c. $ZnS(s) + O_2(g) \overset{heat}{\longrightarrow}$

   d. $Ce(s) + Br_2(g) \longrightarrow$

   e. $Ba(s) + H_2O(l) \longrightarrow$

   f. $Ni(s) + CO(g) \overset{heat}{\longrightarrow}$

   g. $WO_3(s) + H_2(g) \overset{heat}{\longrightarrow}$

   h. $Al_4C_3(s) + H_2O(l) \longrightarrow$

   i. $Tl(s) + H^+(aq) \longrightarrow$

   j. $Ca(s) + Cl_2(g) \longrightarrow$

   k. $Pb(s) + H^+(aq) \longrightarrow$

   l. $V(s) + H_2O(l) \longrightarrow$

   m. $Na(s) + O_2(g) \longrightarrow$

   n. $Al(s) + O_2(g) \overset{heat}{\longrightarrow}$

   o. $Fe(s) + H^+(aq) \longrightarrow$

2. Write the name of each of the following compounds in the space provided:

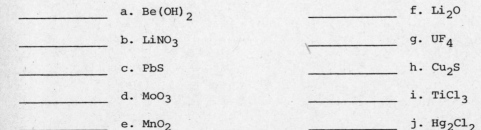

_____ a. $Be(OH)_2$

_____ b. $LiNO_3$

_____ c. $PbS$

_____ d. $MoO_3$

_____ e. $MnO_2$

_____ f. $Li_2O$

_____ g. $UF_4$

_____ h. $Cu_2S$

_____ i. $TiCl_3$

_____ j. $Hg_2Cl_2$

ANSWERS TO
EXERCISES

I. Properties of metals

1. Fe
   [18.6, 18.8, 18.10]

Each element of group I A has an oxidation state of 1+. The 1+ oxidation state is an important oxidation state of the transition metals Cu, Au, Ag, and Hg; compounds of $Cu^+$, $Ag^+$, $Au^+$, and $Hg^{2+}$ are common. Thallium is the only group III A metal that has an oxidation state of 1+.

2. NaCl
   [18.7]

The ions of group II A metals and many compounds containing such ions hydrate readily.

3. Cs
   [18.1, 18.6, 18.7]

The softness of the metals of groups I A and II A increases down the group with increasing atomic number. The metals of group II A are harder than the metals of group I A.

4. Hall process
   [18.4]

5. $3^+$
   [18.9]

6. It is the
   most dense.
   [18.1]

Both barium and strontium are more dense than beryllium.

7. Ru
   [18.8]

An element of the transition series that contains between 4 and 6 electrons in the outer $d$ shell can form a variety of oxidation states.

8. manganese
   [18.8]

Atomic size initially decreases and then increases across a period of transition elements from left to right; the minimum is reached near the center of the period. Atomic size increases with increasing atomic number down a group.

9. titanium
   [18.4]

10. Au [18.2]

11. In [18.1]

12. Sn [18.6, 18.8, 18.11]

13. Be(OH)$_2$  [18.7]

14. Mn  [18.8]

15. Au  [18.8]

### II. Reactions of metals and compounds containing metals

1.  a. $2Al(s) + 3Cl_2(g) \longrightarrow 2AlCl_3(s)$                          [18.10]

    b. $CaO(s) + H_2O(l) \longrightarrow Ca(OH)_2(aq)$                        [18.7]

    c. $2ZnS(s) + 3O_2(g) \xrightarrow{heat} 2ZnO(s) + 2SO_2(g)$            [18.4]

    d. $2Ce(s) + 3Br_2(g) \longrightarrow 2CeBr_3(s)$                        [18.9]

    e. $Ba(s) + 2H_2O(l) \longrightarrow Ba(OH)_2(aq) + H_2(g)$             [18.7]

    f. $Ni(s) + 4CO(g) \xrightarrow{heat} Ni(CO)_4(g)$                       [18.5]

    g. $WO_3(s) + 3H_2(g) \xrightarrow{heat} W(s) + 3H_2O(g)$               [18.4]

    h. $Al_4C_3(s) + 12H_2O(l) \longrightarrow 4Al(OH)_3(s) + 3CH_4(g)$     [18.7]

    i. $2Tl(s) + 2H^+(aq) \longrightarrow 2Tl^+(aq) + H_2(g)$               [18.10]

    j. $Ca(s) + Cl_2(g) \longrightarrow CaCl_2(s)$                          [18.7]

    k. $Pb(s) + 2H^+(aq) \longrightarrow Pb^{2+}(aq) + H_2(g)$              [18.11]

    l. $V(s) + H_2O(l) \longrightarrow NR$                                  [18.8]

    m. $2Na(s) + O_2(g) \longrightarrow Na_2O_2(s)$                         [18.6]

    n. $4Al(s) + 3O_2(g) \xrightarrow{heat} 2Al_2O_3(s)$                    [18.10]

    o. $Fe(s) + 2H^+(aq) \longrightarrow Fe^{2+}(aq) + H_2(g)$             [18.8]

2.  a. beryllium hydroxide  [18.7]

    b. lithium nitrate  [18.6]

    c. lead(II) sulfide  [18.11]

    d. molybdenum(VI) oxide  [18.4, 18.8]

e. manganese(IV) oxide, sometimes called manganese dioxide [18.8]

f. lithium oxide [18.6]

g. uranium(IV) fluoride [18.4]

h. copper(I) sulfide, also called cuprous sulfide [18.4]

i. titanium(III) chloride [18.8]

j. mercury(I) chloride, also called mercurous chloride [18.8]

SELF-TEST

Complete the test in 10 minutes:

1. In the spaces provided, write the formulas of the products obtained from the following reactions:

a. $Mg(s) + H_2O(g) \longrightarrow$ _____ + _____

b. $Na(s) + O_2(g) \longrightarrow$ _____

c. $Li(s) + N_2(g) \longrightarrow$ _____

d. $HgS(s) + O_2(g) \xrightarrow{heat}$ _____ + _____

e. $K(s) + O_2(g) \longrightarrow$ _____

f. $Ba(s) + P_4(l) \xrightarrow{heat}$ _____

g. $Ce(s) + S(l) \xrightarrow{heat}$ _____

# 19

# Complex Compounds

OBJECTIVES

(a) You should be able to demonstrate your knowledge of the following terms by defining them, describing them, or giving specific examples of them:

chelate [19.1]
coordination isomers [19.4]
coordination number [19.1]
crystal field theory [19.5]
degenerate [19.5]
enantiomorph [19.4]
geometric isomers [19.4]
high-spin state [19.5]
hydrate isomers [19.4]
inner complexes [19.5]

ionization isomers  19.4
isomers  19.4
labile [19.2]
ligand [19.1]
ligand field theory [19.5]
linkage isomers [19.4]
low-spin state [19.5]
optical isomers [19.4]
outer complexes [19.5]
racemic [19.4]
stereoisomers  19.4
structural isomers [19.4]
valence bond theory [19.5]

(b) Given the names or chemical formulas of complex compounds, you should be able to write the corresponding chemical formulas or names.

(c) You should be able to draw and identify the various types of isomers.

(d) You should be able to discuss the bonding in complex compounds.

EXERCISES

I. Answer each of the following with *true* or *false*. If the statement is false, correct it.

1. In general, $\Delta_o$ is larger for low-spin octahedral complexes of a metal than for high-spin octahedral complexes of that metal.

2. The ethylenediaminetetraacetate ion, EDTA, is a sexadentate chelate.

3. Inner octahedral complexes of iron (III) have all electrons paired.

4. Outer octahedral complexes of iron (III) have five unpaired electrons.

5. Low-spin and high-spin octahedral complexes only exist for $d^4$, $d^5$, $d^6$, and $d^7$ ions.

6. Tetrahedral complexes never exist as *cis-trans* isomers.

7. Tetrahedral complexes never exist as optical isomers.

_____    8. With the addition of Ag$^+$, two moles of chloride can be precipitated as AgCl from a solution containing one mole of $[Cr(H_2O)_5Cl]Cl_2 \cdot H_2O$.

_____    9. Labile complexes rapidly undergo reactions in which the ligands are replaced.

_____   10. Both chlorophyll and hemoglobin are metal ion complexes of porphyrins.

II. Write the name of each of the following complex compounds in the space provided:

_____ 1. $[Co(NH_3)_6]Cl_3$      _____ 7. $Ir(NH_3)_3Cl_3$

_____ 2. $K_4[Fe(CN)_6]$      _____ 8. $K_3[Fe(CN)_6]$

_____ 3. $[Co(NH_3)_5(SO_4)]Br$      _____ 9. $NH_4[Cr(SCN)_4(NO)_2]$

_____ 4. $[Co(NH_3)_5Br]SO_4$      _____ 10. $Pt(NH_3)_2Cl_2$

_____ 5. $[Pt(NH_3)_4Cl_2]PtCl_4$      _____ 11. $[Ni(H_2O)_5Cl]Cl$

_____ 6. $[Co(H_2O)_6]Cl_2$      _____ 12. $[Cr(NH_3)_6][Cr(CN)_6]$

III. Do the following:

    1. Using Figure 19.1, identify the type of isomers represented by each of the following pairs:

        a. $[Co(en)_2ClBr]Cl$ and $[Co(en)_2Cl_2]Br$

_____    b. $[Pd(dipy)(SCN)_2]$ and $[Pd(dipy)(NCS)_2]$

_____    c.

_____    d. $[Pt(NH_3)_4(OH)_2]SO_4$ and $[Pt(NH_3)_4(SO_4)OH]OH$

_____    e. $[Co(NH_3)_3(H_2O)_2Cl]Br_2$ and

                 $[Co(NH_3)_3(H_2O)ClBr]Br \cdot H_2O$
                    coordination isomers

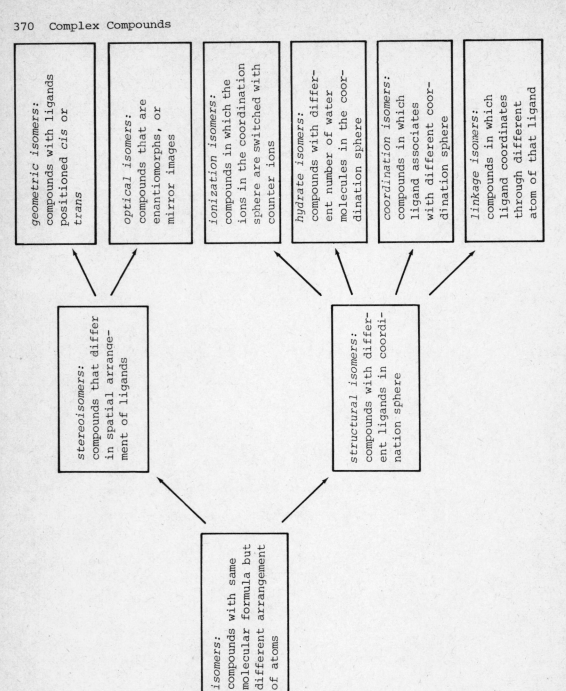

*geometric isomers:*
compounds with ligands
positioned *cis* or
*trans*

*optical isomers:*
compounds that are
enantiomorphs, or
mirror images

*ionization isomers:*
compounds in which the
ions in the coordination
sphere are switched with
counter ions

*hydrate isomers:*
compounds with differ-
ent number of water
molecules in the coor-
dination sphere

*coordination isomers:*
compounds in which
ligand associates
with different coor-
dination sphere

*linkage isomers:*
compounds in which
ligand coordinates
through different
atom of that ligand

*stereoisomers:*
compounds that differ
in spatial arrange-
ment of ligands

*structural isomers:*
compounds with differ-
ent ligands in coordi-
nation sphere

*isomers:*
compounds with same
molecular formula but
different arrangement
of atoms

Figure 19.1 Isomerism in complex compounds

f.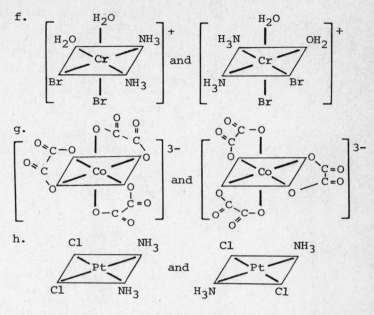

and

g.

and

h.

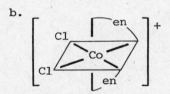

and

2. Which of the following complexes cannot have an optical isomer?

a. *trans*-Pt(NH$_3$)$_2$Cl$_2$, square planar

b.

c.

d.

3. Which of the complexes in problem 2 of this section can exist as *cis-trans* isomers?

4. An elemental analysis of an unknown compound indicates an empirical formula of $CrN_4Cl_3H_{12}$. When dissolved in water, the compound forms two ions, and one mole of chloride ion can be precipitated for each mole of compound. The molecular weight of the compound is found to be 226. The compound exists in two isomeric forms. Draw the isomers.

5. The compound *cis*-$Pt(NH_3)_2Cl_2$ is an active anti-tumor agent. Does it have an optical isomer?

6. A compound when analyzed shows 65.0% Pt, 23.6% Cl, 9.3% N, and 2.0% H. The compound has a molecular weight of 600 and forms two ions in solution. No chloride can be removed from solution by the addition of $Ag^+$, and no ammonia can be evolved. Draw two possible isomers of the compound and identify the type of such isomers.

7. Determine whether each of the following transition metal ions is a $d^1$, $d^2$, $d^3$, ..., $d^9$, or $d^{10}$ ion:

   a. $V^{5+}$

   b. $[Fe(CN)_6]^{4+}$

   c. hexaaquomanganese(III) ion

   d. $[Ni(H_2O)_5Cl]Cl$

   e. $Ni^{3+}$

   f. $NH_4[Cr(SCN)_4(NO)_2]$

IV. Answer each of the following:

1. Using valence bond theory, diagram the electronic arrangement (3d and higher) of $Fe(H_2O)_6^{2+}$, an outer complex, and that of $Fe(CN)_6^{4-}$, an inner complex.

2. Using crystal field theory, diagram the electronic arrangement of iron in each complex ion mentioned in problem 1 of this section.

## ANSWERS TO EXERCISES

I. Properties of complex compounds

1. True
   [19.5]

For a high-spin octahedral complex the electrons are distributed in the $t_{2g}$ and $e_g$ orbitals such that the number of unpaired electrons is a maximum. Such a distribution occurs when the pairing energy is greater than $\Delta_o$. For a low-spin octahedral complex the pairing energy is less than $\Delta_o$ and thus the lower-energy orbitals are filled completely before the higher-energy orbitals. For example, in an octahedral complex the high-spin state of a $d^6$ metal ion complex is

$$\underline{1} \qquad \underline{1} \quad e_g \qquad \updownarrow \; \Delta_o$$
$$\underline{1\!\!\downarrow} \quad \underline{1} \quad \underline{1} \quad t_{2g}$$

and the low-spin state of that $d^6$ metal ion complex is

$$\underline{\quad} \qquad \underline{\quad} \quad e_g \qquad \updownarrow \; \Delta_o$$
$$\underline{1\!\!\downarrow} \quad \underline{1\!\!\downarrow} \quad \underline{1\!\!\downarrow} \quad t_{2g}$$

2. True
   [19.1]

3. False
   [19.5]

The Fe(III) ion is a $d^5$ ion. In inner octahedral complexes of Fe(III) there is one unpaired electron:

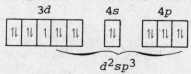

$$3d \qquad\qquad 4s \qquad 4p$$

$$\underbrace{\qquad\qquad\qquad\qquad}_{d^2sp^3}$$

4. True
[19.5]

In outer octahedral complexes of Fe(III) the electron arrangement is

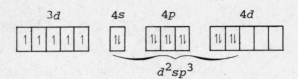

5. True
[19.5]

In an octahedral complex the configuration of a $d^8$ ion is

$$
\begin{array}{ccc}
\underline{\uparrow} & \underline{\uparrow} & e_g \\
\underline{\uparrow\downarrow} \quad \underline{\uparrow\downarrow} \quad \underline{\uparrow\downarrow} & & t_{2g} \\
d^8 & &
\end{array}
$$

and that of a $d^9$ ion is

$$
\begin{array}{ccc}
\underline{\uparrow\downarrow} & \underline{\uparrow} & e_g \\
\underline{\uparrow\downarrow} \quad \underline{\uparrow\downarrow} \quad \underline{\uparrow\downarrow} & & t_{2g} \\
d^9 & &
\end{array}
$$

6. True
[19.4]

Each ligand is 109° from each of the other three ligands; therefore, *cis* and *trans* isomers cannot exist.

7. False
[19.4]

Optical isomers are possible and have been separated for some compounds. For instance, salts of the borosalicylaldehydo complex cannot be superimposed:

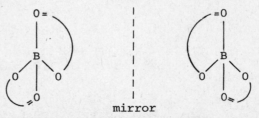

8. True
[19.4]

The two chlorides outside the brackets are not in the first coordination sphere of the metal ion. They exist as ions in solution and can be precipitated as AgCl with the addition of $Ag^+$ to the solution.

9. True
   [19.2]

10. True
    [19.1]

II. Names of complex compounds

Refer to Table 19.1 of the study guide or Section 19.3 of your text if you have any difficulty with this section.  After you have checked your answers, write the formula of each compound in the space provided.

1. Hexaamminecobalt(III) chloride                                   _____

2. Potassium hexacyanoferrate(II)                                   _____

3. Sulfatopentaamminecobalt(III) bromide                            _____

4. Bromopentaamminecobalt(III) sulfate                              _____

5. Dichlorotetraammineplatinum(IV) tetrachloroplatinate(II)         _____

6. Hexaaquocobalt(II) chloride                                      _____

7. Trichlorotriammineiridium(III)                                   _____

8. Potassium hexacyanoferrate(III)                                  _____

9. Ammonium tetrathiocyanatodinitrosylchromate(III)                 _____

10. Dichloroammineplatinum(II)                                      _____

11. Chloropentaaquonickel(II) chloride                              _____

12. Hexaamminechromium(III) hexacyanochromate(III)                  _____

## Table 19.1 Rules for naming complex compounds

1. If the complex compound is a salt, the cation is named first whether or not it is the complex ion.

2. The constituents of the complex are named in the following order: anions, neutral molecules, central metal ion.

3. Anionic ligands are given -o endings; examples are: $OH^-$, hydroxo; $O^{2-}$, oxo; $S^{2-}$, thio; $Cl^-$, chloro; $F^-$, fluoro; $CO_3^{2-}$, carbonato; $CN^-$, cyano; $CNO^-$, cyanato; $C_2O_4^{2-}$, oxalato; $NO_3^-$, nitrato; $NO_2^-$, nitro; $SO_4^{2-}$, sulfato; and $S_2O_3^{2-}$, thiosulfato.

4. The names of neutral ligands are not changed. Exceptions to this rule are: $H_2O$, aquo; $NH_3$, ammine; CO, carbonyl; and NO, nitrosyl.

5. The number of ligands of a particular type is indicated by a prefix: di-, tri-, tetra-, penta-, and hexa- (for two to six). For complicated ligands (such as ethylenediamine), the prefixes bis-, tris, and tetrakis- (two to four) are employed.

6. The oxidation number of the central ion is indicated by a Roman numeral, which is set off by parentheses and placed after the name of the complex.

7. If the complex is an anion, the ending -ate is employed. If the complex is a cation or a neutral molecule, the name is not changed.

---

III. Isomers

See Section 19.3 of your text.

1. a. ionization isomers — The Cl and Br are switched. In the first compound Br is in the first coordination sphere and in the second compound it is the counter ion.

b. linkage isomers — The SCN is bonded to the metal through the S atom in the first compound and through the N atom in the second compound.

c. geometric isomers — The chlorines are *cis* in the first compound and *trans* in the second.

d. ionization isomers

e. hydrate isomers

f. optical isomers

g. optical isomers

h. geometric isomers

2. a, c, and d      Mirror images of these complexes can be superimposed on the originals.

3. a.

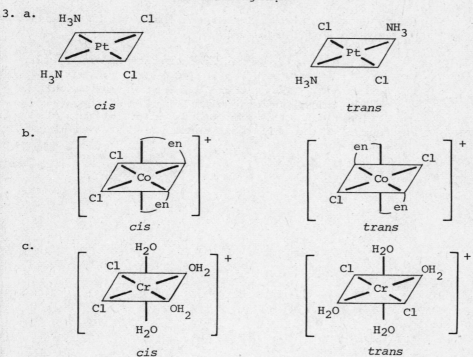

*cis*                    *trans*

b.

*cis*                    *trans*

c.

*cis*                    *trans*

4.

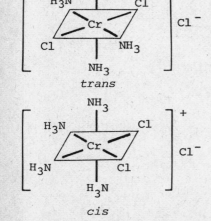

*trans*

*cis*

The nitrogen-hydrogen ratio indicates the possibility of $NH_3$ in the compound and a formula of $Cr(NH_3)_4Cl_3$ for the compound. Chromium usually forms octahedral complexes. Since one mole of chloride ion can be precipitated for each mole of compound, a chloride ion is free in solution, i.e., it is not in the first coordination sphere of Cr. The complex may exist in *cis* and *trans* forms.

5.  No

The compound

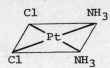

is square planar.  Its mirror image can be super-imposed on the original.

6. $[Pt(NH_3)_4][PtCl_4]$

    and

    $[Pt(NH_3)_3Cl]$
    $[Pt(NH_3)Cl_3]$

From the data on percent composition an empirical formula of $PtCl_2N_2H_6$, or $Pt(NH_3)_2Cl_2$, can be determined.  The compound has a molecular weight of 300; thus, the molecular formula must be $Pt_2(NH_3)_4Cl_4$.  The compound is neutral; thus, the charge of the anion must equal the charge of the cation.

7. a. $d^0$

   b. $d^6$

   c. $d^4$

   d. $d^8$

   e. $d^7$

   f. $d^3$

IV. Bonding

1. [19.5]
   $Fe(H_2O)_6^{2+}$ :

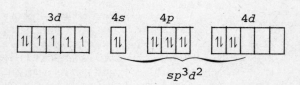

outer complex

$Fe(CN)_6^{4-}$ :

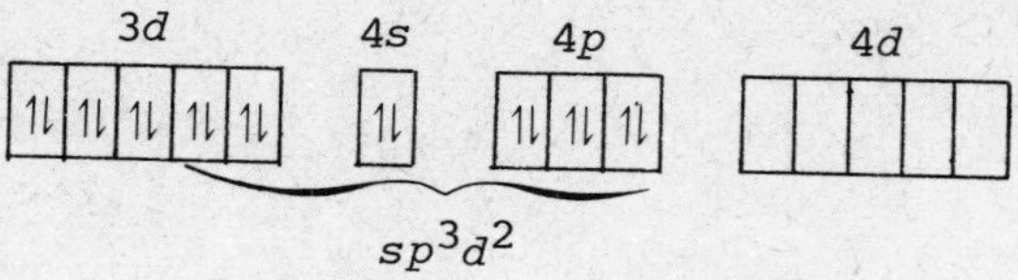

$3d$    $4s$    $4p$    $4d$

$sp^3d^2$

inner complex

2. [19.5]

Iron(II) is a $d^6$ ion. In $Fe(H_2O)_6^{2+}$ the iron(II) is in a high-spin state and in $Fe(CN)_6^{4-}$ it is in a low-spin state:

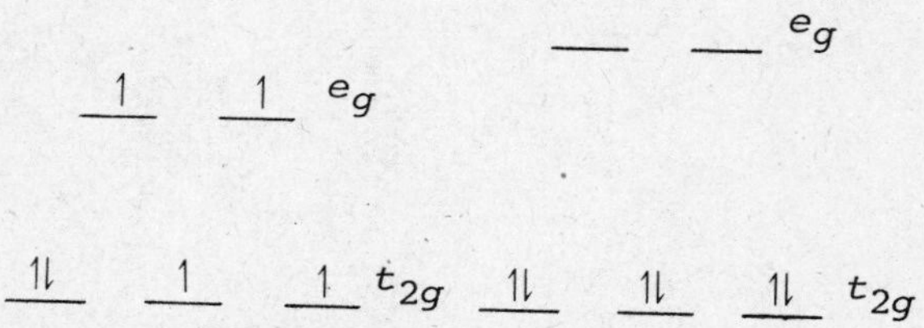

$e_g$

$\underline{1} \quad \underline{1} \quad e_g$

$\underline{1\!\!\downarrow} \quad \underline{1} \quad \underline{1} \; t_{2g}$    $\underline{1\!\!\downarrow} \quad \underline{1\!\!\downarrow} \quad \underline{1\!\!\downarrow} \; t_{2g}$

high-spin state    low-spin state

$Fe(H_2O)_6^{2+}$    $Fe(CN)_6^{4-}$

weak field    strong field

The crystal field approach indicates the unpaired, nonbonding electrons *and* explains the difference in energy of the $d$ orbitals.

## SELF-TEST

Complete the test in 10 minutes:

1. Write the name of each of the following compounds in the space provided:

_____    a. $K_2[CuCl_4]$

_____    b. $K_2[PtCl_4]$

_____    c. $K_4[Fe(CN)_6]$

_____    d. $Pt(NH_3)_2Cl_4$

e. [Co(en)$_2$(SCN)Cl]Cl

f. [Co(NH$_3$)$_4$Br$_2$]Br

g. [Cu(NH3)$_4$]$_3$[CrCl$_6$]$_2$

2. What is the isomeric relationship between the compounds in each of the following pairs?

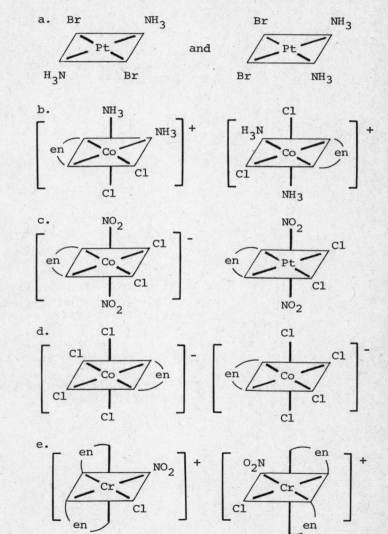

# 20
# Organic Chemistry

OBJECTIVES

(a) You should be able to demonstrate your knowledge of the following terms by defining them, describing them, or giving specific examples of them:

    alcohols [20.6]
    aldehydes [20.7]
    alkanes [20.1]
    alkenes [20.2]
    alkynes [20.3]
    amides [20.9]
    amines [20.9]
    amino acids [20.10]
    aromatic [20.4]
    carbohydrates [20.11]

carbonium ion [20.5]
carbonyl group [20.7]
carboxyl group [20.8]
carboxylic acids [20.8]
chain isomers [20.2]
*cis-* [20.2]
conjugated system [20.12]
copolymer [20.12]
esters [20.8]
ethers [20.6]
functional group isomerism [20.6]
geometric isomers [20.2]
homologous [20.1]
ketone [20.7]
Markovnikov's rule [20.5]
*meta-* [20.4]
*normal-* [20.1]
olefins [20.2]
*ortho-* [20.4]
*para-* [20.4]
peptide linkage [20.10]
polymers [20.12]
position isomers [20.2]
primary [20.6, 20.9]
proteins [20.10]
saponification [20.8]
saturated [20.2]
secondary [20.1, 20.6, 20.9]
stereoisomerism [20.2]
structural isomers [20.1]
tertiary [20.1, 20.6, 20.9]
*trans-* [20.2]
unsaturated [20.2]
zwitter ion [20.10]

(b) Given the structures or names of simple organic compounds, you should be able to write the corresponding names or structures.

(c) You should be familiar with the properties and reactions of the major classes of organic molecules.

(d) You should be able to identify isomers.

EXERCISES

I. Write the name of each of the following compounds in the space provided:

1. $CH_3CH_2CH_2CH_3$

2. 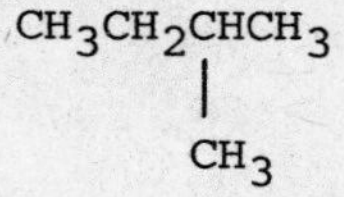 $CH_3CH_2CHCH_3$
        |
       $CH_3$

3.
                    $CH_3$
                     |
                    $CH_2$
                     |
   $CH_3-CH_2-C-CH_2-CH_3$
                     |
                    $CH_3$

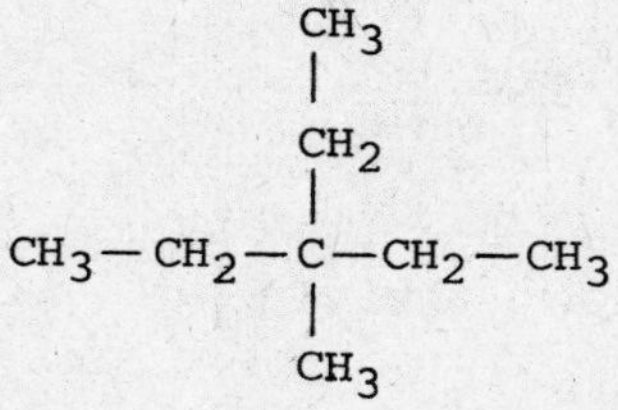

4. 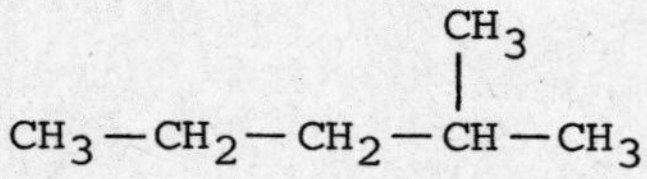
                        $CH_3$
                         |
   $CH_3-CH_2-CH_2-CH-CH_3$

5. 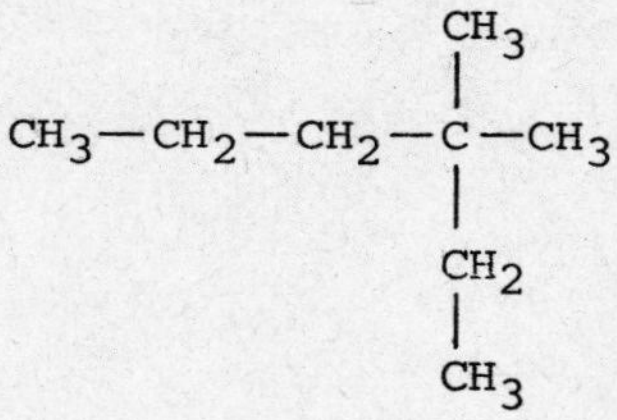
                       $CH_3$
                        |
   $CH_3-CH_2-CH_2-C-CH_3$
                        |
                       $CH_2$
                        |
                       $CH_3$

6. 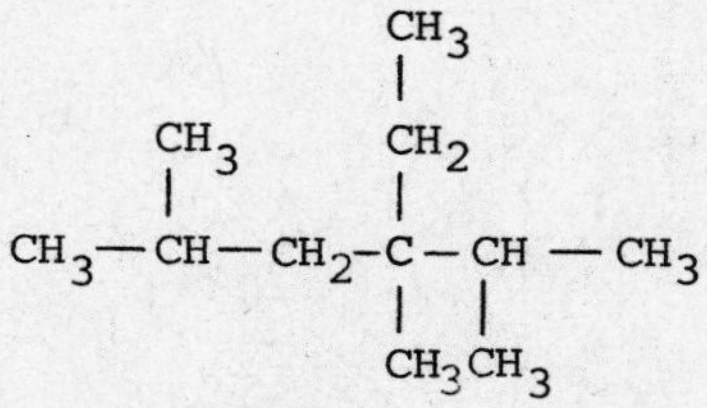
                      $CH_3$
                       |
             $CH_3$   $CH_2$
              |        |
   $CH_3-CH-CH_2-C-CH-CH_3$
                      |   |
                    $CH_3 CH_3$

7. 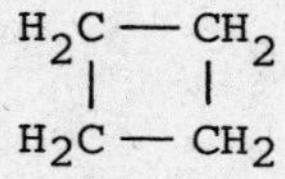 $H_2C-CH_2$
     |      |
   $H_2C-CH_2$

8. 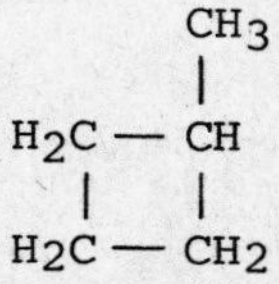
           $CH_3$
            |
   $H_2C-CH$
     |     |
   $H_2C-CH_2$

9. $CH_3-CH = CH-CH_3$

_____  10. $CH_3 - CH_2 - CH = CH - CH_3$

_____  11. 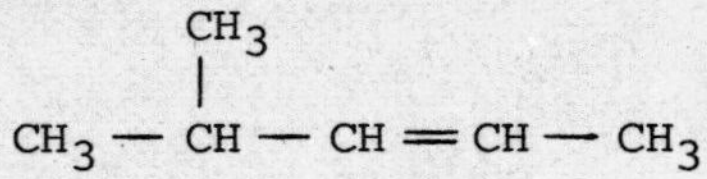
$$CH_3$$
$$|$$
$$CH_3 - CH - CH = CH - CH_3$$

_____  12. 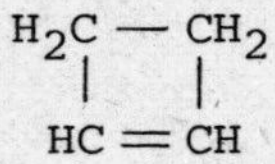 $H_2C - CH_2$
$\quad\quad\quad\quad\;\; | \quad\quad |$
$\quad\quad\quad\;\; HC = CH$

_____  13. 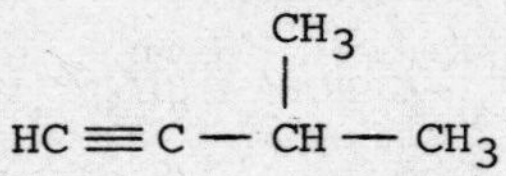
$$CH_3$$
$$|$$
$$HC \equiv C - CH - CH_3$$

_____  14. 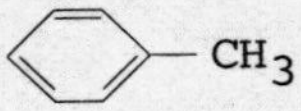 $- CH_3$

_____  15. 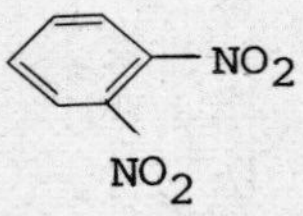 $- NO_2$
$\quad\quad\quad\quad\quad NO_2$

_____  16. 
$Cl -$ $- Cl$

_____  17. 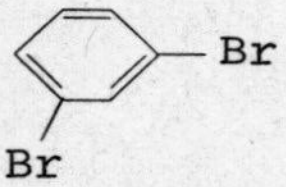 $- Br$
$\; Br$

_____  18. 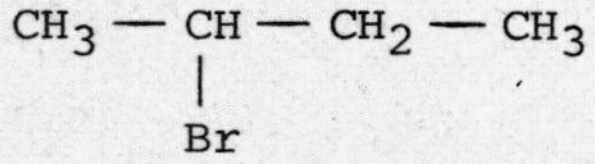 $CH_3 - CH - CH_2 - CH_3$
$\quad\quad\quad\quad\quad |$
$\quad\quad\quad\quad\; Br$

_____  19. 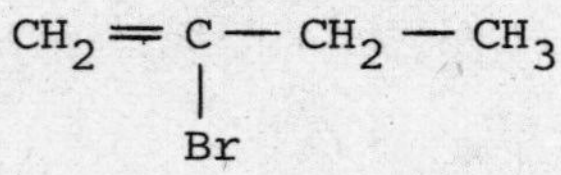 $CH_2 = C - CH_2 - CH_3$
$\quad\quad\quad\quad\quad\; |$
$\quad\quad\quad\quad\; Br$

_____  20. 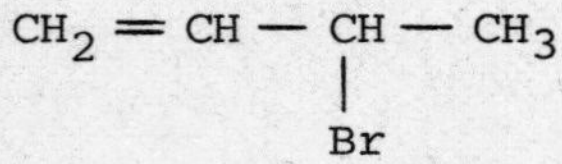 $CH_2 = CH - CH - CH_3$
$\quad\quad\quad\quad\quad\quad\quad |$
$\quad\quad\quad\quad\quad\; Br$

_____  21. $Br_2C = CH - CH_2 - CH_2Br$

_____  22. 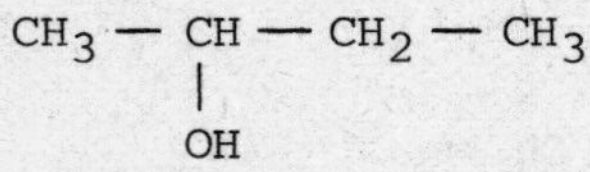 $CH_3 - CH - CH_2 - CH_3$
$\quad\quad\quad\quad\quad |$
$\quad\quad\quad\quad\; OH$

23.

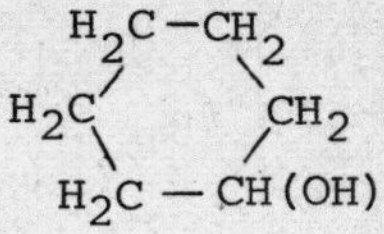

$$H_2C - CH_2$$
$$H_2C \qquad CH_2$$
$$H_2C - CH(OH)$$

24.   $CH_3 - CH - CH - CH_3$
                  $|$         $|$
                 $OH$       $OH$

25. 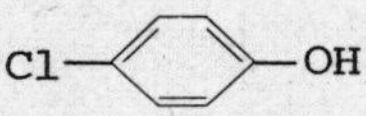   $Cl \underset{\phantom{x}}{\bigcirc} OH$

26.   $CH_3 - CH_2 - CH_2 - \overset{\overset{\displaystyle O}{\|}}{C} - H$

27.
$$\phantom{CH_3 - }Cl \qquad\quad O$$
$$\phantom{CH_3 - }| \qquad\qquad \|$$
$$CH_3 - CH - CH_2 - C - H$$

28.
$$\phantom{CH_3 - }O$$
$$\phantom{CH_3 - }\|$$
$$CH_3 - C - CH_2 - CH_2 - CH_3$$

29.
$$\phantom{CH_3 - CH_2 - CH - }O$$
$$\phantom{CH_3 - CH_2 - CH - }\|$$
$$CH_3 - CH_2 - CH - C - CH_3$$
$$\phantom{CH_3 - CH_2 - }|$$
$$\phantom{CH_3 - CH_2 - }CH_2$$
$$\phantom{CH_3 - CH_2 - }|$$
$$\phantom{CH_3 - CH_2 - }CH_3$$

30.
$$\phantom{CH_3 - CH_2 - }O$$
$$\phantom{CH_3 - CH_2 - }\|$$
$$CH_3 - CH_2 - C - OH$$

31. 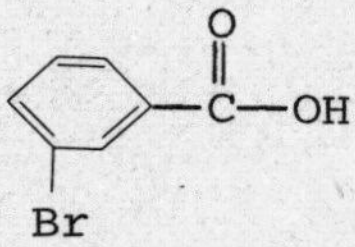   $\bigcirc\!\!-\!\overset{\overset{\displaystyle O}{\|}}{C}\!-\!OH$
       $Br$

32. 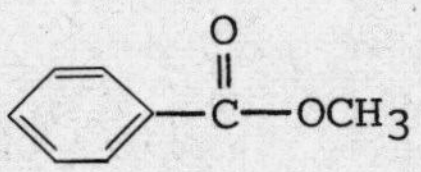   $\bigcirc\!\!-\!\overset{\overset{\displaystyle O}{\|}}{C}\!-\!OCH_3$

33.

$$ClCH_2-\overset{\overset{\displaystyle O}{\|}}{C}-O-CH_2CH_3$$

34.

$$CH_3CH_2-\overset{\overset{\displaystyle H}{|}}{N}-CH_2CH_3$$

35.

36.

37.

$$CH_3-\overset{\overset{\displaystyle O}{\|}}{C}-O-CH_2-\overset{\overset{\displaystyle CH_3}{|}}{CH}-CH_3$$

38.

$$CH_2=\overset{\overset{\displaystyle CH_3}{|}}{\underset{\underset{\displaystyle CH_3}{|}}{C}}-C=\overset{\overset{\displaystyle CH_2}{|}}{\underset{}{C}}-CH_3$$

with $CH_3$ on top of the fourth carbon's $CH_2$

39.

40.

$$CH_3-\overset{\overset{\displaystyle CH_3}{|}}{\underset{\underset{\displaystyle CH_3}{|}}{C}}-C\equiv CH$$

II. Complete the following reactions.  If no reaction occurs, write *NR* in the space provided for products.

1. $CH_3CH_2-CH \equiv CH-CH_3 \xrightarrow{MnO_4^-}$

2. $CH_3CH_2-CH = CH-CH_3 \xrightarrow{H_2/Pt}$

3. $CH_3-CH_2-CH = CH-CH_3 \xrightarrow{Br_2}$

4. 
$$CH_3 \overset{\overset{\textstyle CH_3}{|}}{-CH-CH_2OH} \xrightarrow[Cr_2O_7^{2-}/H+]{mild}$$

5. 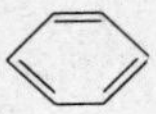 $\xrightarrow{Br_2/FeBr_3}$

6. 
$$CH_3-\overset{\overset{\textstyle O}{||}}{C}-H \xrightarrow{hot\ MnO_4^-}$$

7. 
$$CH_3-CH_2-\overset{\overset{\textstyle O}{||}}{C}-H \xrightarrow{H_2/catalyst}$$

8. 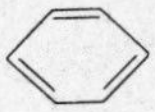 $+ CH_3CH_2Cl \xrightarrow{AlCl_3}$

9. 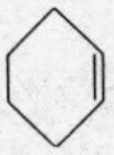 $\xrightarrow{Cr_2O_7^{2-}/H_2SO_4}$

III. Answer each of the following:

1. The molecule

$$CH_3-\overset{\overset{\textstyle O}{||}}{C}-OCH_2-CH_2-\overset{\overset{\textstyle CH_3}{|}}{\underset{\underset{\textstyle CH_3}{|}}{CH}}$$

is an ester that has a banana-like odor.  Write a reaction for the preparation of the ester from a carboxylic acid and an alcohol.  Name the carboxylic acid, alcohol, and ester.

2. Teflon is prepared by polymerizing tetrafluoro-
ethene (also called tetrafluoroethylene). Draw
the structure of teflon.

3. The compound commonly known as DDT has the following structure. Are there any stereoisomers or
optical isomers of the molecule?

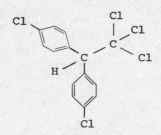

4. Draw and name all structural isomers with the
formula $C_7H_{16}$.

5. Polyvinyl chloride, abbreviated PVC, has the
formula

$$- CH-CH_2-CH-CH_2 -$$

From what single molecule can it be formed?

6. Identify the oxidation states of carbon in each
compound of the following series:

$CH_4$,   $H_3C-OH$,   $H_2C = O$,   $H-C{\underset{\displaystyle OH}{\overset{\displaystyle O}{}}}$,   $O = C = O$

Name each compound.

7. In a certain reaction benzaldehyde, which has the
formula

and an almond-like aroma, is converted to benzoic
acid. Is the reaction an oxidation, reduction,

or displacement reaction?

8. Draw all possible structural isomers of $C_3H_6O$. Name the isomers.  You do not have sufficient information to name all of them.

ANSWERS TO
EXERCISES

I. Names of organic compounds

1. butane
   [20.1]

The names of some straight-chain alkanes are given in Table 20.1 of the study guide.

2. methylbutane
   [20.1]

Substituents on the longest chain are given special radical names (see Table 20.2 of the study guide).

3. 3-ethyl
       3-ethylpentane
   [20.1]

When a substituent can be at one of several places on the longest chain, the positions on the chain are numbered and the substitution position is indicated with the appropriate number.

4. 2-methylpentane
   [20.1]

The numbering of the longest chain is always begun at the end of the chain that will give the lowest number to the first substituted position.

Table 20.1  Names of some straight-chain compounds

| NAME OF COMPOUND | FORMULA |
| --- | --- |
| methane | $CH_4$ |
| ethane | $C_2H_6$ |
| propane | $C_3H_8$ |
| butane | $C_4H_{10}$ |
| pentane | $C_5H_{12}$ |
| hexane | $C_6H_{14}$ |
| heptane | $C_7H_{16}$ |
| octane | $C_8H_{18}$ |
| nonane | $C_9H_{20}$ |
| decane | $C_{10}H_{22}$ |
| hexadecane | $C_{16}H_{34}$ |
| heptadecane | $C_{17}H_{36}$ |

Table 20.2  Names of simple radicals

| FORMULA | NAME |
|---|---|
| $CH_3$— | methyl |
| $CH_3CH_2$— | ethyl |
| $CH_3CH_2CH_2$— | *normal*-propyl or *n*-propyl |
| $CH_3CH$—<br>$\quad\vert$<br>$\quad CH_3$ | isopropyl |
| $CH_3CH_2CH_2CH_2$— | *normal*-butyl or *n*-butyl |
| $CH_3CHCH_2$—<br>$\quad\vert$<br>$\quad CH_3$ | isobutyl |
| $CH_3CH_2CH$—<br>$\qquad\vert$<br>$\qquad CH_3$ | *secondary*-butyl or *sec*-butyl |
| $\quad CH_3$<br>$\quad\ \vert$<br>$CH_3C$—<br>$\quad\ \vert$<br>$\quad CH_3$ | *tertiary*-butyl or *tert*-butyl |
| ⟨phenyl ring⟩— | phenyl |
| Br— | bromo |
| Cl— | chloro |
| $O_2N$— | nitro |

5.  3,3-dimethylhexane   The longest chain is circled and the numbering is
    [20.1]               included:

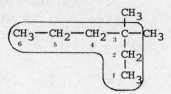

6.  2,3,5-trimethyl-
    3-ethylhexane
    [20.1]

7.  cyclobutane          The names of some cycloalkanes are given in Table
    [20.1]               20.3 of the study guide.

Table 20.3   Names of some cycloalkanes

| FORMULA | COMPOUND | FORMULA | COMPOUND |
|---|---|---|---|
| $CH_2$ / $H_2C$ —— $CH_2$ | cyclopropane | $CH_2$ / $H_2C$ $CH_2$ / $H_2C$ $CH_2$ / $CH_2$ | cyclohexane |
| $H_2C$ —— $CH_2$ / $H_2C$ —— $CH_2$ | cyclobutane | $CH_2$ / $H_2C$ $CH_2$ / $H_2C$ $CH_2$ / $H_2C$ —— $CH_2$ | cycloheptane |

8.  methylcyclobutane    The names of the substituents in Table 20.2 of the
    [20.1]               study guide are used to name such compounds, but the
                         naming becomes more complex when more than one sub-
                         stituent is added because of the possibility of
                         isomers.

9. 2-butene
[20.2]

The name of an alkene is derived from the name of the corresponding alkane by changing the ending from -ane to -ene. When necessary, a number is used to show the double-bond position. Because of the restricted rotation about a double bond, cis and trans isomers exist:

cis-2-butene                      trans-2-butene

10. 2-pentene
[20.2]

The numbering always starts at the end of the chain that gives the lowest numbers to the substituents. Therefore, the compound is 2-pentene, not 3-pentene. For this compound cis and trans isomers exist:

cis-2-pentene                    trans-2-pentene

11. 4-methyl-2-pentene
[20.2]

For this compound cis and trans isomers exist:

cis-4-methyl-2-pentene     trans-4-methyl-2-pentene

12. cyclobutene
[20.2]

The name of a cycloalkene is derived from the name of the corresponding cycloalkane by changing the ending from -ane to -ene.

13. methyl-1-butyne
[20.3]

No number is needed for the methyl group because there is only one possible position. The name of a compound containing a triple bond is derived from the name of the corresponding alkane by changing the ending from -ane to -yne.

14. methylbenzene
[20.4]

Another acceptable name is toluene.

15. *o*-dinitrobenzene
    [20.4]

The prefixes *ortho-* (*o-*), *meta-* (*m-*), and *para-* (*p-*) are used to designate relative positions of two substituents on a ring.

16. *p*-dichlorobenzene
    [20.4]

17. *m*-dibromobenzene
    [20.4]

18. 2-bromobutane
    [20.1]

19. 2-bromo-1-butene
    [20.2]

20. 3-bromo-1-butene
    [20.2]

21. 1,1,4-tribromo-
    1-butene
    [20.2]

22. 2-butanol
    [20.6]

A hydrocarbon containing an OH functional group is an alcohol. The name of an alcohol is derived from the name of the corresponding alkane by changing the ending from *-ane* to *-anol*. When necessary, a number is used to indicate the position of the OH group.

23. cyclohexanol
    [20.6]

The name of a cycloalcohol is derived from the name of the corresponding cycloalkane by changing the ending *-ane* to *-anol*.

24. 2,3-butanediol
    [20.6]

A polyhydroxy alcohol contains more than one OH group. The number of OH groups in a polyhydroxy alcohol is indicated in the name by addition of the appropriate ending to the name of the corresponding alkane: *diol, triol,* etc. The position of an OH group is indicated in the name by a number.

25. *p*-chlorophenol
    [20.4, 20.6]

26. butanal
    [20.7]

The name of an aldehyde is derived from the name of the corresponding alkane by changing the ending *-ane* to *-al*.

27. 3-chlorobutanal     For aldehydes the numbering of carbon atoms always
    [20.7]              starts with the carbon atom that is double bonded
                        to oxygen.

28. 2-pentanone        The name of a ketone is derived from the name of the
    [20.7]             corresponding alkane by changing the ending -ane
                       to -one.  The numbering of carbon atoms begins at the
                       end of the chain that gives the lowest number to
                       the carbon of the carbonyl group, $C = O$.

29. 3-ethyl-2-pentanone
    [20.7]

30. propanoic acid     The name of a carboxylic acid is derived from the
    [20.8]             parent hydrocarbon by elision of the final -e,
                       addition of the ending -oic, and addition of the
                       separate word acid.  This compound is also called
                       propionic acid.

31. m-bromobenzoic acid
    [20.8]

32. methyl benzoate    The name of an ester reflects the alcohol and acid
    [20.8]             from which the ester is derived:  the name of the
                       hydrocarbon radical attached to the OH of the
                       alcohol (see Table 20.2 of the study guide) is used
                       to indicate the parent alcohol, and the ending -ate
                       added to the base of the name of the parent acid is
                       used to indicate the parent acid.

33. ethyl chloroacetate
    [20.8]

34. diethylamine       The name of an amine is formed by the addition of the
    [20.9]             names of the radicals (see Table 20.2 of the study
                       guide) that are attached to the N atom to the word
                       amine.

35. m-chloroaniline    The molecule ⬡— $NH_2$ is called aniline.
    [20.9]

36. cyclopentanone
    [20.7]

37. isopropyl ethanoate
    [20.8]

38. 2, 3, 4-trimethyl-
    1, 3-hexadiene     [20.2]

39. triphenylamine
    [20.9]

40. 3,3-dimethyl-1-
            butyne
       [20.3]

## II. Reactions of organic compounds

1. [20.8]   $CH_3CH_2-CH = CH-CH_3 \xrightarrow[\text{heat}]{MnO_4^-}$   $CH_3CH_2-\overset{\overset{\displaystyle O}{\|}}{C}-OH$ + $CH_3-\overset{\overset{\displaystyle O}{\|}}{C}-OH$
                                                    propanoic acid    ethanoic acid

2. [20.5]   $\cdot CH_3CH_2-CH = CH-CH_3 \xrightarrow{H_2/Pt}$   $CH_3(CH_2)_3CH_3$
                                                        pentane

3. [20.5]   $CH_3-CH_2-CH = CH-CH_3 \xrightarrow{Br_2}$   $CH_3-CH_2-\overset{\overset{\displaystyle Br}{|}}{CH}-\overset{\overset{\displaystyle Br}{|}}{CH}-CH_3$
                                                          2,3 dibromopentane

4. [20.7]   $CH_3-\overset{\overset{\displaystyle CH_3}{|}}{CH}-CH_2OH \xrightarrow[]{\text{mild} \atop Cr_2O_7^{2-}/H+}$   $CH_3-\overset{\overset{\displaystyle CH_3}{|}}{CH}-\overset{\overset{\displaystyle O}{\|}}{C}-H$
                                                          methylpropanal

5. [20.5]   ⬡ $\xrightarrow{Br_2/FeBr_3}$ ⬡ $-Br$
                       bromobenzene

6. [20.8]   $CH_3-\overset{\overset{\displaystyle O}{\|}}{C}-H \xrightarrow{\text{hot } MnO_4^-}$ $CH_3-\overset{\overset{\displaystyle O}{\|}}{C}-OH$
                                          ethanoic acid
                                          (acetic acid)

7. [20.7]   $CH_3-CH_2-\overset{\overset{\displaystyle O}{\|}}{C}-H \xrightarrow{H_2/\text{catalyst}}$ $CH_3-CH_2-\overset{\overset{\displaystyle OH}{|}}{CH_2}$
                                                      1-propanol

8. [20.5]   ⬡ + $CH_3CH_2Cl \xrightarrow{AlCl_3}$ ⬡ $-CH_2CH_3$ + HCl

9. [20.8]   ⬡ $\xrightarrow{Cr_2O_7^{2-}/H_2SO_4}$   $\overset{\overset{\displaystyle O}{\|}}{C}-OH$
                                                    $|$
                                                    $\overset{\displaystyle C}{\underset{\displaystyle O}{\|}}-OH$
            
            or drawn differently
            
            $HO-\overset{\overset{\displaystyle O}{\|}}{C}-\overset{\overset{\displaystyle H}{|}}{\underset{\displaystyle H}{C}}-\overset{\overset{\displaystyle H}{|}}{\underset{\displaystyle H}{C}}-\overset{\overset{\displaystyle H}{|}}{\underset{\displaystyle H}{C}}-\overset{\overset{\displaystyle H}{|}}{\underset{\displaystyle H}{C}}-\overset{\overset{\displaystyle O}{\|}}{C}-OH$

## III. Properties of organic compounds

1.  20.8

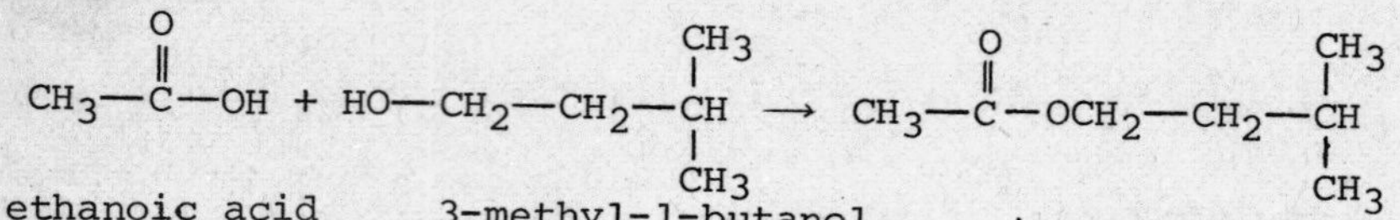

| | | |
|---|---|---|
| ethanoic acid (acetic acid) | 3-methyl-1-butanol | isopentylacetate |

2.

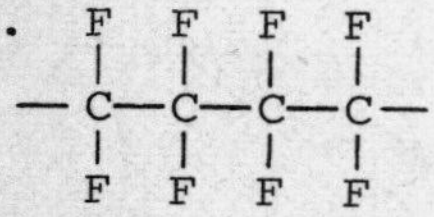

The compound has the structure of polyethene, but the H atoms are replaced by F atoms.

   20.12

3. no
   20.2

4.  20.1

| | |
|---|---|
| $CH_3-CH_2-CH_2-CH_2-CH_2-CH_2-CH_3$ | heptane |
| $CH_3-CH-CH_2-CH_2-CH_2-CH_3$ with $CH_3$ | 2-methylhexane |
| $CH_3-CH_2-CH-CH_2-CH_2-CH_3$ with $CH_3$ | 3-methylhexane |

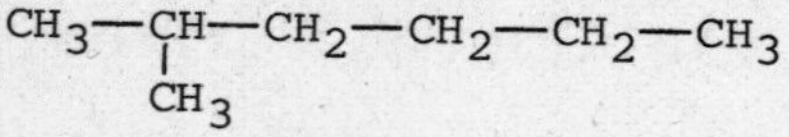

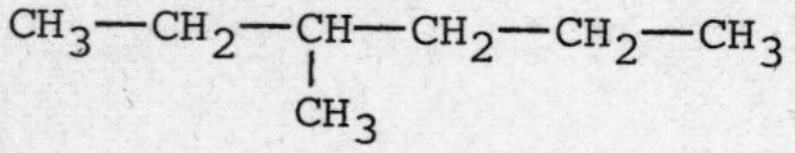

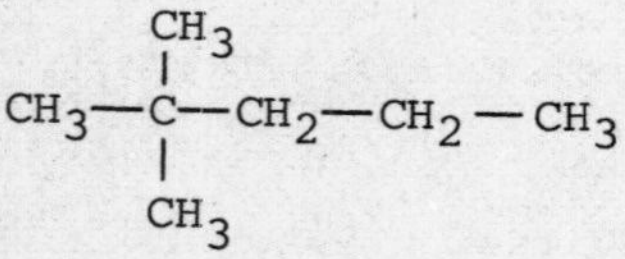

2,2-dimethylpentane

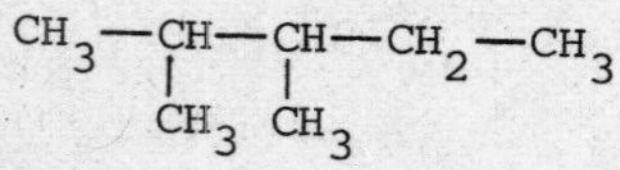

2,3-dimethylpentane

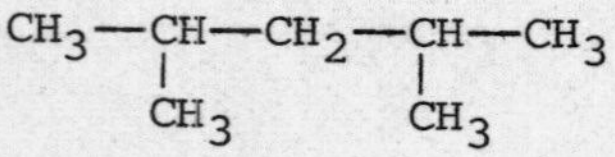

2,4-dimethylpentane

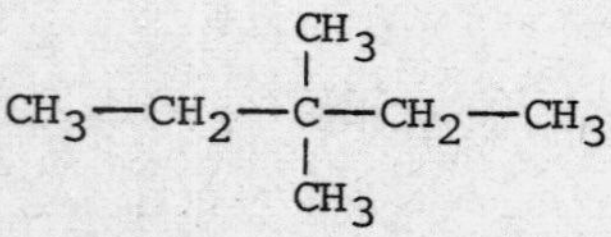

3,3-dimethylpentane

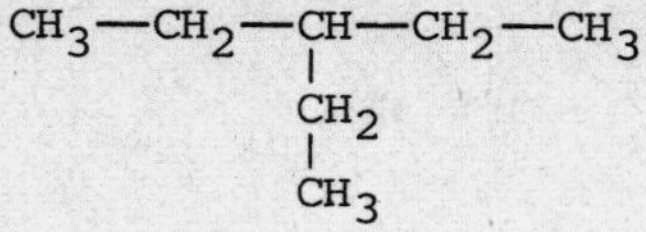

$$CH_3-CH_2-CH-CH_2-CH_3$$

with branch:

$$CH_2$$
$$CH_3$$

ethylpentane

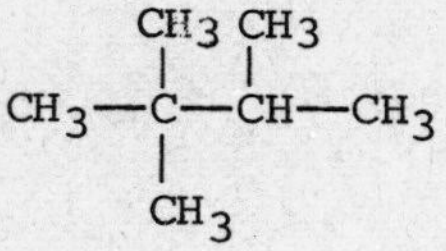

$$CH_3-C-CH-CH_3$$ with $CH_3$ $CH_3$ top and $CH_3$ bottom

2,2,3-trimethylbutane

5. 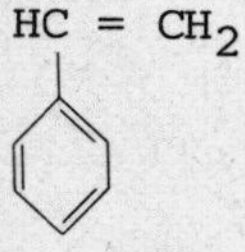 $HC = CH_2$

[benzene ring structure]

[20.12]

6. [3.10]

| methane | methanol | formaldehyde | formic acid | carbon dioxide |
|---------|----------|--------------|-------------|----------------|
| 4- | 2- | 0 | 2+ | 4+ |
| $CH_4$ | $H_3C-OH$ | $H_2C=O$ | 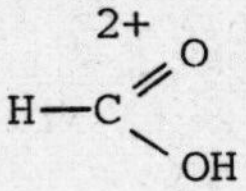 $H-C$ with $=O$ and $OH$ | $O=C=O$ |

7. oxidation
   [20.8]

Note the change in the oxidation state of the carbon atom that is double bonded to the oxygen:

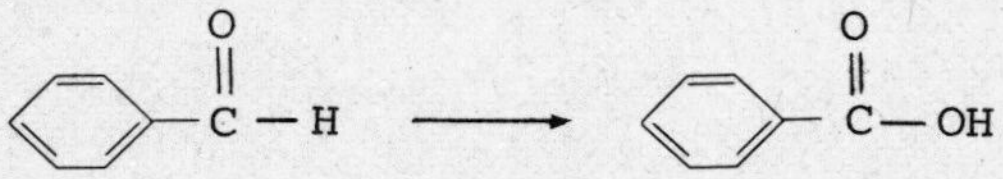

[benzene ring]$-C-H$ with $=O$ above and $0$ below $\longrightarrow$ [benzene ring]$-C-OH$ with $=O$ above and $2+$ below

8. [20.6, 20.7]

You should have been able to name the compounds in the left column, but you may not have been able to name those in the right column:

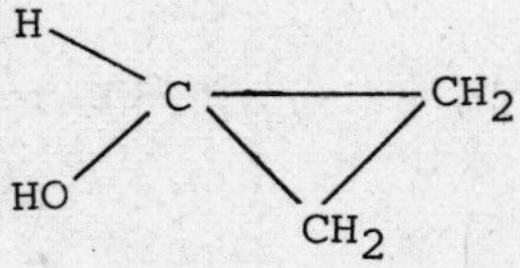

$$\begin{array}{c} H \\ HO \end{array} C \underset{CH_2}{\overset{CH_2}{\diagup\!\!\!\diagdown}}$$

$$OH$$
$$CH_3-CH=CH$$

cyclopropanol

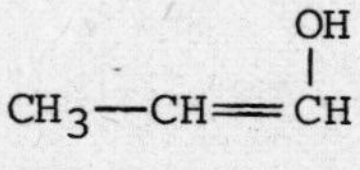

1-propen-1-ol

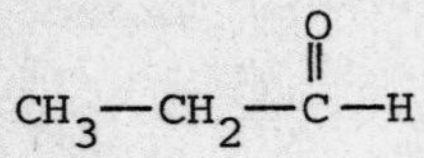

$$CH_3—CH_2—\overset{\overset{O}{\|}}{C}—H$$

propanal

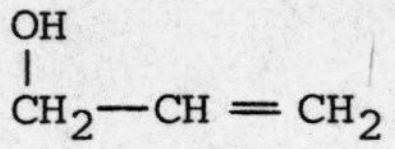

$$CH_3—\overset{\overset{O}{\|}}{C}—CH_3$$

propanone or acetone

$$\overset{\overset{OH}{|}}{CH_2}—CH = CH_2$$

2-propen-1-ol

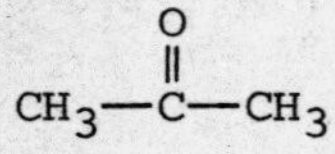

$$CH_3—\overset{\overset{OH}{|}}{C} = CH_2$$

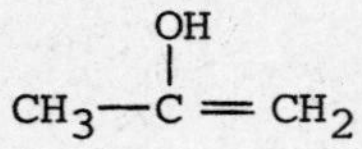

1-propen-2-ol

$$CH_3—O—CH = CH_2$$

3-oxa-1-butene or methylvinyl ether

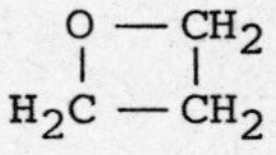

$$\overset{O — CH_2}{\underset{H_2C — CH_2}{|\quad\quad|}}$$

trimethylene oxide

SELF-TEST

I. Complete the test in 20 minutes:

1. Write the name of each of the following compounds in the space provided:

_____

a.

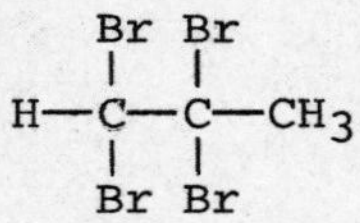

$$\overset{\overset{Br\ \ Br}{|\ \ \ |}}{H—C—C—CH_3}\underset{\overset{|\ \ \ |}{Br\ Br}}{}$$

_____

b.

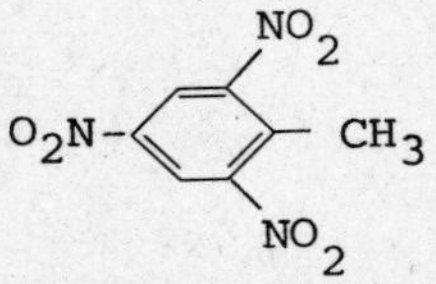

$$O_2N-\overset{NO_2}{\underset{NO_2}{\bigcirc}}-CH_3$$

_____

c.   $CH_3—CH_2—C \equiv CH$

_____

d. 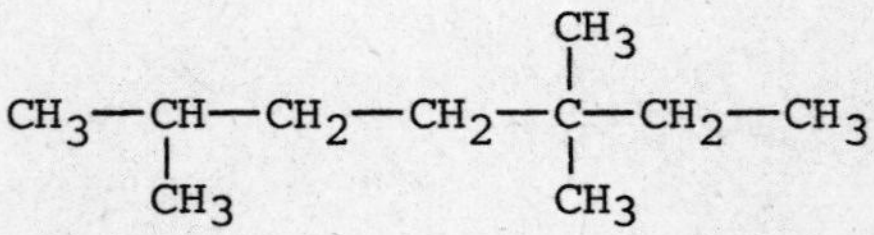   $$CH_3—\overset{\overset{}{|}}{\underset{\overset{|}{CH_3}}{CH}}—CH_2—CH_2—\overset{\overset{CH_3}{|}}{\underset{\overset{|}{CH_3}}{C}}—CH_2—CH_3$$

_____

e. 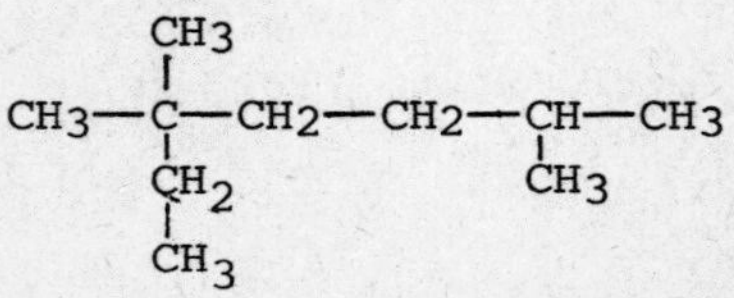   $$CH_3—\overset{\overset{CH_3}{|}}{\underset{\overset{|}{CH_2}}{C}}—CH_2—CH_2—\overset{\overset{}{}}{\underset{\overset{|}{CH_3}}{CH}}—CH_3$$
$$\underset{\overset{|}{CH_3}}{}$$

_____

f.

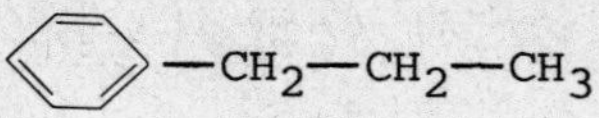

⬡—$CH_2$—$CH_2$—$CH_3$

g.  $CH_2 = CH—CH_2—CH = CH_2$

h.  $CH_3—CH_2$ $\quad$ H

$C = C$

H $\qquad$ $CH—CH_3$

$\qquad\qquad$ $CH_3$

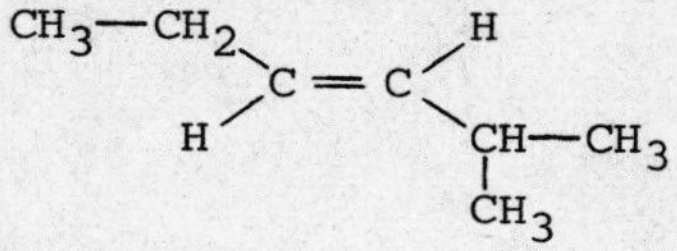

i.
$\qquad$ H
$\qquad$ |
$CH_3—CH_2—N—CH_3$

j.
$\qquad\qquad$ O
$\qquad\qquad$ ||
$CH_3—CH_2—C—O—CH_2—CH_3$

k.  $CH_3—CH_2—CH2—O—C—CH_3$
$\qquad\qquad\qquad\qquad\quad$ ||
$\qquad\qquad\qquad\qquad\quad$ O

l.
$\qquad\qquad\qquad\qquad\qquad\qquad$ O
$\qquad\qquad\qquad\qquad\qquad\qquad$ ||
$CH_3—CH_2—CH_2—CH_2—CH_2—C—H$

2. Write the formulas of the products obtained from each of the following reactions.  If no reaction occurs, write *NR*.

a.
$CH_3CH_2—OH$ $\xrightarrow[\text{heat}]{\text{Cu, } O_2}$

b.
$CH_2 = CH_2$ $\xrightarrow{\text{HCN}}$

c.
$CH_3—CH_2—C \equiv N$ $\xrightarrow[\text{catalyst}]{H_2}$

3. Draw all structural isomers of $C_3H_6$ and name each.

4. Draw all isomers of $C_3H_8O$ and name each.

# 21

# Nuclear Chemistry

OBJECTIVES

(a) You should be able to demonstrate your knowledge of the following terms by defining them, describing them, or giving specific examples of them:

    alpha particle [21.2]
    beta particle [21.2]
    binding energy [21.6]
    critical mass [21.6]
    curie [21.3]
    disintegration series [21.4]
    electron capture [21.2]
    fission [21.6]
    fusion [21.6]
    gamma radiation [21.2]

Geiger-Müller counter [21.3]
neutron capture [21.5]
nucleons [21.1]
radioactivity [21.2]
radiocarbon dating [21.3]
scintillation counter [21.3]
transuranium elements [21.5]

(b) You should be able to write equations for nuclear reactions and account for the mass and energy of such reactions.

(c) You should be able to determine the value of the half-life of a radioactive element from experimental data and use the value to determine the amount of radioactive sample present at any time, $t$.

EXERCISES

I. Complete and balance the following nuclear reactions:

1.  $^{218}_{84}Po \longrightarrow ^{4}_{2}He +$

2.  $^{254}_{102}No \longrightarrow ^{4}_{2}He +$

3.  $^{238}_{92}U + ^{1}_{0}n \longrightarrow \gamma +$

4.  $^{238}_{92}U + ^{12}_{6}C \longrightarrow 6 \, ^{1}_{0}n +$

5.  $^{207}_{84}Po \longrightarrow ^{0}_{1}e +$

6.  $^{228}_{88}Ra \longrightarrow ^{0}_{-1}e +$

7.  $^{14}_{6}C \longrightarrow ^{0}_{-1}e$

8.  $^{238}_{92}U \longrightarrow ^{4}_{2}He +$

9.  $^{87}_{36}Kr \longrightarrow ^{0}_{-1}e +$

10.  $^{87}_{36}Kr \longrightarrow ^{1}_{0}n +$

11.  $^{13}_{7}N \longrightarrow ^{13}_{6}C +$

12. $_{19}^{39}K + _0^1n \longrightarrow 2\ _0^1n +$

13. $_{92}^{238}U \longrightarrow _{90}^{234}Th +$

14. $_{92}^{235}U + _0^1n \longrightarrow _{56}^{139}Ba + 3\ _0^1n +$

15. $_1^2H + _1^3H \longrightarrow _2^4He +$

16. $_{-1}^0e + _{26}^{55}Fe \longrightarrow$

II. Work the following problems:

1. In the electron capture reaction

$$_{-1}^0e + _4^7Be \longrightarrow _3^7Li$$

how much energy in MeV is released? The mass of $_4^7Be$ is 7.0169 u and that of $_3^7Li$ is 7.0160 u.

2. Radon in a tube is often used in cervical cancer therapy. The half-life of radioactive radon is 3.8 days. If 11 micrograms of radon are sealed in a tube, how many micrograms remain after 21 days?

3. How long will it take before the amount of radon in the tube described in problem 2 of this section is reduced to one millionth of a microgram?

4. The $_6^{14}C$ activity of a fiber from an Egyptian mummy shroud is 7.50 disintegrations per minute per gram of carbon. How old is the fiber? The half-life of $_6^{14}C$ is 5770 years and, the $_6^{14}C$ activity of a piece of wood from the outer layer of a freshly cut tree is 15.2 disintegrations per minute per gram of carbon.

5. Use information from problem 4 of this section to determine the age of the oldest sample that can be dated by the $_6^{14}C$ technique. Assume that less than 1 disintegration per minute per gram of carbon cannot be detected with the equipment used with this technique.

6. A sample of $^{147}_{59}$Pr is prepared and placed in a scintillation counter.  The initial counting rate is 200/min.  After 36 min the rate is 25 counts/min.  What is the half-life of $^{147}_{59}$Pr?

7. How many grams of $^{60}_{27}$Co will give 75 x 10$^{-3}$ curies, or 75 millicuries, of radiation? The half-life of $^{60}_{27}$Co is 5.2 years.

**ANSWERS TO EXERCISES**

I. Balancing nuclear equations.
See Section 21.5 of your text.

1. $^{214}_{82}$Pb

An alpha particle is a helium nucleus, $^{4}_{2}$He.  Thus, when an alpha particle is emitted, two protons and

2. $^{250}_{100}$Fm

two neutrons are lost from the nucleus of the atom.

3. $^{239}_{92}$U

4. $^{244}_{98}$CF

The capture of a neutron, $^{1}_{0}$n, by an atom increases the mass number of that atom by 1 unit.  The loss of a neutron by an atom decreases the mass number of that atom by 1 unit.

5. $^{207}_{83}$Bi

A positron, $^{0}_{1}$e, is the product of a transformation of a nuclear proton into a nuclear neutron.

6. $^{228}_{89}$Ac

The beta particle, $^{0}_{-1}$e, is a negatively charged, low-mass particle.  It is a product of the transformation of a nuclear neutron into a nuclear proton.

7. $^{14}_{7}$N

8. $^{234}_{90}$Th

9. $^{87}_{37}$Rb

10. $^{86}_{36}$Kr

11. $^{0}_{1}$e

12. $^{38}_{19}$K

13.  $^{4}_{2}He$

14.  $^{94}_{36}Kr$          This is the $^{235}U$ atomic bomb reaction.

15.  $^{1}_{0}n$          This is a typical fusion reaction.

16.  $^{55}_{25}Mn$

II. Radioactive decay problems

1. 0.8 MeV          We determine the loss of mass:

mass of reactant $^{7}_{4}Be$ = 7.0169 u
mass of product  $^{7}_{3}Li$ = 7.0160 u
_____
loss of mass = 0.0009 u

We calculate the energy equivalent of this mass difference by means of Einstein's equation:

$E = mc^2$

  = 0.0009 u (930MeV/u)

  = 0.8 MeV

Thus, the energy released is 0.8 MeV.

2. 0.24 µg          We first calculate the rate constant for the radio-
   [21.3]           active decay of radon:

$k = \dfrac{0.693}{t^{\frac{1}{2}}}$

  $= \dfrac{0.693}{3.8 \text{ days}}$

  $= 0.182 \text{ day}^{-1}$

We then calculate the fraction of radon remaining after 21 days:

$$\log\left(\frac{N_0}{N}\right) = \frac{kt}{2.303}$$

$$= \frac{(0.182 \text{ day}^{-1})(21 \text{ days})}{2.30}$$

$$= 1.66$$

and

$$\frac{N_0}{N} = 10^{1.66}$$

$$= 10^{0.66} \times 10^1$$

$$= 4.6 \times 10^1, \text{ or } 46$$

Since $N_0 = 11$ μg, the amount of radon remaining after 21 days is

$$\frac{N_0}{N} = 46$$

$$N = \frac{N_0}{46}$$

$$= \frac{11 \text{ μg}}{46}$$

$$= 0.24 \text{ μg}$$

3. 90 days
   [21.3]

Since $N_0 = 11$ μg and $N = 1 \times 10^{-6}$ μg, the length of time, $t$, can be computed directly:

$$\log\left(\frac{N_0}{N}\right) = \frac{kt}{2.303}$$

$$t = \log\left(\frac{N_0}{N}\right)\left(\frac{2.303}{k}\right)$$

$$= \log\left(\frac{1.1}{1 \times 10^{-6}}\right)\left(\frac{2.3}{0.18 \text{ day}^{-1}}\right)$$

$$= 9 \times 10^1 \text{ days, or 90 days}$$

4. $6.45 \times 10^3$ years
   [21.3]

We can determine the value of $k$ from the half-life:

$$k = \frac{0.693}{t_{\frac{1}{2}}}$$

$$= \frac{0.693}{5770 \text{ years}}$$

$$= 1.20 \times 10^{-4}/\text{year}$$

The number of disintegrations per minute is proportional to the number of atoms present; therefore, we can substitute the values for the number of disintegrations per minute into the fraction $N_0/N$:

$$\log\left(\frac{N_0}{N}\right) = \frac{kt}{2.303}$$

$$t = \left(\frac{2.303}{k}\right) \log\left(\frac{N_0}{N}\right)$$

$$= \left(\frac{2.303}{1.20 \times 10^{-4}/\text{year}}\right) \log\left(\frac{15.2 \text{ disint./min}}{7.00 \text{ disint./min}}\right)$$

$$= 6.45 \times 10^3 \text{ years}$$

5. $2 \times 10^4$ years     We determine the time, $t$:

$$\log\left(\frac{N_0}{N}\right) = \frac{kt}{2.303}$$

$$t = \left(\frac{2.303}{k}\right) \log\left(\frac{N_0}{N}\right)$$

$$= \left(\frac{2.3}{1.2 \times 10^{-4}/\text{year}}\right) \log\left(\frac{15 \text{ disint./min}}{1 \text{ disint./min}}\right)$$

$$= 2 \times 10^4 \text{ years}$$

The technique is actually limited to less than 20,000 years by other restrictions.

6. 12 min     In one half-life, half the original material disintegrates, and the rate drops to 100 counts per minute. During the second half-life, the rate drops to 50 counts per minute, and during the third half-life it drops to 25 counts per minute. Therefore, 3 half-lives elapse between the initial and final counts. Thus, the half-life is 1/3 of the elapsed time, or 36 min/3 = 12 min.

7. $6.6 \times 10^{-5}$ g $^{60}_{27}Co$    We convert the activity from millicuries to disintegrations/sec:

$$activity = (75mc) \left( \frac{3.70 \times 10^7 \text{ disint/sec}}{1c} \right) \left( \frac{1c}{1000mc} \right)$$

$$= 2.78 \times 10^9 \text{ disint/sec, or}$$

$$2.78 \times 10^9 \text{ atom/sec}$$

Since the rate constant is expressed in years$^{-1}$, we convert the activity from atoms/sec to atoms/year:

$$activity = \left( \frac{2.78 \times 10^9 \text{ atoms}}{1 \text{ sec}} \right) \left( \frac{60 \text{ sec}}{1 \text{ min}} \right) \left( \frac{60 \text{ min}}{1 \text{ hr}} \right) \left( \frac{24 \text{ hr}}{1 \text{ day}} \right) \left( \frac{365 \text{ days}}{1 \text{ year}} \right)$$

$$= 8.77 \times 10^{16} \text{ atoms/year}$$

We then calculate the rate constant, $k$:

$$k = \frac{0.693}{t^{\frac{1}{2}}}$$

$$= \frac{0.693}{5.2 \text{ years}}$$

$$= 0.133 \text{ year}^{-1}$$

Since activity $= kN$, we can calculate the number of atoms, $N$:

$$activity = kN$$

$$N = \frac{activity}{k}$$

$$= \frac{8.77 \times 10^{16} \text{ atoms/year}}{0.133 \text{ year}}$$

$$= 6.59 \times 10^{17} \text{ atoms}$$

Finally we calculate the number of grams:

$$? \text{ g } ^{60}_{27}Co = 6.59 \times 10^{17} \text{ atoms } ^{60}_{27}Co \left( \frac{1 \text{ mol } ^{60}_{27}Co}{6.022 \times 10^{23} \text{ atoms}} \right) \left( \frac{60 \text{ g } ^{60}_{27}Co}{1 \text{ mol } ^{60}_{27}Co} \right)$$

$$= 6.6 \times 10^{-5} \text{ g } ^{60}_{27}Co$$

SELF-TEST          Complete the test in 10 minutes:

I. Answer the following:

1. A sample contains $^{35}_{16}$S as the only radioactive
   species.  How many moles of $^{35}_{16}$S are in a sample
   that has an activity of $1.01 \times 10^3$ disintegrations per minute.  The half-life of $^{35}_{16}$S is 86.6
   days.

2. The half-life of $^{31}_{14}$Si is 2.6 hours.  How many
   grams of $^{31}_{14}$Si must be prepared if $1.0 \times 10^{-12}$
   g  will be needed in an experiment 13 hours later?

3. Calculate the binding energy of an atom of $^{208}_{82}$Pb,
   which has a mass of 208.060 u.  The masses of the
   proton, neutron, and electron are 1.007277 u,
   1.008665 u, and 0.0005486 u, respectively.

# Answers to Self-Tests

**CHAPTER 1**
1. c. 2. d. 3. c. 4. d. 5. d. 6. a. 7. c. 8. b.
9. c. 10. d. 11. b. 12. c. 13. a. 14. b. 15. b.

**CHAPTER 2**
I. 1. energy. 2. electron, proton, neutron. 3. neutron.
4. isotopes. 5. 18. 6. diamagnetic. 7. $1s^2\ 2s^1$, 3.

II. 1. The atomic mass unit is defined as one-twelfth the mass of the nuclide $^{12}C$. 2. $n = 5$, $l = 2$, $m = +1$, $s = -\frac{1}{2}$. 3. Ir : $1s^2\ 2s^2\ 2p^6\ 3s^2\ 3p^6\ 3d^{10}\ 4s^2\ 4p^6$ $4d^{10}\ 4f^{14}\ 5s^2\ 5p^6\ 5d^7\ 6s^2$ and $Ir^{2+}$ : $1s^2\ 2s^2\ 2p^6$ $3s^2\ 3p^6\ 3d^{10}\ 4s^2\ 4p^6\ 4d^{10}\ 4f^{14}\ 5s^2\ 5p^6\ 5d^7$.
4. 6.94 u.

III. 1. c. 2. d. 3. d. 4. c. 5. c. 6. a. 7. a. 8. b.

9. b.   10. a.   11. d.   12. d.   13. d.   14. a.   15. b.

## CHAPTER 3

I. See the following table:

| FORMULA OF COMPOUND | NAME OF COMPOUND | SYMBOL OF ELEMENT | OXIDATION NUMBER OF ELEMENT |
|---|---|---|---|
| $SnCl_4$ | tin(IV) chloride or stannic chloride | Sn | 4+ |
| $As_2O_3$ | arsenic(III) oxide or arsenous oxide | As | 3+ |
| $Cu(NO_3)_2$ | copper(II) nitrate or cupric nitrate | N | 5+ |
| $MnO_2$ | manganese(IV) oxide | Mn | 4+ |
| $Cr_2O_3$ | chromium(III) oxide | Cr | 3+ |
| $NaC_2H_3O$ | sodium acetate | Na | 1+ |
| $N_2O_4$ | dinitrogen tetroxide | N | 4+ |
| $H_2SO_4$ | sulfuric acid | S | 6+ |
| $NaClO_2$ | sodium chlorite | Cl | 3+ |

II.  1. d    2. a    3. b    4. d    5. d    6. c
     7. c    8. b    9. a    10. d

III.

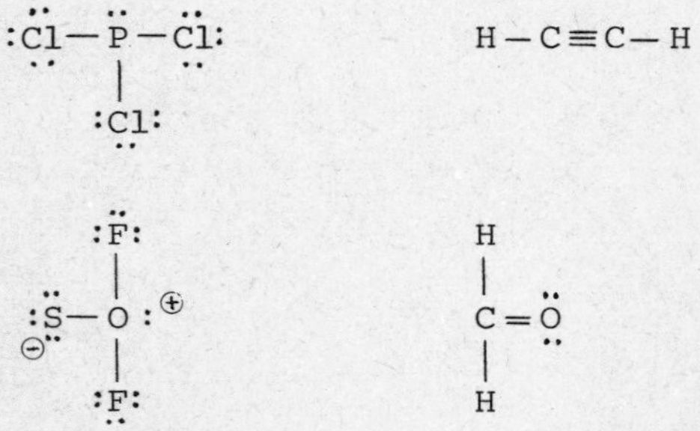

:Cl—P—Cl:    H—C≡C—H
    |
   :Cl:


   :F:          H
    |           |
:S—O:⊕        C=O:
⊖   |           |
   :F:          H

# CHAPTER 4

**1.**

| FORMULA | NUMBER OF ELECTRON PAIRS | | SHAPE OF MOLECULE OR ION |
|---------|--------------------------|-----------|--------------------------|
| | BONDING | NONBONDING | |
| $SCl_2$ | 2 | 2 | angular |
| $XeF_4$ | 4 | 2 | square planar |
| $AlH_4^-$ | 4 | 0 | tetrahedral |
| $TeCl_4$ | 4 | 1 | trigonal pyramidal or distorted tetrahedral |
| $SeF_5^-$ | 5 | 1 | square pyramidal |

**2.**

| | HYBRID ORBITALS | GEOMETRIC SHAPE |
|--|-----------------|-----------------|
| $PCl_5$ | $dsp^3$ | trigonal bipyramidal |
| $PCl_4^+$ | $sp^3$ | tetrahedral |
| $PCl_6^-$ | $d^2sp^3$ | octahedral |

**3.**

| MOLECULE | TOTAL NUMBER OF ELECTRONS IN | | | | BOND ORDER | NUMBER OF UNPAIRED ELECTRONS |
|----------|------------|-------------|------------|-------------|------------|------------------------------|
| | $\sigma$ ORBITALS | $\sigma^*$ ORBITALS | $\pi$ ORBITALS | $\pi^*$ ORBITALS | | |
| $Be_2$ | 2 | 2 | 0 | 0 | 0 | 0 |
| $B_2$ | 2 | 2 | 2 | 0 | 1 | 2 |
| $N_2$ | 4 | 2 | 4 | 0 | 3 | 0 |
| $O_2$ | 4 | 2 | 4 | 2 | 2 | 2 |
| $NO^+$ | 4 | 2 | 4 | 0 | 3 | 0 |

**4.** $\left[ \ddot{S}\!=\!C\!=\!\ddot{N} \right]^- \longleftrightarrow \left[ :S\!\equiv\!C\!-\!\overset{\oplus}{\ddot{N}}: \right]^- \longleftrightarrow \left[ :\overset{\ominus}{\ddot{S}}\!-\!C\!\equiv\!N: \right]^-$

# CHAPTER 5

1. $3.03 \times 10^{22}$ atoms.   2. CaO.   3. +21.0 kcal/mol [or +87.86 kJ/mol].   4. $C_6H_5OCl_2$.   5. 7.44 g.   6. $\Delta H = -63.2$ kcal [or -264.4 kJ].

# CHAPTER 6

I.  1. 16.0 g/mol.   2. 105 g/mol.   3. 12 mm.   4. 800 mm.
5. 0.900 g/l.   6. 2 liter $NH_3$, 0 liter $O_2$, 8 liter NO, 12 liter $H_2O$.   7. 43.9 g/mol.

II. 1. $H_2$, 6.   2. same, same.   3. 1/17.   4. intermolecular attractive forces, molecular volume.

III.

a. pressure vs. volume

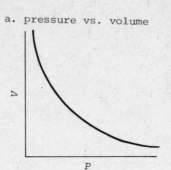

c. absolute temper-
ature vs. volume

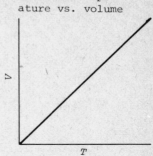

b. pressure vs. the
   product of pres-
   sure and volume

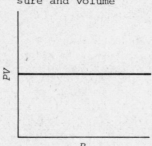

d. energy distribu-
   tion of molecules

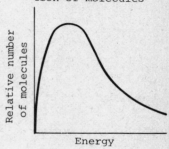

CHAPTER 7    1.    218 atm

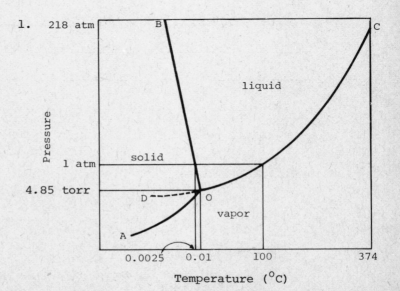

2. 1,2.   3. $334^\circ K$.   4. 58.7 g/mol.

CHAPTER 8

1. $2ClO_2 + 2OH^- \rightarrow ClO_2^- + ClO_3^- + H_2O$

$IO_4^- + H_2AsO_3^- \rightarrow IO_3^- + H_2AsO_4^-$

$4Sb + 4NO_3^- + 4H+ \rightarrow Sb_4O_6 + 4NO + 2H_2O$

$O_3 + 6I^- + 6H^+ \rightarrow 3H_2O + 3I_2$

$2NO_3^- + 3H_2S + 2H^+ \rightarrow 2NO + 3S + 4H_2O$

$Cr_2O_7^{2-} + 6Cl^- + 14H^+ \rightarrow 2Cr^{3+} + 3Cl_2 + 7H_2O$

2. 67.4 g $ClO_2$/eq, 95.4 g $IO_4^-$/eq, 62.4 g $H_2AsO_3^-$/eq, 40.6 g Sb/eq, 20.7 g $NO_3^-$/eq, 8.00 g $O_3$/eq, 126.9 g $I^-$/eq, 20.7 g $NO_3^-$/eq, 17.0 g $H_2S$/eq, 36.0 g $Cr_2O_7^{2-}$/ eq, 35.4 g $Cl^-$/eq

3. $Mg(s) + H_2O(g) \rightarrow MgO(s) + H_2(g)$

$H_2(g) + Cl_2(g) \rightarrow 2HCl(g)$

$2Na(s) + 2H_2O(l) \rightarrow 2Na^+(aq) + 2OH^-(aq) + H_2(g)$

CHAPTER 9

1. 123 g/eq.   2. 4.30%.   3. 6.00$N$.   4. 16.4 ml.
5. 125 g/mol.

CHAPTER 10

I. 1. +1.23 V  2. 0.755 g Mg.   3. a. $Cu^{2+}$; b. $Cu^{2+}$/Cu;
c. positive; d. yes; e. negative.  4. $4.5 \times 10^{-3}$.

II. 1. b.   2. b.   3. a.   4. c.   5. d.   6. c.

CHAPTER 11

1. a. $2OH^-(aq) + Cl_2(aq) \rightarrow ClO^-(aq) + Cl^-(aq) + H_2O(l)$

b. $FeS(s) + 2H^+(aq) + 2Cl^-(aq) \rightarrow$
$$Fe^{2+}(aq) + 2Cl^-(aq) + H_2S(g)$$

c. $2Br_2(g) + Ag_2O(s) + H_2O(l) \rightarrow 2AgBr(s) + 2HOBr(aq)$

d. $PCl_3(l) + 3H_2O(l) \rightarrow 3HCl(aq) + H_3PO_3(aq)$

e. $Cl^-(aq) + 3H_2O(l) \xrightarrow[\text{heat}]{\text{electrolysis}} ClO_3^-(aq) + 3H_2(g)$

f. $Cl^-(aq) + H^+(aq) + H_2O(l) \rightarrow NR$

g. $Cl_2(g) + 2I^-(aq) \rightarrow 2Cl^-(aq) + I_2(s)$

h. $Ba^{2+}(aq) + SO_4^{2-}(aq) \rightarrow BaSO_4(s)$

i. $C_{12}H_{22}O_{11}(s) + 11H_2SO_4(conc) \rightarrow 11H_2SO_4 \cdot H_2O +$
sucrose
$12C(s)$

or

$$C_{12}H_{22}O_{11}(s) + 6H_2SO_4 \rightarrow$$
$$5H_2SO_4 \cdot 2H_2O + H_2SO_4 \cdot H_2O + 12C$$

or any other balanced equation containing hydrated sulfuric acid species and 12 carbon atoms
as products.

j. $S_2O_3^{2-}(aq) + 2H^+(aq) \rightarrow S(s) + SO_2(g) + H_2O(l)$

**CHAPTER 12**

1.

2. a. $2Sb_2S_3(s) + 9O_2(g) \xrightarrow{heat} Sb_4O_6(g) + 6SO_2(g)$

b. $CaO(s) + 3C(s) \xrightarrow{heat} CaC_2(s) + CO(g)$

c. $Ca_3N_2(s) + 6H_2O(l) \rightarrow$
$$3Ca^{2+}(aq) + 6OH^-(aq) + 2NH_3(g)$$

d. $As_4O_6(s) + 6C(s) \xrightarrow{heat} As_4(g) + 6CO(g)$

e. $PBr_3(l) + 3H_2O(l) \rightarrow H_3PO_3(aq) + 3HBr(aq)$

f. $2PBr_3(l) + O_2(g) \rightarrow 2POBr_3(s)$

g. $PI_3(s) + I_2(s) \rightarrow NR$

3. monoprotic acid

**CHAPTER 13**

1. order of $HgCl_2$ = 1, order of $C_2O_4^{2-}$ = 2, k = 4.7
x $10^{-3}M^{-2}$ min$^{-1}$.

2. a. remains the same;  b. increases;  c. remains the
same;  d. decreases;  e. increases.

3. a. increase;  b. no change;  c. no change;  d. increase;  e. increase.

4. increase temperature, decrease pressure, remove $SO_2$.

5. $K_p$ = 133 atm.$^{-1}$

6. 6.25 x $10^{-3}M$.

**CHAPTER 14**

1. -59.5 kcal/mol [or -249 kJ/mol].  2. $K_p$ = 1.03 x
$10^{16}$.  3. reaction is spontaneous, $K_p$ = 9.84 x $10^4$.

**CHAPTER 15**

1. $NH_3$, $H^-$.   2. $H_2S$.   3. $SOCl_2$ ($SO^{2+}$ is the acid ion).
4. $F^-$.   5. $\ddot{\underset{..}{S}}{:}$.   6. $HS^-$, $HC_2H_3O_2$.   7. $HPO_4^{2-}$.

**CHAPTER 16**

1. 12.9   2. $2.5 \times 10^{-10}$.   3. $2.0 \times 10^{-6}$.   4. $0.18M$.

**CHAPTER 17**

1. $1.2 \times 10^{-4}$ g $Ag^+$/250 ml.   2. 8.8   3. 5.2.
4. 0.0075%.

**CHAPTER 18**

1. a. $MgO(s)$, $H_2(g)$;   b. $Na_2O_2(s)$;   c. $Li_3N(s)$;
d. $Hg(g)$, $SO_2(g)$;   e. $KO_2(s)$;   f. $Ba_3P_2(s)$;
g. $Ce_2S_3(s)$.

**CHAPTER 19**

1. a. potassium tetrachlorocuprate(II)
   b. potassium tetrachloroplatinate(II)
   c. potassium hexacyanoferrate(II)
   d. tetrachlorodiammineplatinum(IV)
   e. chlorothiocyantobis(ethylenediamine)cobalt(III) chloride
   f. dibromotetraamminecobalt(III) bromide
   g. tetraamminecopper(II) hexachlorochromate(III)

2. a. geometric isomers
   b. same compound
   c. not isomers, different central metal atom
   d. same compound
   e. optical isomers

**CHAPTER 20**

I. 1. a. 1,1,2,2-tetrabromopropane
     b. 1,3,5-trinitrotoluene
     c. 1-butyne
     d. 2,5,5-trimethylheptane
     e. 2,5,5-trimethylheptane
     f. n-propylbenzene
     g. 1,4-pentadiene
     h. *trans*-2-methyl-3-hexene
     i. methylethylamine
     j. ethyl propionate
     k. *n*-propyl acetate
     l. hexanal

2. a.
$$CH_3-\overset{\displaystyle O}{\overset{\|}{C}}-H$$

b. $CH_3-CH_2-C\equiv N$

c. $CH_3-CH_2-CH_2-NH_2$

3.

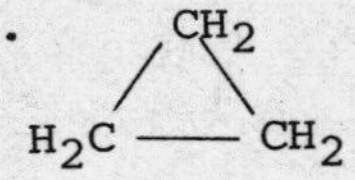

$$\begin{array}{c} CH_2 \\ H_2C \underline{\quad\quad} CH_2 \end{array}$$

cyclopropane

$CH_2-CH=CH_2$

propene

4. 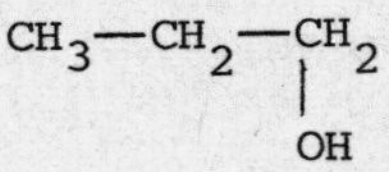 $CH_3-CH_2-CH_2$
     |
     $OH$

1-propanol

$CH_3-CH-CH_3$
        |
        $OH$

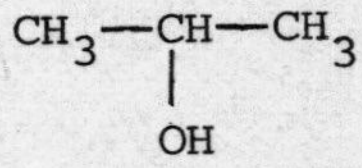

2-propanol

$CH_3-O-CH_2-CH$

ethylmethylether

CHAPTER 21     1. $3.0 \times 10^{-16}$ mol.   2. $3.2 \times 10^{-11}$ g.   3. 7.84 MeV.

# PERIODIC CLASSIFICATION

| I A | II A | III B | IV B | V B | VI B | VII B | VIII B | | |
|---|---|---|---|---|---|---|---|---|---|
| 1<br>**H**<br>Hydrogen<br>1.0079 | | | | | | | | | |
| 3<br>**Li**<br>Lithium<br>6.941 | 4<br>**Be**<br>Beryllium<br>9.01218 | | | | | | | | |
| 11<br>**Na**<br>Sodium<br>22.98977 | 12<br>**Mg**<br>Magnesium<br>24.305 | | | | | | | | |
| 19<br>**K**<br>Potassium<br>39.098 | 20<br>**Ca**<br>Calcium<br>40.08 | 21<br>**Sc**<br>Scandium<br>44.9559 | 22<br>**Ti**<br>Titanium<br>47.90 | 23<br>**V**<br>Vanadium<br>50.9414 | 24<br>**Cr**<br>Chromium<br>51.996 | 25<br>**Mn**<br>Manganese<br>54.9380 | 26<br>**Fe**<br>Iron<br>55.847 | 27<br>**Co**<br>Cobalt<br>58.9332 | |
| 37<br>**Rb**<br>Rubidium<br>85.4678 | 38<br>**Sr**<br>Strontium<br>87.62 | 39<br>**Y**<br>Yttrium<br>88.9059 | 40<br>**Zr**<br>Zirconium<br>91.22 | 41<br>**Nb**<br>Niobium<br>92.9064 | 42<br>**Mo**<br>Molybdenum<br>95.94 | 43<br>**Tc**<br>Technetium<br>98.9062b | 44<br>**Ru**<br>Ruthenium<br>101.07 | 45<br>**Rh**<br>Rhodium<br>102.9055 | |
| 55<br>**Cs**<br>Cesium<br>132.9054 | 56<br>**Ba**<br>Barium<br>137.34 | 57*<br>**La**<br>Lanthanum<br>138.9055 | 72<br>**Hf**<br>Hafnium<br>178.49 | 73<br>**Ta**<br>Tantalum<br>180.9479 | 74<br>**W**<br>Tungsten<br>183.85 | 75<br>**Re**<br>Rhenium<br>186.2 | 76<br>**Os**<br>Osmium<br>190.2 | 77<br>**Ir**<br>Iridium<br>192.22 | |
| 87<br>**Fr**<br>Francium<br>(223)a | 88<br>**Ra**<br>Radium<br>226.0254b | 89**<br>**Ac**<br>Actinium<br>(227)a | 104<br>(260)a | 105<br>(260)a | | | | | |

| * | 58<br>**Ce**<br>Cerium<br>140.12 | 59<br>**Pr**<br>Praseo-<br>dymium<br>140.9077 | 60<br>**Nd**<br>Neodymium<br>144.24 | 61<br>**Pm**<br>Promethium<br>(145)a | 62<br>**Sm**<br>Samarium<br>150.4 |
|---|---|---|---|---|---|
| ** | 90<br>**Th**<br>Thorium<br>232.0381b | 91<br>**Pa**<br>Protactnium<br>231.0359b | 92<br>**U**<br>Uranium<br>238.029 | 93<br>**Np**<br>Neptunium<br>237.0482b | 94<br>**Pu**<br>Plutonium<br>(242)a |

a Mass number of most stable or best known isotope.

b Mass of most commonly available, long-lived isotope.

# OF THE ELEMENTS

| | | | | | 0 |
|---|---|---|---|---|---|
| | | | | | **2**<br>**He**<br>Helium<br>4.00260 |

| III A | IV A | V A | VI A | VII A | |
|---|---|---|---|---|---|
| 5<br>**B**<br>Boron<br>10.81 | 6<br>**C**<br>Carbon<br>12.011 | 7<br>**N**<br>Nitrogen<br>14.0067 | 8<br>**O**<br>Oxygen<br>15.9994 | 9<br>**F**<br>Fluorine<br>18.99840 | 10<br>**Ne**<br>Neon<br>20.179 |
| 13<br>**Al**<br>Aluminum<br>26.98154 | 14<br>**Si**<br>Silicon<br>28.086 | 15<br>**P**<br>Phosphorus<br>30.97376 | 16<br>**S**<br>Sulfur<br>32.06 | 17<br>**Cl**<br>Chlorine<br>35.453 | 18<br>**Ar**<br>Argon<br>39.948 |

| I B | II B | | | | | | | |
|---|---|---|---|---|---|---|---|---|

| 28<br>**Ni**<br>Nickel<br>58.71 | 29<br>**Cu**<br>Copper<br>63.546 | 30<br>**Zn**<br>Zinc<br>65.38 | 31<br>**Ga**<br>Gallium<br>69.72 | 32<br>**Ge**<br>Germanium<br>72.59 | 33<br>**As**<br>Arsenic<br>74.9216 | 34<br>**Se**<br>Selenium<br>78.96 | 35<br>**Br**<br>Bromine<br>79.904 | 36<br>**Kr**<br>Krypton<br>83.80 |
| 46<br>**Pd**<br>Palladium<br>106.4 | 47<br>**Ag**<br>Silver<br>107.868 | 48<br>**Cd**<br>Cadmium<br>112.40 | 49<br>**In**<br>Indium<br>114.82 | 50<br>**Sn**<br>Tin<br>118.69 | 51<br>**Sb**<br>Antimony<br>121.75 | 52<br>**Te**<br>Tellurium<br>127.60 | 53<br>**I**<br>Iodine<br>126.9045 | 54<br>**Xe**<br>Xenon<br>131.30 |
| 78<br>**Pt**<br>Platinum<br>195.09 | 79<br>**Au**<br>Gold<br>196.9665 | 80<br>**Hg**<br>Mercury<br>200.59 | 81<br>**Tl**<br>Thallium<br>204.37 | 82<br>**Pb**<br>Lead<br>207.2 | 83<br>**Bi**<br>Bismuth<br>208.9804 | 84<br>**Po**<br>Polonium<br>(210)a | 85<br>**At**<br>Astatine<br>(210)a | 86<br>**Rn**<br>Radon<br>(222)a |

metals ← → nonmetals

| 63<br>**Eu**<br>Europium<br>151.96 | 64<br>**Gd**<br>Gadolinium<br>157.25 | 65<br>**Tb**<br>Terbium<br>158.9254 | 66<br>**Dy**<br>Dysprosium<br>162.50 | 67<br>**Ho**<br>Holmium<br>164.9304 | 68<br>**Er**<br>Erbium<br>167.26 | 69<br>**Tm**<br>Thulium<br>168.9342 | 70<br>**Yb**<br>Ytterbium<br>173.04 | 71<br>**Lu**<br>Lutetium<br>174.97 |
|---|---|---|---|---|---|---|---|---|
| 95<br>**Am**<br>Americium<br>(243)a | 96<br>**Cm**<br>Curium<br>(247)a | 97<br>**Bk**<br>Berkelium<br>(249)a | 98<br>**Cf**<br>Californium<br>(251)a | 99<br>**Es**<br>Einsteinium<br>(254)a | 100<br>**Fm**<br>Fermium<br>(253)a | 101<br>**Md**<br>Mendelevium<br>(256)a | 102<br>**No**<br>Nobelium<br>(254)a | 103<br>**Lr**<br>Lawrencium<br>(257)a |